THE SPECIFICITY
OF
SEROLOGICAL REACTIONS

Karl Landsteiner

1868 - 1943

THE SPECIFICITY OF SEROLOGICAL REACTIONS

Revised Edition

BY

KARL LANDSTEINER, M.D.

With a Chapter on Molecular Structure and
Intermolecular Forces by
LINUS PAULING

and with a Bibliography of Dr. Landsteiner's Works

and a New Preface by
MERRILL W. CHASE
The Rockefeller Institute

DOVER PUBLICATIONS, INC.
New York   New York

This new Dover edition, first published in 1962, is
an unabridged republication of the Second English
edition, published by the Harvard University Press
in 1945, to which has been added the following
material:
A new Preface especially prepared for this edition
by Merrill W. Chase, The Rockefeller Institute.
A Bibliography of Dr. Landsteiner's Works, reprinted with permission from the *Journal of Immunology*, Vol. 48, No. 1, January, 1944.
This edition is published by special arrangement
with the Harvard University Press.

International Standard Book Number: 0-486-66203-9

Manufactured in the United States of America

Dover Publications, Inc.
31 East 2nd Street
Mineola, New York 11501

PREFACE TO DOVER EDITION

THE legacy of a lifetime's pondering on serological specificity and the interaction of antibody with antigens becomes available once again with the reprinting of this unique volume by the late Karl Landsteiner. Its particular character lies in the flow of thought and the fashioning — from vast acquaintance with the literature — of principles that serve to appraise and link diverse contributions of so many workers. Throughout, there runs as *Leitmotiv* the author's personal experiences in the laboratory from 1897 to his death in 1943; but by no means "light" were the areas that he subjected to continuing laboratory examination — the multiplicity of erythrocytic antigens, the serological specificity of split products of native proteins, the searching for evidences of nascent specificity as short polypeptide chains were synthesized, his demonstration that new specificities could arise when certain chemical groupings ("haptens") were attached to carrier protein, original and fundamental studies on hypersensitivity to chemical allergens, and so on and on. His definitive bibliography was printed in the *Journal of Immunology* (January, 1944), and over it there appeared appropriately words of William James: "The great use of a life is to spend it for something that outlasts it."

A volume of this nature has seldom received so much searching concern by an author. The writer had the privilege of listening to the animated luncheon conversations during the final fashioning of the first version (*Die Spezifizität der serologischen Reaktionen*, Berlin, Springer, 1933, 123 pp.), the second version (1936, published by C. C. Thomas as *The Specificity of Serological Reactions*, Springfield, Illinois, 178 pages), and this matured third version published posthumously in 1945 from the completed manuscript by the Harvard University Press, 310 pages. Each of the editions was redone completely; some of the data presented in tabular form from the author's own laboratory files that seemed at variance with later reports were re-checked by repeating the experiments. A most unusual instance is presented by Table 18, in which interpretations of Berger and

Erlenmeyer caused the author to re-examine chemically the same bottles that had been used in 1918 for the original publication, and indeed a few (but significant) errors in the commercial labels were uncovered and corrected. With the third version (the second edition printed in English), the citations underlying the conclusions appeared to the author to be embarrassingly large, and accordingly code abbreviations for journals were introduced with the idea of establishing a better balance in space devoted to text and bibliography, and papers meriting special attention because of excellent bibliographies were signalized by appending the letter (B) to the cited paper.

Winner of the Nobel Prize in Medicine in 1930 for discovery of the human blood groups and for elaboration of knowledge of other antigenic determinants on human red blood cells, Landsteiner, although deriving great personal satisfaction from the award, believed that the sheer accident of discovering the human blood groups could have fallen to anyone's lot, and that his own greater contribution to immunology was the concept of haptens — artificial, or synthesized, chemical radicals that, attached to carrier protein, impose new immunological specificities. Passage of the years has justified his self-appraisal. Modern immuno-chemistry has leaned heavily on Landsteiner's demonstration that, in reacting systems of hapten-protein antigens with their corresponding antiserums, the addition of simple chemical derivatives of such hapten *inhibits* the reaction: it competes with those sites on the antibody that would result in formation of a precipitate upon binding with the full hapten-protein antigen. This basic principle has allowed studies on the binding forces between antibody and dialyzable hapten by equilibrium dialysis, quantitative appraisal of the effect of varying chain length, and the like.

One may ask whether the content of this volume is not superseded by more recent studies. Indeed we have grown in understanding and factual knowledge, chapter by chapter, and one would live impoverished were he unaware of newer knowledge in antibody structure, in "homotransplantation antigens," in new methods for separating proteins and for detecting antibodies, and many another. Yet the beginner and the advanced worker alike will find in no other place the sweep of thought here given, basic to a grasp of immunology.

The pattern of thought of a mighty mind can be appreciated again with this reprinting. The new will be found only to embellish, not to contradict the content, and the reader will have his ample reward.

A feature added to this printing is the reprinting of the definitive bibliography of Karl Landsteiner by courtesy of the Editorial Board of the *Journal of Immunology*.

New York	MERRILL W. CHASE
September, 1962	*The Rockefeller Institute*

FOREWORD

AT THE TIME of Dr. Karl Landsteiner's death in June of 1943, this revised edition of his book, "The Specificity of Serological Reactions," had been completed textually, though actual publication had not yet been started. It has been a long and arduous task to continue the work from the point where it was interrupted by the author's death. In general, the manuscript has been left as it was except for occasional minor changes in sentence structure. In thus appearing the book must lack the final touch and critical eye of the author.

I wish to express my very sincere appreciation to Dr. Merrill W. Chase and Dr. Alexander Wiener for their assistance in the problems and details of publication. Dr. Chase, through his association with the author, has been able to supply much necessary information and technical advice; Dr. Wiener has also given assistance of great value in matters pertaining to the text and index.

ERNEST K. LANDSTEINER

Boston, Massachusetts
July, 1944

AUTHOR'S PREFACE

THE scope of this book was stated in the former edition as follows: "In the preparation of this review it was primarily the author's intention to give an account of the experiments on antigens and serological reactions with simple compounds carried out by himself and his colleagues and concurrently to discuss the phenomena of serological specificity, not yet fully explained, and certain related topics. It was chiefly the chemical aspects of the immunological reactions that were considered, and the material was selected according to its bearing upon basic questions. Within these limits the writer has attempted to include the salient facts and to offer a bibliography comprehensive enough for the use of workers in the field. On the other hand, explanations of elementary concepts and phenomena of serology are provided for readers not acquainted with the subject."

The accumulation of new material comprising significant advances in the study of antibodies, bacterial antigens, haptens and toxins, was an incentive for venturing upon a revision. At the same time the opportunity was taken to expand several sections and to include a number of subjects, formerly omitted, in order to give an outline of the field of immunochemistry. The general plan of the book and the emphasis on matters of theoretical and biological significance have been maintained. In the allotment of space preference was given to subjects on which other treatises have left something to be added; on a similar consideration, in the new chapter dealing mainly with physico-chemical investigations, treated fully and with clarity in well-known reviews, only a synoptic presentation was attempted, sufficient, it is hoped, to provide an introduction and to supply the main experimental findings.

Arbitrariness could not be avoided in handling the vast literature; reasonable completeness was aimed at, however, with subjects on which adequate compilations are not available. Regrettably, the most recent European literature, other than British, was not available.

As before, the sequence of the chapters corresponds broadly to the progress of serological research. Natural antigens and antibodies are first discussed, then artificial conjugated antigens and the reactivity of simple chemical compounds (including a summary of the experimental work on "drug allergy"), next the chemistry of specific non-protein cell substances, and in conclusion the developments in our knowledge of serological reactions from a physico-chemical approach.

That Dr. Linus Pauling accepted the invitation to contribute a chapter on molecular structure and intermolecular forces is highly appreciated by the writer and will, no doubt, be welcome to many.

The author again expresses his gratitude to Dr. H. Lampl and J. van der Scheer, and to the late Dr. E. Prášek, for their most valuable collaboration in the investigations on antigens, and to those who have assisted in other ways. In particular he is greatly indebted to Dr. M. W. Chase who with critical and constructive advice rendered assistance throughout the composition of the manuscript, and to Miss E. H. Tetschner for her invaluable help in preparing the manuscript.

K. L.

CONTENTS

CONTENTS

KEY TO ABBREVIATIONS

(References with the designation B should be consulted for
papers not listed)

AA	Arch. für Augenheilkunde	BJ	Biochem. J.
AC	Annalen der Chemie	BJH	Johns Hopkins Hospital Bull.
ADS	Arch. für Dermatologie und	BK	Berliner Klin. Wschr.
	Syphilis	BP	Bull. Inst. Pasteur
ADV	Acta dermatol. venereologia	BPA	Beitr. z. path. Anatomie
AE	Advances in Enzymology	BPh	Ber. f.·d. ges. Physiol.
AEP	Arch. für exp. Pathologie u.	Br	Brit. J. Exp. Pathology
	Pharmakol.	BZ	Biochem. Ztschr.
AH	Arch. für Hygiene		
AI	Arch. Int. Med. Exp.	C	Chem. Centralblatt
AIF	Arb. Inst. exp. Therapie,	CA	Chem. Abstracts
	Frankfurt	CAA	The Chemistry of Antigens and
AIM	Arch. Internal Medicine		Antibodies, Med. Res. Coun-
AJ	Amer. J. Public Health		cil, London, 1938, Spec. Rep.
AJA	Amer. J. Phys. Anthropology		Ser. No. 230
AJB	Amer. J. Botany	CAI	The Chemical Aspects of Im-
AJC	Amer. J. Cancer		munity, Chem. Catalog Co.,
AJE	Austral. J. Exp. Biology		New York, 1929
AJH	Amer. J. Hygiene	CaR	Cancer Research
AJM	Amer. J. Med. Science	CB	Centralblatt für Bakt.
AJP	Amer. J. Physiology	ChR	Chem. Reviews
AM	Ann. Int. Med.	CIY	Carnegie Inst. Yearbook
AMO	Arb. Med. Fakultät Okayama	CJP	Chin. J. Physiology
AN	Amer. Naturalist	CP	Centralblatt für allg. Pathol.
ANY	Ann. New York Acad. Science	CR	S. R. Acad. Sci.
AP	Ann. Inst. Pasteur	CSH	Cold Spring Harbor Symp.
APS	Acta Pathologica Scandinavia		
ARB	Annual Rev. Biochemistry	DA	Deutsch. Arch. für klin. Med.
ArP	Arch. of Pathology	DM	Deutsche Med. Wochenschr.
ART	Amer. Rev. Tuberculosis	DW	Dermatol. Wochenschr.
AS	Arch. of Dermatology and Syph.	DZG	Deutsche Ztschr. f. d. ges.
AT	Arch. für Tierheilkunde		gerichtl. Medizin
BaR	Bacteriological Review	E	Endocrinology
BB	Biological Bulletin	EE	Erg. der Enzymforschg.
BC	Ber. d. deutsch. chem. Ges.	EH	Erg. der Hygiene
BCB	Bull. Soc. Chim. Biologie		
BCP	Beitr. z. chem. Physiol. und	F	Fermentforschung
	Path.	FI	Fundamentals of Immunology,
BDB	Ber. d. deutsch. botan. Ges.	FP	Interscience, New York, 1943
BiR	Biological Rev.		Federation Proceedings

G	Genetics	N	Nature
		NEM	New England J. Med.
H	Helvetica Chimica Acta	NK	Newer Knowledge of Bacteriol-ogy, University of Chicago Press, 1928
HL	Harvey Lectures		
HP	Handb. d. norm. und path. Physiol.		
		NT	Nederl. Tijdschr. voor Geneesk.
HPM	Handb. d. pathogenen Mikro-organismen	Nw	Naturwissensch.
		OH	Oppenheimer's Handb. der Bio-chemie
IC	3rd Intern. Congr. of Micro-biology, New York, 1939.		
		PB	The Principles of Bacteriology and Immunity, Wood, Balti-more, 1937
JA	J. Allergy		
JAC	J. Am. Chem. Soc.		
JAM	J. Am. Med. Ass.	PH	Physiol. Rev.
JB	J. Biochemistry	PNA	Proc. Nat. Acad. Sci.
JBa	J. Bacteriology	Pr	Proc. Soc. Exp. Biol. and Med.
JBC	J. Biol. Chem.	PZ	Physiological Zoology
JC	J. Soc. Chem. and Ind.		
JCI	J. Clin. Investig.	Q	Quarterly Rev. of Biology
JCS	J. Chem. Soc.		
JD	J. Invest. Dermatology	RI	Revue d'Immunologie
JE	J. Endocrinol.	RS	Proc. Royal Society
JEB	J. Exp. Biol.		
JEM	J. Exp. Med.	S	Science
JG	J. Genetics	SB	C. R. Soc. Biol.
JGP	J. Gen. Physiol.	SM	Standard Methods, etc., Wil-liams and Wilkins, Baltimore, 1939
JH	J. Hygiene		
JI	J. Immunology		
JID	J. Inf. Dis.	SMW	Schweiz. Med. Wochenschr.
JJ	Jap. J. Med. Science	SystB	A System of Bacteriology, Great Britain Med. Res. Council
JLC	J. Lab. and Clin. Med.		
JM	J. Med. Res.		
JP	J. Path. and Bact.	WK	Wiener Klin. Wschr.
JPC	J. Physical Chem.		
		YJB	Yale J. Biol. and Med.
K	Klin. Wochenschr.		
KB	Kolloid Beihefte	Z	Zoologica
KZ	Kolloid Ztschr.	ZH	Ztschr. f. Hygiene
		ZI	" " Immunitätsforschung
		ZK	" " Krebsforschung
L	Lancet	ZP	" " physik. Chem.
LC	C. R. Lab. Carlsberg	ZPC	" " physiol. Chem.
		ZR	" " Rassenphysiol.
MK	Med. Klinik	ZZ	" " Züchtung
MM	Münchener Med. Wochenschr.		

THE SPECIFICITY
OF
SEROLOGICAL REACTIONS

I

INTRODUCTION

THE morphological characteristics of plant and animal species form the chief subject of the descriptive natural sciences and are the criteria for classification. But not until recent times and the advent of serology has it been clearly recognized that in living organisms, as in the realm of crystals, chemical differences parallel the variation in structure. This conclusion was arrived at indirectly, not as the result of studies made with that aim in view. The idea of specificity originated in the knowledge that after recovery from an infectious disease there remains an immunity for that particular disease, a fact which found its first practical application in Jenner's vaccination against smallpox. The search for the explanation of this remarkable phenomenon led to the discovery of a peculiar sort of substances in the blood serum, the so-called antibodies, some of them protecting against infectious agents (bacteria and viruses), or neutralizing toxins. These substances, now definitely recognized as modified globulins, are formed not only as a result of infection but also in consequence of the administration of certain poisons of large molecular size (toxins from bacilli,[1] higher plants[2] and animals[3]), or of dead bacilli.[4] A new line of serological research, and the separation of serology from the original close connection with the question of immunity to disease, began with the discovery that the immunization against microbes and toxins is only a special instance of a general principle and that the same mechanism is in play when innocuous materials, such as cells or proteins derived from a foreign species[5] are injected into animals. In these cases, likewise, there appear in the serum antibodies causing the agglomeration or destruction of cells or precipitation of proteins.

[1] Roux, Behring, Kitasato.
[2] Ehrlich.
[3] Calmette, Phisalix and Bertrand.
[4] Pfeiffer, Metchnikoff, Gruber and Durham, Kraus.
[5] Tschistowitch, Bordet, Belfanti and Carbone, von Dungern, Landsteiner, Uhlenhuth.

3

The immune antibodies all have in common the property of specificity, that is, they react as a rule only with the antigens that were used for immunizing and with closely similar ones, for instance with proteins or blood cells of one species, or particular bacteria, and related varieties.

Hence a general method for differentiating proteins, distinguishable only with difficulty or not at all by the chemical methods available, was furnished, and it was found that particular proteins characterize every species of animals and plants.

The specificity of antibodies, whose range of activity was later found to extend far beyond the proteins and to include simple chemical substances, underlies the practical applications of serology and constitutes one of the two chief theoretical problems, the other being the formation of antibodies. The term "specificity" is often used to imply that a certain immune serum reacts with only one of many biologically similar substances, as tetanus antitoxin with no other toxin but that produced by B. tetani. It is known, however, that snake antivenins neutralize not only the venom used for immunization but neutralize also to some degree venoms from other snakes, or scorpions (1) and, as has already been pointed out, the selectivity is not absolute when proteins or cells of related origin are tested with an immune serum; indeed it will be seen later that, using chemically well defined compounds, overlapping reactions occur regularly, provided the substances are sufficiently similar in chemical constitution. The word "specificity" then signifies that the reaction with one of the antigens, namely, that used for immunization, is stronger than with all others. Yet even this definition is not comprehensive enough, since it does not take into consideration a group of phenomena which resemble antibody reactions in all essentials.

A case in point is that of plant hemagglutinins.[6] In the seeds of Abrus precatorius and Ricinus communis, there are, along with toxins (abrin, ricin), substances that clump blood corpuscles, quite like the hemagglutinins of animal sera. Agglutinins of this type have also been found in numerous non-poisonous plants, particularly in Papilionaceae (phasins). Many of these substances, which are proteins and antigenic, act in very high dilutions and upon practically all

[6] Kobert (2), Landsteiner and Raubitschek (3), Schneider (4), Eisler (5) (56), Schiff (6), Jacoby (7). Agglutination takes place with animal cells other than erythrocytes, not with bacteria. Whether the clumping of bacteria and precipitation of proteins by plant extracts, occasionally recorded [Kritschewsky (7a), Wilenko (7b)], are also due to antibody-like substances is undecided.

sorts of blood. However, when solutions — for example, of abrin and ricin — which contain such agglutinins, are mixed with the blood of different animals, it will be found that the reactions differ in strength; thus one of two sorts of blood may be agglutinated more intensely by the abrin solution, the other by ricin. Even more striking differences are demonstrable with crotin, a substance from the seeds of Croton tiglium which has hemolyzing and hemagglutinating properties, and with hemolysins derived from certain bacteria and animals. These lysins act strongly on the erythrocytes of numerous species that are in no way related, but have little or no effect on others. For example, arachnolysin,[7] contained in the spider Eperia diadema, reacts strongly with the blood of rabbit and man but has practically no effect on guinea pig or horse erythrocytes, while the latter blood is very sensitive to the lysins produced by tetanus bacilli. Natural antibodies, in the author's opinion, belong in the same category.

The action of plant agglutinins,[8] not limited to a single substrate yet to some extent selective, has not commanded much attention in spite of its theoretical interest — the agglutinins are occasionally referred to in the literature as non-specific — and for this reason as well as the scarcity of reliable data, the following experiment is presented. The highest dilutions were determined in which solutions prepared from seeds still agglutinated suspensions of red blood corpuscles. The titers after a given time are shown in Table 1. It is of

TABLE 1

	Blood Rabbit	Pigeon		Blood Horse	Pigeon
Bean extract	125	2000	Abrin	128	256
Lentil extract	160	0	Ricin	4	512

importance that corresponding to the variations in the sensitivity of blood cells, there are also distinct differences in the binding capacity for the agglutinins.

The reactions just described may properly be termed specific and, accordingly, it seems adequate to define serological specificity as

[7] Sachs (8).
[8] Recently influenza and vaccinia virus were found to cause hemagglutination, likewise to some degree selective [Hirst (9), McClelland et al. (10), Nagler (11)]; the agglutination is inhibited by viral antibodies.

the disproportional action of a number of similar agents on a variety of related substrata. Depending upon the number of substances acted upon and the relative strength of the reactions caused by one reagent, one can distinguish differences in the range of activity and the degree of specificity. The definition includes as limiting case the specificity of the immune antibodies, highly though not completely selective, which, because of their origin, are uniquely related to one substance, and — although it is applicable to many chemical reactions — it serves to distinguish the serological effects from apparently similar ones.

Thus, there are various substances which agglutinate blood cells,[9] such as, salts of heavy metals, inorganic colloidal acids and bases, and basic proteins (protamines,[10] histones). The hemagglutinating and (with the aid of complement) hemolyzing action of some of these substances, as colloidal silicic acid,[11] which is detectable in concentrations as low as 0.001 per cent by agglutination of blood, or tannin,[12] parallel the serological phenomena sufficiently to serve as non-specific models thereof and to yield information concerning their mechanism (Reiner). On the other hand, these agents do not possess the characteristic property of disproportional action and specific absorption. From the results of a few experiments, the specificity ascribed to hemagglutination by metallic salts (19) seems doubtful; at any rate, it is necessary in such tests to consider the influence of the hydrogen ion concentration. A slight degree of specificity, in the sense defined above, could be demonstrated in the hemolysis produced by saponins.[13]

Finally, it should be mentioned that antibodies act by combining with the inciting substances, as first shown by Ehrlich (22) with an antibody prepared to the plant agglutinins mentioned above; trivial as this statement now sounds, in those days the opinion was shared by some authorities that the protection by immune sera is due rather to an action on the animal body. Subsequent to the specific union there occur visible effects constituting the "second stage" of antigen-antibody reactions (p. 252). This second stage, of subordinate significance for the problem of specificity, depends on the properties of the substrate and subsidiary conditions.

[9] On non-specific agglutination of bacteria, v. Schiff (6).
[10] Thompson (12).
[11] Landsteiner (13), Browning (14).
[12] Reiner et al. (15), Kruyt (16), Freund (17); v. (18).
[13] Kofler (20), Ponder (21).

The investigations to be discussed on antigens and antibodies and their specificity have developed into a special branch of biochemistry, peculiar in regard to materials, methods and the nature of the reactions, which depend on affinities different from those that are involved in the covalent bonds of organic chemistry. These studies will bear upon other biological phenomena in which selective reactivity of the agents is a prominent feature, such as enzymatic, pharmacological, hormonal and chemotherapeutic effects. Other than that, as will be seen, they have already served to bring many biological substances, discovered serologically, within the scope of chemical research.

PRINCIPAL IMMUNOLOGICAL PHENOMENA AND NOMENCLATURE

Substances inciting the formation of and reacting with antibodies are named antigens.[14] Sera that contain antibodies as the result of the injection of antigens [15] are called "immune sera" (antisera); the designation "normal" or "natural" antibodies is applied to substances found in the serum of untreated animals, which are similar in their effects to the antibodies developed by immunization. A rather detailed nomenclature has been built up around the diverse manifestations of antigen-antibody reactions, the antigens and antibodies being described with reference to the kind of reaction observed; but it is necessary to state at the outset that the same antibodies can act in various capacities, e.g., as agglutinins and precipitins (see below).

Poisonous antigens characterized by high toxicity and their capacity of being fully neutralized, even in high multiples of the lethal dose, by their antibodies are termed toxins (as the toxin of the diphtheria bacillus) [16] or exotoxins (an expression used in distinction to endotoxins, which are less toxic, firmly attached to the cell and not readily neutralized by antisera). Toxins include substances which destroy cells, e.g., the hemotoxin (or hemolysin) of tetanus

[14] For the words "hapten," "conjugated antigen," v. (pp. 156, 76, 110).

[15] In amplification of its original meaning the term immunization is commonly used also when antigens are not harmful and the antibodies which are formed have no curative or protective role.

[16] Hypothetical chemical groupings of toxins responsible for toxicity and combination with antibodies were called by Ehrlich toxophore and haptophore groups respectively; in the sense of the "side chain theory" of Ehrlich lytic antibodies, too, were supposed to contain a cytophilic and a complementophilic group and were called amboceptors.

bacilli, the leucotoxin (or leucocidin) of staphylococci, etc. The neutralizing antibody is called antitoxin. Toxoids are modified toxins which are atoxic but still antigenic and capable of combining with antibodies. The clumping of cells is known as agglutination, and the antigens and antibodies involved are called agglutinogens and agglutinins (hemagglutinins, bacterial agglutinins) respectively. Similarly, the antibodies causing disruption of cells (lysis) are designated as lysins (bacteriolysins, hemolysins), precipitins those which cause precipitation when mixed with the inciting soluble antigens (precipitinogens), while tropins (opsonins) [17] are antibodies that render cells susceptible to ingestion by leucocytes (phagocytosis). Conglutinin is a colloidal substance, occurring especially in beef serum, that combines with cells after they have absorbed antibody and complement, and enhances lysis and agglutination.[18] Bactericidal substances other than those mentioned above and, contrariwise, active towards gram-positive bacteria are to be found in serum (β-lysins) and in leucocyte extracts (leukins); these lysins are not increased by immunization.[19]

Antibodies produced with material taken from the animal selected for immunization, or from other individuals of the same species, are referred to as auto- or isoantibodies respectively. The expression passive immunization, in contrast to active immunization, signifies the temporary protection conferred upon an animal by the administration of immune sera. Reactions of an antibody with the corresponding antigen are said to be homologous, while heterologous reactions, known also as overlapping, cross or group reactions, are those taking place with substances other than the inciting antigen.

Referring to the ability of antigens to induce a state of hypersensitivity and subsequently to produce the symptom-complex known as anaphylaxis the term anaphylactogen is in use. The anaphylactic state is induced by parenteral administration of a protein; to describe a typical experiment, after 0.01 cc [20] of ox serum has been injected into a guinea pig (and much smaller doses may be effective, as 0.0001 cc or less), and an interval of 10 days or longer has elapsed, a second (intravenous) injection of, say, 0.1 cc produces shock and

---

[17] v. Ward and Enders (23).
[18] (24–26); v. (27).
[19] v. Ledingham (28), Pettersson (29).
[20] The minimal doses for sensitization are much smaller than those for eliciting shock.

kills the animal in a few minutes under characteristic and violent symptoms (choking, convulsions, dyspnoea); or, a rabbit previously sensitized would show intense local inflammatory and necrotic reactions on subcutaneous reinjection of the same antigen (Arthus phenomenon). Furthermore, anaphylactic reactions, rendered manifest by contraction of smooth muscle, can be elicited in isolated organs (uterus, intestine, etc.) of sensitized animals (Schultz-Dale reaction). When anaphylactically sensitive animals are injected with a sufficient amount of antigen in such manner that they survive (sublethal doses, subcutaneous injection), a temporary refractoriness to shock ensues (desensitization). The anaphylactic condition can be transferred with the serum of sensitized animals to normal ones (passive anaphylaxis) and, fundamentally, anaphylactic shock has the significance of an antigen-antibody reaction *in vivo* and can be used in place of a reaction *in vitro*. The opinion that anaphylactic reactions generally exhibit greater specificity than *in vitro* tests, would require confirmation.[21]

The terms allergen and atopen are used to connote the ability of certain materials to induce specific manifestations of hypersensitivity in man (hay fever, asthma), and the associated special antibodies in the serum of such patients are known as reagins, a name also given to some other agents occurring in serum (Wassermann reagins).

Remarkably similar but hardly related to anaphylaxis are nonspecific hemorrhagic reactions (Shwartzman phenomenon) which occur in a skin site injected with certain bacterial culture filtrates when a day later the same or also certain other culture filtrates have been injected intravenously.

COMPLEMENT, COMPLEMENT FIXATION. — Lysis of red cells or bacteria by antisera requires the aid of a special agent, present in normal sera and not augmented upon immunization, which is called complement or alexin [22] [Bordet (33)]; the amount necessary varies inversely with that of the lytic antibody employed. It deteriorates slowly on storing and is inactivated by heating, e.g., within half to one hour at 54–60°C.[23] Complement is readily taken up by various adsorbents and is inactivated by organic solvents,[24] sufficient con-

---

[21] In special cases the anaphylaxis experiment has proved of advantage (p. 23).

[22] The subject is presented at length by Osborn (30), Browning (31), Sachs (32).

[23] Cf. Ecker, Pillemer and Kuehn (34). As in denaturation of proteins there is no critical temperature.   [24] v. Ecker et al. (35).

centrations of salts, particularly those containing the anions SCN and I,[25] and by oxidation (this being reversible,[26] if not too drastic), as some enzymes and bacterial toxins; it does not act in high dilution, unlike most antibodies. Because of the lability of complement, unless immune sera are fresh (and not much diluted), fresh normal serum, usually taken from guinea pigs, must be added to the antibody solutions in order to produce lytic effects, the amount necessary depending on the concentration of the antibody; or cells are allowed to absorb lytic antibody, are centrifuged off, and complement is added to these "sensitized" cells.

Separation of serum by dilution and slight acidification into two fractions, both thermolabile — the one precipitated containing the euglobulin and called mid-piece, the other remaining in solution called end-piece — revealed a complex constitution of complement, the fractions alone being inactive but active when recombined [Ferrata (39)]. The existence of other more heat stable constituents has been shown, namely a third component which was indicated by inactivation upon treatment with yeast or zymin,[27] and a fourth component destructible by ammonia and certain other bases, announced by Gordon, Whitehead and Wormall (42). Pillemer, Seifter and Ecker (43, 44) suggest that the inactivation by ammonia involves the carbonyl group of a carbohydrate.

More detailed information on the substances associated with complement activity has accrued through important recent work of Pillemer, Ecker, Oncley and Cohn (44), who characterized by their electrophoretic behavior two protein fractions, one a euglobulin amounting to 0.7% of the serum protein, the other, rich in carbohydrate, representing 0.2% of the total serum protein, which comprises the end-piece and the fourth component.[28]

Whether the activity of complement is enzymatic, as may well be surmised, or otherwise, and what its physiological function may be is still to be learned. A mutant strain of guinea pigs, deficient in complement activity, showed no evident abnormality in behavior but was less resistant to un-

[25] Gordon and Thompson (36).

[26] v. Pillemer, Seifter and Ecker (37). Experiments by Bordet and Ehrlich on the question whether a serum contains several kinds of complement are reviewed by Zinsser-Enders-Fothergill (38); the protein constituents of complement are doubtless not the same in different species [cf. (38)].

[27] Von Dungern; Coca; Gordon, Whitehead and Wormall (40), Pillemer and Ecker (41).

[28] For mid-piece, end-piece, third and fourth component the symbols C_1, C_2, C_3, C_4, respectively, have been proposed [Pillemer et al. (45), Heidelberger (46)].

favorable living conditions; to which of the components of complement the deficiency is referable was not definitely ascertained.[29]

The characteristic of complement to be bound [30] by the aggregates (precipitates, sensitized cells) formed through the interaction of antigens and antibodies is the basis for a frequently used serological test. In this "complement fixation reaction," introduced by Bordet and Gengou (50), antigen and antibody to be tested are mixed with fresh normal serum and, after incubation, hemolytic immune serum and corresponding red cells are added as indicator for the presence of complement. If an immunological reaction takes place in the first stage, complement is fixed and removed from the solution and hemolysis is prevented, completely or in part, according to the intensity of the reaction. In general, the reactions run parallel to precipitin reactions but complement fixation, in certain cases, gives positive results in the absence of visible precipitation. In the fixation tests the antibodies may react in considerable dilution.

Because of the fixation phenomenon it is evident that precipitins prepared to a serum will inactivate the complement contained therein independent of any special anticomplement; antisera against the separated complement constituents have not been produced.

The inhibition of bacteriolysins by an excess of bacteriolytic antibody (Neisser-Wechsberg's complement deviation) has been put down to precipitation of dissolved antigen with consequent fixation of complement [Gay (51).]

IMMUNIZATION. — Precipitins may be produced, against serum proteins for instance, by injecting rabbits intravenously three to four times at weekly intervals with one or two cc. of serum, or daily with 0.1 cc followed by rest periods on alternate weeks. The serum is usually drawn 7–10 days after the last injection. Similarly, hemagglutinins and hemolysins may be obtained upon injecting the washed cells of, say, two to three cc blood intravenously a few times at intervals of several days. Various techniques for preparing precipitins are given in the texts of Sherwood (52), Kolmer (53), and Boyd (54); the production of anti-bacterial and antitoxic sera is described in "Standard Methods" by Wadsworth (55).

---

[29] v. Coca; Hyde; Whitehead et al. (47).
[30] Cf. Dean; Goldsworthy; (54). This was established also by nitrogen analyses [Heidelberger (46); v. Haurowitz et al. (48)]. A study of the fixation of complement and its separate constituents by antigen-antibody combinations under various experimental conditions, and by adsorbents has been made by Pillemer, Seifter and Ecker (49).

To present a picture of the development of immunochemistry a list of significant steps is here appended.

Hemagglutinins and hemolysins (1875), and bactericidal substances (1888–1889) in normal serum

Discovery of bacterial toxins (1889)

Antitoxins, and antibacterial sera (1890–1896)

Demonstration of complement (1896)

Antibodies for proteins and animal cells; serological species (and individual) specificity (1898–1901)

Chemically modified proteins as antigens (1902–1906)

Reaction of syphilis sera with non-protein substances from tissues, extracted with alcohol (1907)

Heterogenetic antigens (1911)

Simple chemicals rendered antigenic by attachment to proteins (1918)

Separation from animal tissues of substances that are practically non-antigenic but react specifically with antibodies (haptens) (1918–1921)

Specific reactions of antibodies with simple chemical compounds (1920)

Discovery of serologically reactive bacterial polysaccharides (1923)

Systematic application of quantitative methods to antigen-antibody reactions (from 1929 on)

Complex bacterial antigens containing polysaccharides and lipids (1934)

Characterization of antibodies by physico-chemical methods (ultracentrifugation and electrophoresis). Crystallization of diphtheria antitoxin (1936–1941)

BIBLIOGRAPHY *

(1) Kraus et al: Giftschlangen, Fischer, Jena, 1931. — (2) Kobert: Beitr. z. Kenntn. der vegetabl. Hämagglutinine, Landwirtsch. Versuchsstat. 79, 1, 1913, Berlin, Parey. — (3) Landsteiner et al: CB 45, 660, 1907. — (4) Schneider: JBC 11, 47, 1912. — (5) Eisler et al: ZI 47, 59, 1926 (B); CB 66, 309, 1912. — (6) Schiff: OH 3, 346, 1924. — (7) Jacoby: HPM 3, 107, 1930. — (7a) Kritschewsky: ZI 22, 381, 1914. — (7b) Wilenko: ZI 5, 91, 1910. — (8) Sachs: BCP 2, 125, 1902. — (9) Hirst: JEM 75, 49, 76, 195, 1942. — (10) McClelland et al: Canad. Publ. Health J. 32, 530, 1941. — (11) Nagler: Med. J. Australia 1, 281, 1942. — (12) Thompson: ZPC 29, 1, 1900. — (13) Landsteiner et al: MM 1904, p. 1185; v. ZI 14, 21, 1912. — (14) Browning: Immunochemical Studies, London, Constable 1925. — (15) Reiner et al: ZI 61, 317, 397, 459, 1929. — (16) Kruyt: KZ 31, 338, 1922. — (17) Freund: JI 21, 127, 1931; Pr 28, 1010, 1931. — (18) Gordon et al: Br 18, 390, 1937. — (19) Hirschfeld: AH 63, 237, 1907. — (20) Kofler: Die Saponine, Wien, Springer, 1927. — (21) Ponder et al: BJ 24, 805, 1930; v. AH 70, 1, 1908. — (22) Ehrlich: Fortschr. Med. 15, 41, 1897. — (23) Ward and Enders: JEM 57, 527, 1933. — (24) Muir et al: JH 6, 20, 1906. — (25) Streng: CB 50, 47, 1909. — (26) Leschly: ZI 25,

* For abbreviations see p. xii.

219, 1916. — **(27)** Dean: RS *B*, *84*, 416, 1911. — **(28)** Ledingham: SystB *6*, 31, 1931. — **(29)** Pettersson: Die Serum-β-Lysine, Fischer, Jena, 1934. — **(30)** Osborn: Complement or Alexin, Oxford University Press, London, 1937. — **(31)** Browning: SystB *6*, 332, 1931. — **(32)** Sachs: HPM *2*, 779, 1929. — **(33)** Bordet: AP *10*, 193, 1896. — **(34)** Ecker et al: Pr *45*, 115, 1940. — **(35)** Ecker et al: Pr *38*, 318, 1938. — **(36)** Gordon et al: Br *14*, 33, 277, 1933. — **(37)** Pillemer et al: JI *40*, 97, 1941 (B); v. RI *4*, 528, 1938. — **(38)** Zinsser-Enders-Fothergill: Immunity, MacMillan, New York, 1939. — **(39)** Ferrata: BK *44*, 366, 1907. — **(40)** Gordon et al: BJ *19*, 618, 1925. — **(41)** Pillemer et al: JBC *137*, 139, 1941. — **(42)** Gordon et al: BJ *20*, 1028, 1036, 1926. — **(43)** Pillemer et al: JI *40*, 89, 1941. — **(44)** Pillemer et al: JEM *74*, 297, 1941; ANY *43*, 63, 1942 (B). — **(45)** Pillemer et al: S *94*, 437, 1941. — **(46)** Heidelberger: JEM *73*, 681, 1941, *75*, 285, 1942. — **(47)** Whitehead et al: JI *13*, 439, 1927. — **(48)** Haurowitz et al: ZI *95*, 478, 1939. — **(49)** Pillemer et al: JEM *75*, 421, 1942 (B), *76*, 93, 1942 (B); JI *45*, 51, 1942, *40*, 81, 1941. — **(50)** Bordet et al: AP *15*, 289, 1901. — **(51)** Gay: AP *19*, 593, 1905. — **(52)** Sherwood: Immunology, Mosby, St. Louis, 1941. — **(53)** Kolmer: Infection, Immunity, etc., Saunders, Philadelphia, 1925. — **(54)** Boyd: FI. — **(55)** Wadsworth: SM. — **(56)** Eisler: Handb. d. Pflanzenanalyse, Springer, Wien, *4*, 987, 1933.

II

THE SEROLOGICAL SPECIFICITY OF PROTEINS

SPECIES SPECIFICITY. — While species specificity is a general attribute of plant proteins as well as those of animal origin, the principal investigations on species differences of proteins have been made with precipitins obtained by injecting animals, usually rabbits, with blood serum from other species [Tschistovitch, Bordet (1)]. Owing to technical complications, tissue proteins have been much less thoroughly examined, and, for the same reason, whole serum has been mostly employed instead of the use, however preferable, of purified serum proteins. Nevertheless, in this way an important and general law was revealed by the work of several authors, especially Nuttall (1a) who tested the blood of more than 500 animal species with about 30 immune sera prepared in rabbits.[1] The material at his disposal was scant in some instances and not always well preserved, and the tests could not be performed simultaneously, but his careful experiments were entirely sufficient to prove that immune sera act most

---

[1] Immunization of chickens is recommended by Wolfe (1b). On precipitin-production in guinea pigs and rats v. (1c-f).

intensely with the kind of serum used for the immunization and, in addition, with sera of related animals, the intensity of the reactions in general being in proportion to the degree of zoological relationship.[2] In consequence, it would be possible to outline broadly the genealogical tree of animals on the basis of serum reactions alone if the data were extensive enough.

For illustration, Nuttall's reactions with two precipitins produced by injecting human serum are given in Table 2. They show that the intensity of precipitation diminishes in the order: anthropoid apes, Old World monkeys, American monkeys. The figures indicate the

TABLE 2

	Immune Serum	
	1	2
Man	100	100
Orang-Utan	47	80
Cynocephalus mormon	30	50
Cercopithecus petaurista	30	50
Ateles vellerosus	22	25

volumes of the precipitates formed in the serum of the different species in comparison with the volume (100) of the precipitate with human serum.

It will be in place to consider at this point the methods used in assessing the strength of precipitin reactions.[3] One procedure consists of the determination of the highest dilutions (usually successively doubled) of antigen that precipitate with a given quantity of immune serum; the tests are made either by mixing the immune serum and the antigen solution, or as interfacial reactions [4] ("ring test") by overlayering the serum with the antigen solution in narrow tubes. These methods do not really measure the antibody content [5] — this being obvious in the case of agglutinin reactions —

[2] Cf. the reviews by Erhardt (2) and Boyden (3). For the most part mammals and birds have been examined (1a), (4); (4a) (Bovidae and Cervidae); (4b) (birds). Concerning lower animals v. (4), (5) (reptiles); (6) (amphibians); (7) (fishes); (8–10), (2) (crustaceans); (9), (11–12) (molluscs); (13) (worms); (p. 81) (insects). Wilhelmi (14) deduces from precipitin tests a relationship between prochordates and echinoderms and, on the other hand, annelids and arthropods.

[3] (15–18); (19) (weighing of precipitates); (20–23). Technical details are described by Kolmer (24), Boyd (513).

[4] This procedure is useful for detecting weak reactions. Similar tests can be made with antigens solidified by gelatin. (See 24a.)

[5] (16–18), (25–27), (36). Antigen titration is, however, of value for assess-

and sera of unlike potency, yielding quite different amounts of precipitate, may give the same end titer. (Indeed, when a serum is diluted several fold the antigen titer may remain undiminished.) [6] Incidentally, this shows, and it will become still more evident in the following, that an immune serum is not to be characterized, even apart from specificity differences, merely by the amount of antibodies which it contains.

More rational is titration of a constant amount of antigen with increasing dilutions of antiserum to the point where no precipitation (or complement fixation) is discernible.[7] An inexpedience of this technique, namely, that ordinarily precipitins cannot be much diluted, has been overcome by adsorbing the antigen to collodion particles or killed bacteria and conducting the titration as an agglutination test.[8] Dilution titers of one part in several hundred are then obtained with potent sera and very weak precipitins can be detected. A highly sensitive method devised to magnify the reactions has been proposed by Goodner; in it collodion particles are added after the reagents have been mixed (30).

Frequently used for comparing the strength of precipitin sera is the method of optimal proportions recommended by Dean and Webb.[9] It determines in a series of tubes containing a constant amount of immune serum and increasing dilutions of antigen the one which first shows flocculation.[10] The ratio of antigen and antibody in this tube, the optimal proportion, was found to be fairly independent of the antibody concentration used and therefore to be characteristic for a given serum. Often, not invariably,[11] this ratio practically coincides with that at which both antibody and antigen are completely precipitated (equivalence point or zone) and are barely or not at all detectable in the supernatant fluid. Hence, broadly speaking, the greater the amount of antigen at the optimum, the higher will be the potency of the serum, the method being comparable to the titration of antitoxin by its capacity to neutralize toxin. Similar in principle is the establishment of the equivalence point by directly locating the tube in the series in which addition of neither antigen nor antibody to the supernatant fluid produces a precipitate.[12]

---

ing the content of an antigen, in samples of unknown composition, for instance as an aid during purification procedures.

[6] Satoh (17).

[7] Kister et al.; Culbertson (18). It is possible that the relative titers of two sera vary when determined with different antigen concentrations, if the reaction curves are not parallel throughout.

[8] Cannon et al.; Lowell (28), Roberts et al. (29).

[9] (31), Taylor et al. (32); v. Opie (33), Boyd (92) (use for determination of species relationships).

[10] The occurrence of several flocculation maxima is referable to a plurality of antigens. Such observations are described by Goldsworthy and Rudd (34) who tested antisera to horse serum containing antiglobulin and antialbumin. A single zone is no proof for homogeneity [Taylor and Adair (35)].

[11] v. (35a), (1c).   [12] (18), (32), (36a).

Determination of the flocculation optimum when varied amounts of anti-body are added to constant antigen is the procedure prescribed by Ramon and Richou (36).

When the antigen is free of nitrogen or can be estimated independently, it is possible to determine accurately the absolute amount of antibody in a serum [13] by nitrogen analyses (micro-Kjeldahl) on the precipitate in the zone where all antibody is carried down; the method has been adapted also to common protein-antiprotein systems.

A rough estimation of antibody may be made volumetrically by cen-trifuging the sediment in narrow graduated tubes after complete precipita-tion for, with the exception of antigens of very high molecular weight, the bulk of specific precipitates is made up of antibody protein with which is combined a relatively small quantity of antigen; and in common practice the quantity and speed of formation of the precipitates, apparent upon simple inspection, is used for judging the potency of an immune serum. In fact most of the data on specificity to be presented were obtained in qualitative and semi-quantitative fashion, and when only definite differ-ences are taken into account such results are quite dependable for com-paring the strength of reactions. Nevertheless, a great advance was made by the development of accurate quantitative techniques, due mainly to Heidelberger and his colleagues, and these methods are indispensable for precise evaluation, as of antibody content or cross reactions, and investiga-tions along physico-chemical lines.

The majority of studies on serological species differences have been carried out with precipitins [14] and the method of antigen titra-tion which, although allowing the demonstration of species differences, is, as stated, not unobjectionable in principle. Results of this sort are given in Table 3.

The following determinations were made with purified ovalbumins and a fair number of antisera to crystalline hen ovalbumin; the pre-cipitates were measured volumetrically. Despite considerable varia-tion in the strength of the cross reactions, the sequence of the species arranged according to the amounts of precipitates was, with minor exceptions, the same with all sera, the more distant Anseriformes giving, as would be expected, lower values than the Galliformes species (Table 4). On the other hand, titration of these sera with progressive antigen dilutions failed to demonstrate clearly the dif-

[13] Wu et al. (37), Heidelberger and Kendall (38); v. (597). Density measure-ments on the supernatant fluids have been suggested for the purpose (39). On optical measurement of precipitates v. (39a, b, c).

[14] According to some reports, finer differentiation is attained in tests for species specificity with complement fixation than with precipitation (40).

ferences between the five proteins [15] and, consequently, the method must be in error in so far as a distinction is intended.

In allied species the protein differences are often too minute to be

TABLE 3

[after Boyden (41) and Satoh (17)]

| | Rabbit Immune Sera for Ox Protein | | |
	No. 1	No. 2	No. 3
Ox	100	100	100
Sheep	66	50	40
Goat	50	50	40
Pig	16	8	0
Horse	16	8	<1
Dog	16	8	0
Man	8	8	0
Wild rat	<1	<1	

The homologous titers of the sera (antigen dilution of 5000 or more) are taken as 100. The decimal fractions given in the original paper are omitted. Other tabulations of this sort are found in papers by Boyden (41) and Wolfe (42); cf. (2).

TABLE 4

[after Landsteiner and van der Scheer (44)]

| Ovalbumin Immune Serum No. | Galliformes | | Anseriformes | |
	Turkey	Guinea hen	Duck	Goose
1	35	26	9	9
3	67	57	42	31
9	61	51	30	18
13	43	53	25	20
21	19	14	14	12
Mean of 22 sera and standard error	50±2.4	42±2.3	25±1.7	19±1.4

Volumetric measurement of precipitates secured with hen egg albumin immune sera by complete precipitation with different egg albumins; the value for chicken ovalbumin is taken as 100. Serum 21 was exceptional among 22 sera prepared in giving weak cross reactions of about equal strength.

detectable by direct precipitin tests.[16] A method for distinguishing such proteins consists of first absorbing the antibodies reacting with the heterologous antigen by adding this protein and centrifuging off the precipitate formed. Tests are then made with the supernatant fluid. The same purpose may be served by partial absorption in the animal body (partial desensitization of sensitized animals) with

[15] v. (3), (4b), (43).
[16] v. (15), (45), (46).

heterologous antigens, as demonstrated by Dakin and Dale (47), and Wells and Osborne (48). Clear-cut results have often been obtained with the above technique of partial precipitation,[17] but it is not regularly successful.[18] In experiments of the author small differences were detectable between the serum proteins of horse and donkey and of man and chimpanzee,[19] yet the distinction was not much sharper than with direct precipitation. The discrepancies between the various reports probably find their explanation in peculiarities of individual immune sera and the methods used.[20] Examples of actual experiments and the intricate relations prevailing in the absorption of precipitin sera with several cross-reacting antigens will be described in a later section (pp. 54, 55; 270).

Some statements, based upon absorption tests, to the effect that, in the evolution of races or species, proteins may acquire higher serological complexity are difficult to comprehend from the viewpoint of chemical constitution (61a, 61b).

Another way for the differentiation of proteins from closely related species, less widely applicable for technical reasons, is the principle of cross-immunization devised by Uhlenhuth, who was able by injection of rabbits with the serum of hares to prepare immune sera precipitating hare serum but without action on rabbit serum.[21] In this method the formation of antibodies, which would react with the protein of the immunized animal, is suppressed, or, if such antibodies are formed, they must be neutralized in the body, the procedure then being equivalent to an absorption experiment *in vivo*.

A considerable number of data having been accumulated, the question arises to what extent the differences between species found in serological tests can be taken as a measure of zoological relationships.

---

[17] (11), (42), (49), (50), (52–57). The differentiation of the sera of man and lower monkeys, for which Fujiwara (58) recommends the absorption method, can also be carried out without absorption by comparing the strength of the reactions.

[18] (59–61).

[19] v. (3), (42), (51).

[20] In such tests the possible presence of multiple antigens should be borne in mind [v. (20)] and the fact that precipitin reactions are inhibited by an excess of the homologous or a related antigen.

[21] (62). In other cases cited by Uhlenhuth (man-lower apes, pigeon-chicken) the species are not very closely related and the procedure is not required. The principle was applied in the preparation of an immune serum by injecting a chimpanzee with human serum (63), and in the sensitization of Cavia porcellus with serum of Cavia rufescens (64).

To some degree this appears to be the case, and even application to undecided questions of taxonomy would seem possible. But, apart from the differences in the various quantitative methods which can lead to divergent results, it should be pointed out that immune sera vary in the range and strength of the cross reactions when produced in individuals of the same species, or obtained from the same animal at different times (p. 145).[22] The significance of the results can be improved by employing a number of sera and by preparing immune sera for each of the antigens to be compared (41). The extent of interspecies cross reactions is not the same with diverse types of proteins (e.g., serum globulins, thyroglobulins), but there is no evidence for disproportional variation which would give different relations according to the protein selected.

Among the factors which modify the quality of antibodies is the mode of immunization. The amount of antibody becomes greater on continued injection and, at the same time, there is an increase in the number and intensity of the reactions. That this is not always a qualitative difference was shown by Satoh (17) who, in examining several such sera rich in antibody, found that upon dilution the cross reactions [23] diminished or disappeared and the diluted sera resembled in specificity those with low antibody content, a fact which may be taken advantage of in practical work [Kister et al.; v. (18)].

It would seem that also the degree of relationship between the animal species furnishing the antigen and the one used for immunization has influence on the results. Thus, several authors concluded from tests with rabbit immune sera that rats and mice are widely distant while dissimilar species of birds, like chicken, pigeon, goose, seemed closely related.[24] Probably this is, so to speak, a case of faulty perspective, and the somewhat paradoxical results are to be explained in accordance with the principle illustrated by Uhlenhuth's "cross immunization." From this it may be understood why rabbit immune sera are very useful for revealing dissimilarities in the proteins of other rodents, while in the case of birds, if rabbit sera are employed, the lesser differences may be hidden by the predominant structural similarities of bird proteins.

The differentiation of mouse and of rat species has been attained — in keeping with morphological classification — by direct precipitin tests with

[22] Manteufel and Beger (65), Levine and Moody (66).
[23] (27), (42), (49), (50), (67).
[24] v. (68), (69).

rabbit antisera,[25] and in anaphylactic (Schultz-Dale) tests which, Moody asserts, permit somewhat sharper distinction [26] when the highest effective antigen dilution is taken as criterion.

Besides the regular cross reactions of proteins derived from related animals, others occurring with unrelated species have repeatedly been reported [27] and sera were found which, owing to antibodies of low specificity, reacted to some degree with all mammalian sera tested (v. pp. 43, 44).[28] At any rate, these exceptions have little significance for the general notion of protein specificity which, despite occasional claims (81), has been firmly established with the aid of highly specific precipitins.

The skin reactions observed in allergic human beings, extending sometimes unexpectedly to all mammalian sera tested,[29] can be due to antibodies of wide reaction range whose formation is perhaps favored by long continued sensitization; multiple sensitization may be a contributory factor.[30]

As already seen, species specificity is not restricted to serum proteins. Thus, precipitins can be prepared (Leblanc, Ide, Demees) which distinguish the hemoglobins [31] (and globins) [32] and hemocyanins [33] of various kinds of animals. One can safely assert that the differences depend upon the globin, the prosthetic group probably being the same in all hemoglobins; in accord is the precipitation of methemoglobin, CO- and CN- hemoglobins by antihemoglobin sera. In inhibition reactions observed with heterologous hemoglobins (83), the heme present in all hemoglobins may play a part; the small number of prosthetic groups could, as Marrack suggests, explain why there is no significant cross-precipitation between hemoglobins of distant species. With hemoglobin it was possible to demonstrate species differences by several methods other than immunological,

---

[25] (66), (69), (70).   [26] v. Roesli (71).
[27] Cf. (65), (72); for bibliography v. (15).
[28] (1a), (73-79). In certain instances anomalous reactions are conceivably caused by substances in the serum other than the serum proteins. Schiff (80) obtained group-specific precipitin reactions in human sera of group A with immune sera produced with blood cells of group A.
[29] Simon (82).   [30] v. (71).
[31] Heidelberger and Landsteiner (83), Higashi; Fujiwara (84), Hektoen and Boor (85); (86).
[32] Gay and Robertson (87), Ottensooser and Strauss (88), Hektoen and Schulhof; Johnson and Bradley (89), Browning and Wilson (90). Globin antisera react with hemoglobin as well, and Went et al. (91) recommend these sera for the forensic examination of blood stains.
[33] Hooker and Boyd (92).

For a long time it has been known that the crystals of various hemoglobins differ in shape, and Reichert and Brown (93) have carried out systematic investigations on hemoglobin crystals and found, in conformity with the serological results, that the shapes and angles are characteristic for each species, and that the differences stand in relation to the distance between the species in the zoological system.[34] The validity of these conclusions is unquestionable, even though the measurements cannot claim a high degree of accuracy, and the data are not detailed enough to estimate the errors in measurement or the degree of variation in individuals or species. Species differences of hemoglobins, furthermore, were established by other methods, e.g., absorption spectra [35] and solubility determinations. The latter investigations [36] were based on the principle that the solubility of a given substance is not affected by the presence of other substances provided they do not react with each other. Accordingly, the solubility of one hemoglobin in a saturated, aqueous solution of another hemoglobin ought to be the same as in water, and preliminary investigations actually were in agreement with this expectation, except in the case of two closely related animals (horse-donkey) where the formation of mixed crystals can be assumed. Other animal proteins found to be more or less species specific are muscle proteins,[37] fibrinogen,[38] seromucoid (109), and the proteins of eggs [39] and of milk (15) [as ovomucoid (109), casein (112), (113)], and to these the tissue proteins, as yet incompletely investigated, will surely be added.[40] Thus from the mass of evidence one may derive with reasonable confidence the rule that proteins, with some exceptions perhaps, are different even in related species, though the variations can be too small to be detected by present methods, a qualification of this statement being that it may not hold where relationships are so close as to make difficult a taxonomic distinction between species and varieties or races. Failure to find differences under some given conditions obviously cannot be treated as conclusive proof for the identity of proteins, as is sometimes alleged.

In agreement with the general principle are the few electrophoretic experiments so far performed which showed a difference between the ovalbumins

---

[34] With regard to some inconsistencies v. (94).

[35] (95–97). On molecular weight, refractive index, v. (p. 65), (98), (99); on hemocyanins (100).

[36] Landsteiner and Heidelberger (101).

[37] (15), (102), (103).

[38] (104–108).

[39] (15), (110), (111).

[40] v. (p. 80); Witebsky (114).

of chicken and goose and to a slight degree even between chicken and guinea hen (115). That this method can be less sensitive than serological tests is seen from the fact that ovalbumins (chicken, turkey) which are readily distinguishable by precipitin tests showed little or no distinction in the electrophoresis apparatus of Tiselius.

Some hemoglobins of unrelated mammals tested at a certain pH displayed very similar electrophoretic mobility. Species differences in hemocyanins were found electrophoretically and with regard to isoelectric points by Pedersen (115a).

Keratins [41] and lens proteins are often cited as lacking species specificity. It has been shown, however, that the keratins of wool and human hair, brought in solution by thioglycolate (Goddard and Michaelis), can easily be differentiated in precipitin tests; [42] whether hair keratins of less distant species are distinguishable is still to be investigated. The proteins of the optic lens are usually quoted as a prototype ever since Uhlenhuth (118) made the interesting observation that precipitins prepared with the lens of one animal react with lens substances of most diverse origin. [43] But here also there are differences which may be considerable, as between mammals, birds and fish, and a higher differentiation is indicated by the difficulty in immunizing or sensitizing animals with lens of their own kind. According to Witebsky, [44] the similarity of the lens substances is attributable, in part at least, to the presence of a special common lipid, which masks existing protein differences. Such a condition was found by Witebsky and Steinfeld with brain tissue (p. 81). In the latter case, species specificity of the proteins can be presumed because attempts to produce antibodies in rabbits with unaltered rabbit brain were unsuccessful, while antibodies were easily obtained with material from other species, which reacted with rabbit brain also (v.p. 103). [45]

Marked overlapping reactions were observed with thyroglobulin, [46] and it may be thought that the presence of thyroxine and diiodotyro-

---

[41] Several examinations of hair and dander were prompted by their role as allergic agents (v. Longcope et al.; Forster (115b)).

[42] Pillemer, Ecker and Wells (116); (117).

[43] (119–129). Serological differentiation of several protein constituents of the lens is dealt with in some of these papers. Cross reactions with other tissues are described by Gotoh (125).

[44] (128); v. (129).

[45] (130); Landsteiner and van der Scheer (unpublished experiments).

[46] Hektoen et al. (131).

sine residues in all thyroglobulins increases the tendency for cross reactions.[47] Definite species differences in keeping with the zoological relationships were demonstrated, however, by Stokinger and Heidelberger (134).

Rather extensive investigations [48] were carried out aimed at utilizing serological reactions in the field of systematic botany, but the results have been controversial (140), owing to intrinsic difficulties and, in part, to the use by a group of workers of a probably fictitious method, namely tests with so-called "Kunstsera" (alleged to be artificial antisera).[49] The conditions are less favorable with botanical material than in the investigation of animal proteins where the task was facilitated by the convenient use of whole blood serum. Blood serum, even when derived from various animal species, contains proteins of similar types, without other substances which seriously interfere with the reading of the specific reactions. On the other hand, in testing plant extracts trouble arises because of the nature of the material, which often consists of quite different sorts of proteins, along with other substances, as lipids,[50] organic acids, tannin, and the like, apt to cause errors through the formation of precipitates; for this reason the anaphylactic method may be advisable.[51] For reliable conclusions it would seem best to compare purified proteins of the same chemical type, derived from different plants, and such tests, for instance, on prolamines of the wheat and corn group,[52] and on legumes, gave results analogous to those obtained with animal proteins as to the parallelism of serological and systematic relationship. On the whole, this principle is borne out also by the work with crude plant extracts.

PROTEINS WITHIN A SPECIES. — The diverse proteins in one species, different in their composition and chemical properties, in part discussed above, in general are quite different serologically,[53]

---

[47] From experiments of Adant et al. (132) and Snapper (133) there are (with materials from different species) no group reactions between thyroglobulin and artificial iodoproteins (p. 194) which also contain diiodotyrosine (134).

[48] A review by Chester (135) gives a compilation of the voluminous literature; v. also (136–139).

[49] v. (141).

[50] v. (142).

[51] (143), (144).

[52] Lewis and Wells (145).

[53] With respect to enzymes with organ- or species specific activity, described by Abderhalden, which lie outside the scope of this review, the reader is referred to (146).

and, in consequence, the serum reactions reveal a twofold specificity, that of the particular protein and, for each one, that of the species. In blood serum, the most thoroughly investigated material, a great variety of proteins has now been demonstrated [54] by the methods of fractional precipitation, electrophoresis and ultracentrifugation.[55] Electrophoretic examination revealed in serum globulin three molecular species, a, β, γ-globulins, further separable into euglobulin and pseudoglobulin fractions. From seemingly homogeneous crystallized serum albumin, two crystalline albumins have been isolated, one with much carbohydrate, the other practically carbohydrate-free, and the fraction not precipitated by half saturation with ammonium sulfate was found to contain a globulin (globoglycoid), seroglycoid and seromucoid (possibly related to seroglycoid), all rich in carbohydrate, and a small amount of haemocuprein, a copper containing protein; plasma contains in addition fibrinogen.

Most of the proteins in this list, which is not complete and hardly final,[56] have been subjected to serological analysis. Serum globulin and albumin from one species behave like different antigens,[57] much more so than globulins or albumins, respectively, of closely related species which, as may be inferred from the data given above on whole sera, show a high degree of overlapping. Hence, it has been possible by means of precipitin tests to verify the questioned chemical individuality of albumins and globulins and to disprove the reports on artificial preparation of globulin from albumin.[58] Besides, fibrinogen, three globulins,[59] two albumins, seromucoid, globoglycoid and sero-

---

[54] v. Cohn (147).

[55] It is not certain whether all fractions that have been separated are different chemical individuals in the sense of having a distinctive amino acid make-up, since differences can result from the state of aggregation or from association with proteins or other substances (lipids, fatty acids, etc.) [cf. Marrack (148) Went et al. (149), Tiselius (150)]. The distribution of lipids (and carbohydrates) in various serum proteins has been examined by Kleczkowski (151) and Blix et al. (152); a and β-globulins were found to have a much higher lipid (and carbohydrate) content than albumin and γ-globulin.

[56] Actually two special protein fractions present in serum in small quantities have been recognized by Ecker and Pillemer as components of complement (p. 43 ff.). Cf. also (153), (154).

[57] Leblanc (155), Michaelis (156), Dale and Hartley (157), Doerr and Berger (158), Hektoen and Welker (159); (160–162), (68).

[58] Fanconi; Hooker and Boyd (163).

[59] Kendall (164); v. Harris and Eagle (165). Because of the cross reactions of Kendall's sera, the differentiation had to be made with immune sera rendered more specific by absorption. According to Hewitt (166) [v. (149)] the difference in the properties of euglobulin and pseudoglobulin is not, as was once assumed (167) attributable merely to the high lipid content of the former.

glycoid have been shown to be serologically distinct.[60] As for the differentiation of related species by antisera, the method of partial absorption is frequently used in this work. The question of actual cross reactions between various serum proteins, other than those due to imperfect purification, has been systematically investigated by Kendall [v. Haurowitz (170)]. Likewise, in eggs [61] and milk [62] a number of different antigens were demonstrable.[63] Casein, fractionated by chemical means,[64] was shown in the ultracentrifuge to consist of three components with the molecular weights of 98000, 188000 and 375000, and the heterogeneity of casein could be established furthermore by electrophoretic examination which gave evidence of three components differing in phosphorus content [Mellander (180)]. The three fractions of different molecular weight were found to be distinguishable by precipitin tests.[65, 66]

Upon injection of whole serum, antibodies for globulins and albumins are produced and can be separated by partial precipitation. In such an experiment, after addition of a sufficient amount of horse globulin to an anti-horse immune serum possessing precipitins for both globulin and albumin, the supernatant fluid obtained by centrifuging no longer acted on globulin, though still with undiminished intensity on albumin.[67] These, and results of Harris and Eagle, and Kendall, with globulin fractions support the view that, whether free or associated,[68] the serologically distinct proteins are not artifacts resulting from the fractionation procedures [69] — which is consonant with evidence from such mild separation methods as electrophoresis and ultracentrifugation.

[60] Kekwick, Gell and Yuill (168), Hewitt (169).
[61] Hektoen and Cole (171), Jukes and Kay (172), Roepke (173).
[62] Bauer (112), Wells and Osborne (174).
[63] Cf. Longsworth, Cannan and McInnes (175), Pedersen (176), Woodman (177). Since ovalbumin is often used in serological studies, it should be remarked that even several times recrystallized ovalbumin is not entirely homogeneous [Longsworth; Pappenheimer (177a)].
[64] Linderström-Lang (178); v. Carpenter (179).
[65] Carpenter and Hucker (181); v. Demanez (182).
[66] Other proteins found to be inhomogeneous are hemoglobin (182a), (182b); clupein (182c); pepsin (182d); globin (598); hemocyanin (92).
[67] (68), (158).
[68] Sörensen (183); v. Reiner (183a).
[69] This is not contradicted by the finding of Marrack and Duff (184) that the water-soluble and insoluble fractions of globulin failed in completely exhausting sera prepared to whole globulin, their conclusion being that these fractions do not exist as such in whole serum.

Animal proteins not yet considered, which have been examined serologically, are glucoproteins (p. 27), silk,[70] collagen,[71] ferritin,[72] and the pathological substances amyloid,[73] and Bence Jones proteins of which at least two varieties could be distinguished in serum tests.[74] Other observations concern peculiar proteins in pathological conditions, two of them myeloma cases.[75]

A protein having the peculiar property of precipitating the C polysaccharide of pneumococci was discovered by MacLeod and Avery (193) in the albumin fraction of sera of patients suffering from various infectious diseases [v. (193a)].

Tests for proteins in urine are mentioned in references 194–196, saliva in 197.

Proteins of worms are discussed by Campbell;[76] for immunity reactions to animal parasites the monographs by Taliaferro (204) and by Culbertson (205) may be consulted.

Evidence that constant serological properties characterize the proteins of a species has been obtained on so large a scale that one is naturally skeptical as to changes in their composition by external conditions. Dirr's far-reaching claim that on administration of arginine, tyrosine, or histidine, serum proteins are formed which contain an increased proportion of these amino acids has been refuted,[77] and the assertions made regarding the modification of proteins by diverse diets or other environmental factors are not convincing in view of the inhomogeneity of the protein preparations examined, and are contradicted by experiments of Abderhalden and others (208). Likewise on technical grounds a number of reports on pathological changes are to be regarded as questionable.[78]

PLANT PROTEINS. — Important work on plant proteins, using chiefly the anaphylaxis method, was carried out by Wells, Osborne

---

[70] Silk antisera are obtainable, not easily, upon injecting silk brought into solution by means of HCl and adsorbed to charcoal (185). Fell (185a) injected silk dissolved by lithium bromide; the sera produced are said to precipitate only after the tests have been kept for hours.

[71] Loiseleur and Urbain (186).

[72] Michaelis and Granick (186a). (Ferritin is species-specific but is identical in various organs.)

[73] Raubitschek (187).

[74] v. Hektoen and Welker (188), Bayne-Jones and Wilson (189).

[75] Everett, Bayne-Jones and Wilson (190), Welker and Hektoen (191), Packalén (192).

[76] (198); v. Taliaferro (199); (200–202). Precipitin tests with extracts of malaria parasites were demonstrated by Dulaney and House (203).

[77] Leipert and Loucoupoulos (206), Block (207).

[78] v. (209), (209a).

and coworkers [79] with the idea of correlating serological and chemical properties. The investigations are of importance because of the careful purification of the substances tested, in which respect certain types of plant proteins offer especial advantages, such as ready crystallization (edestin and similar globulins), or the solubility in dilute alcohol characteristic of prolamines (gliadin, hordein, zein). Quite a number of proteins, including globulins, legumins, prolamines and proteoses were examined. As with animals, proteins derived from the same plant (e.g., two sorts of globulins in leguminous seeds) could be differentiated, and distinct as well as overlapping reactions were observed with chemically similar proteins from various species.[80]

In the search for the substances causing allergic symptoms in asthma and hay fever patients, fractions, mostly regarded as proteins, have been prepared from animal and plant materials,[81] in particular from pollens,[82] that were effective in eliciting skin reactions in allergic individuals. The protein nature of the active agents in pollens has been disputed mainly on account of the resistance to enzymatic digestion,[83] but discrepant results are on record. Abramson and coworkers (220), who examined ragweed pollen extracts by electrophoresis, ultracentrifugation and diffusion, estimated the molecular weight of the active components to be about 5000, and these authors and Rockwell (221) believe the substances to be polypeptides, found by the latter workers to contain a large percentage of basic amino acids. Of interest but not yet explained are observations on skin reactivity, seemingly of low degree, to nucleic acids, nucleotides and purines, of individuals hypersensitive to ragweed.[84]

GLYCOPROTEINS. — The low antigenic power of most mucoids and mucins makes it difficult to obtain immune sera, and the sera may give fallacious reactions because of antibodies formed to contaminating antigens of higher activity. Yet it has been possible, upon prolonged immunization, to secure antisera in a small proportion of

[79] (210); Jones and Gersdorff (211).
[80] A discordant result concerning inter-reactions of gliadin and glutenin from wheat would deserve retesting by reactions in vitro.
[81] e.g., grains (212); cottonseed (212a), (212b); housedust allergens (212c).
[82] Review by Newell (213) (considers also carbohydrates); cf. (214), (214a), (215) (differentiation of various pollens); (216–218); (218a) (cross reactions of pollens).
[83] Kammen; Grove and Coca; Rockwell; Harsh and Huber (219).
[84] Winkenwerder et al. (222), Sherman (222a).

the animals injected.[85] In Goodner's experiments with carefully purified preparations the reactions with beef chondromucoid, beef submaxillary mucin and hog stomach mucin were specific but cross reactions occurred between ovomucin and ovomucoid; ovomucin gave better immunization results than the other glycoproteins.[86]

Apart from glycoproteins in the usual sense of the term, many proteins such as egg albumins and certain serum proteins [87] contain carbohydrates as part of the molecule and the idea has been advanced, suggested evidently by the discovery of bacterial polysaccharides, that the specificity of proteins is determined by these carbohydrates [Bierry et al. (228)]; in fact, Bierry claims to have found differences between the carbohydrates in serum proteins of various animal species. This hypothesis cannot be of general validity in view of the existence of species specific proteins which contain no carbohydrate, e.g., hemoglobin. Another point is that carbohydrates, apparently identical in structure, have been isolated from proteins of widely diverse origin. A carbohydrate, prepared from ovalbumin, was found to be made up from units containing one molecule of glucosamine and two of mannose,[88] and Rimington [89] obtained a polysaccharide of the same composition from horse serum and ox serum proteins.[90] This indicates the wide distribution of the substance and is an argument against the significance of carbohydrates for the species specificity of proteins. If it is true that only a few different prosthetic carbohydrate groups exist in mucins, the same conclusion would hold for these glycoproteins.[91]

It will be desirable at any rate to isolate and study the carbohydrates from the serum of a number of animal species in order to settle definitely the question of species differences in these substances. A certain serological significance of the carbohydrate groups in proteins is suggested by some overlapping between ovomucoid and seromucoid (109), and by experiments of Sevag and Seastone (240) in

[85] Elliott (223), Goodner (224); v. (225).
[86] On mucin of frog eggs v. Uhlenhuth and Wurm (226).
[87] A high proportion of carbohydrates was found in proteins of body fluids of lower animals (227), (228).
[88] Levene, Mori and Rothen (229), Hewitt (230); v. Neuberger (231), Sörensen (232); (233).
[89] (234); v. (235).
[90] v. (232). According to Hewitt (236), [v. (228)], the carbohydrate of horse proteins contains mannose, acetylhexosamine and galactose, which were also recovered from ovomucoid (237).
[91] (228); v. (239).

which anaphylactic shock could be elicited by a polysaccharide preparation (containing 11% N) from egg white in guinea pigs sensitized to egg proteins. The carbohydrate did not sensitize guinea pigs; the quantities necessary for shocking were about 0.5 mg., i.e., considerably more than a shocking dose of egg white. Ferry and Levy (241) on the contrary were unable to induce shock with the carbohydrate, in guinea pigs passively sensitized to egg white, and found that it neither gave precipitation with immune sera for egg white nor inhibited the precipitation of egg white by such antisera. Similar negative results were obtained with the carbohydrates of serum globulin, ovalbumin and ovomucoid, prepared with precautions for avoiding chemical alteration (as deacetylation), which confirms the conclusion that species specificity of proteins is not a function of their carbohydrate groups.[92]

Toxins.[93] — The discovery of antitoxins gave the first proof of the existence of specific antibodies and therefore toxins hold, historically, a prominent place among antigens. That toxins are of protein nature had been rather generally believed and well established for toxins of plant origin (ricin, abrin),[94] but with other toxins clear-cut chemical results have been attained only in recent years. One animal toxin, crotoxin,[95] from the rattlesnake Crotalus terrificus, has now been crystallized by Slotta and Fraenkel-Conrat (249) in the form of well defined quadratic platelets. It is a protein that in the ultracentrifuge and by diffusion measurements appeared to be a homogeneous substance of a molecular weight of 30000.[96] The toxin has the high S content of 4%, cystine and methionine are present in the proportion of 6:1, and there is probably in addition an unknown sulphur compound in small quantity. Cysteine destroys the toxicity, an indication that S-S linkages participate in the composition of the active groups. Inasmuch as the purified toxin seemed to be uniform and possessed both the toxic and hemolytic activity of the original venom, and since the latter is referable to a lecithinase (v. pp. 228, 39), it was submitted that the neurotoxic effect, too, is caused by an enzymatic attack on the nervous tissues.[97] The uniformity of the

[92] Coghill and Creighton (242), Neuberger (243); v. (244).
[93] On endotoxins v. (p. 7, 222).
[94] Osborne et al. (245); Karrer et al. (246) (crotin).
[95] For more complete information on venoms v. Kellaway (247), Kraus and Werner (248).
[96] Gralén and Svedberg (250).
[97] A toxin produced by Cl. welchii was found to act, like venom hemolysins,

crystallized crotoxin is contested by Ghosh and De who reported that they were able to effect a partial separation of the hemolysin and the neurotoxin (253).

Purification of another snake toxin, cobra venom,[98] yielded a preparation with 14.7% N and as much as 5.5% S. The carbon content of 45.2% is significantly low, but otherwise the substance has all distinctive protein properties. By the rate of dialysis through cellophane membranes the molecular weight was estimated to be 2500–4000; if this is correct and the substance is antigenic, as might be supposed, the result would have importance in establishing that so low a molecular size suffices for antigenicity. A low molecular weight assigned to scarlet fever toxin (257) has been questioned (257a).

Scorpion toxin was likewise found to have a high percentage of S (258) and bee poison is very rich in tryptophane (8%).[99] Thus some of the analytical data seem to point to peculiarities in the chemical composition of toxins but, in general, the question is not definitely decided whether the activity of toxins depends merely on a special combination of the amino acids occurring in ordinary proteins, or whether there exist toxins whose reactivity is due to prosthetic groups of a different nature, as in some enzymes, e.g., riboflavin phosphoric acid in the yellow respiratory enzyme.[100]

The above studies did not confirm the results of Faust who had claimed that venoms are saponins, but an N-free preparation, probably not uniform, has been described as the toxic constituent of Bothrops venom.[101]

Considerable progress has lately been made in the isolation and chemical characterization of a bacterial toxin,[102] diphtheria toxin,

---

as lecithinase [Macfarlane et al.; van Heyningen (251)], and enzymatic activity has been suggested to explain the effect of toxins in general [Herbert and Todd (251a)]. Conversely, it has been conjectured that toxins may act by affecting enzymes in the body; actually, inhibition of an enzyme by venom was described by Chain (252).

[98] Micheel et al. (254); v. (255), (256). Remarkable is the high content up to 5–6% of zinc in venoms (256a). While toxins are, as a rule, heat labile, cobra venom was found to be practically unimpaired by heating in slightly acid solution for an hour in a water bath; and the heat resistance of the venom has been utilized for liberating it from the combination with antitoxin [Calmette, Morgenroth (256b)]. Another exception is the relatively heat resistant toxin of Cl. botulinum. See also (256c) (partial restitution of bacterial hemotoxins after inactivation by heat).

[99] (259).   [100] v. (260).
[101] Klobusitzky (261).
[102] See the review on bacterial toxins by Eaton (262). The early work on purification of toxins is reviewed in (471). The preparation of bacterial toxins for

aided by the use of improved, simplified culture media and control of substances which in small quantities promote or inhibit growth and toxin production.[103] By adsorption of impurities and precipitation with ammonium sulfate or heavy metal salts Eaton, and Pappenheimer,[104] succeeded in separating diphtheria toxin in practically pure condition,[105] having the typical attributes of a protein (with about 9% tyrosine), and a molecular weight of about 74000.[106] The toxin is heat coagulable and easily destroyed by acid; the preparation contained only a small amount of atoxic protein and the properties and composition were not changed by further fractionation. For guinea pigs of 250 gm. the lethal dose was 10^{-4} mg. The figures for some purified toxins are given in Table 5. Although it is seen that several toxins possess the same activity as diphtheria toxin (or even a higher

TABLE 5

Toxin	Animal	Minimal lethal dose for 1 kg. weight
		mg
Diphtheria	Guinea pig	4×10^{-4}
Ricin	Rabbit	1×10^{-3} to 5×10^{-4}
Botulinus	Mouse	2×10^{-4}
Tetanus	Mouse	4×10^{-4}
Tetanus	Guinea pig	1.5×10^{-4}

one), it cannot be said that other bacterial toxins have been isolated in a state of approximate purity, with the probable exception of streptolysin (251a) which has been well purified and appears to be a protein; statements concerning the preparation of protein-free bacterial toxins [107] have in part been contradicted (257a).

Diphtheria anatoxin (see below) purified by Pope [108] contained 1500 flocculation units (Lf) per mg. N which compares fairly well with the values for purified toxins.

practical use in immunization and animal experiments is described in Wadsworth's "Standard Methods" (262a); v. (262b) (methods for concentration).

[103] J. H. Mueller (263).

[104] (264), (265), (270); v. Wadsworth (266), Linggood (267).

[105] On criteria for purity v. Eaton (264).

[106] Petermann and Pappenheimer (268).

[107] On tetanus toxin v. (269), (270); dysentery toxin (271); scarlet fever toxin (272).

[108] (273); v. Theorell (274), Schmidt and Hansen (274a), Reiner (274b); (262b).

Among alterations of toxins by chemical agents,[109] the treatment with formaldehyde [110] is of practical importance since toxoids made in this manner (anatoxins) which are not toxic but still antigenic and reactive with immune sera — a behaviour ascribed by Ehrlich to blocking of special toxophore groups — are widely used for immunization purposes. According to Velluz (278) the reaction of tryptophane groups with formaldehyde leading to ring closure is the significant chemical change in the transformation of toxins into atoxic products. Carbon disulfide detoxicates tetanus toxin, not diphtheria toxin, the difference being ascribed to the absence of SH groups in the latter.[111]

Characteristic of certain toxins, e.g., hemolysins of streptococci, Cl. welchii, is the reversible inactivation by oxygen [Todd (280)]; the participation of SH groups in the process is discussed by Herbert and Todd (251a); (v. 467).

In connection with the chemical data it should be noted that, apart from the production of multiple toxins, including hemotoxins and leucocidins, by a given bacterium, various strains of the same bacterial species may produce different toxins, e.g. streptococci,[112] Cl. botulinum,[113] C. diphtheriae,[114] Cl. tetani,[115] Cl. welchii.[116] The existence of "toxons" with neurotropic affinity in diphtheria toxin is refuted by Prigge (286). A further instance of complexity is Forssman's observation on hemotoxins of staphylococci having unlike effects on various sorts of erythrocytes.[117]

The reactions of toxins with immune sera are carried out as neutralization tests, or by precipitation as in the case of venoms and Ramon's flocculation test (288) for diphtheria toxin and toxoid, which is used also for other toxins (scarlatina, dysentery, etc.).[118] Cross tests with the limited choice of bacterial toxins have been made

---

[109] v. (275), (599), (275a); (275b) (formation of toxoid by peptic digestion). The inactivation by acids is, under certain conditions, reversible [Morgenroth, Doerr, Hallauer (275c)].
[110] Eaton (262), (276), Pappenheimer (265), Hewitt (277), Haas (277a); v. (p. 46). On acetylation with ketene, treatment with phenylisocyanate and isothiocyanates etc., v. (278), (278a), (265).
[111] (279), (265); v. (p. 48).
[112] (280), (281).
[113] (282).
[114] Behring; (283), (284).
[115] (284a).
[116] (285).
[117] (287); v. (287a).
[118] Methods for the assay of antitoxins are presented in Otto and Hetsch (289).

rather to substantiate the general principle of antibody specificity, while venoms from numerous species have been examined [119] for overlapping reactions, since cross neutralization is desired in anti-venin therapy when, as frequently happens, the species of the attacking snake is not known. Sharp distinction was found between the venoms of the families Colubridae and Viperidae, which, however, is no proof of species differences in one type of protein, because the chief toxic agents of cobra venom and those of other Colubridae are neurotoxins, while in the venoms of Viperidae (Crotalus) hemorrhagins are effective components. (In general, the secretion of a given species often contains several sorts of toxic substances such as neurotoxins, hemorrhagins, coagulins, hemolysins, and in different venoms these components are present in unlike proportions.) Nevertheless, besides cross reactions of various degrees, true species differences are commonly found in corresponding toxins (e.g., neurotoxins, or hemotoxins) even of closely allied snakes. The investigations are not systematic enough to allow one to form an accurate opinion on the extent of parallelism between zoological and serological relationships — as is the rule with proteins and would be presumed for venoms too — but discrepancies have been recorded.[120]

An interesting specificity phenomenon is exhibited by bacterial hemolysins, namely their unequal activities when tested with various kinds of red cells.[121] In experiments of Neisser and Wechsberg, and Todd, rabbit cells, for example, were found to be rather resistant to a lysin of B. megatherium but were readily hemolyzed by staphylolysin; human blood was sensitive to megatheriolysin and little affected by staphylolysin (v. p. 32).

BACTERIAL PROTEINS.[122] — Although coagulable proteins were demonstrated in the "press juice" of bacteria long ago [Buchner (301)], chemical and serological work has been done to a large part on nucleoproteins,[123] prominent constituents of bacteria and responsible, it is reasonably assumed, for their affinity, like that of cell

---

[119] (290–295).
[120] v. (295).
[121] (296), (297), (287), (600); (297a) (review). References on bacterial hemolysins and hemagglutinins (of low potency but perhaps of pathological significance) are quoted in (297b); v. (297c). On leucocidins (leucoagglutinins) v. (297d).
[122] Data on proteins of bacteria and yeasts and molds are given in Buchanan et al. (298); v. (299), (300).
[123] v. (302), (307).

nuclei, for basic dyes. As regards the method of preparation, of importance for the properties of the products, the customary acid precipitation of nucleoproteins from alkaline extracts has been replaced in recent studies by procedures designed to avoid alteration — grinding and extraction in the cold,[124] sonic disintegration,[125] enzymatic autolysis.[126]

While some of the preparations separated are considered to represent salt-like combinations between proteins and nucleic acid [Zittle et al. (308)], evidence for more firmly bound nucleic acid has been produced.[127] In pneumococci nucleic acid of the ribose type has been identified by Thompson and Dubos (307); besides, thymonucleic acid has been recognized in bacteria.[128] The protein portion of pneumococcus nucleoprotein was found to be probably of low molecular weight, not heat-coagulable, and deficient in aromatic amino acids (Thompson and Dubos).[129]

Detailed data which cannot well be summarized briefly have been recorded on the composition and serological properties of proteins and nucleoproteins, particularly of tubercle bacilli [130] and streptococci,[131] including Lancefield's type specific, weakly antigenic M-protein, characteristic of hemolytic streptococci of group A, which has been well purified and separated from its original combination with ribose nucleic acid [Zittle (310)]. As an illustration of the complicated conditions encountered, the fractionation of tubercle bacilli proteins by Menzel and Heidelberger may be cited (318). Despite careful fractionation the isolation of homogeneous proteins was not achieved with certainty; the presence of at least three antigenic components could be demonstrated. Corresponding fractions of avian tubercle and timothy-grass bacilli were serologically different from one another and from those of human or bovine strains, and proteins of human and bovine tubercle bacilli, too, were differentiable.[132]

The proteins of V. cholerae and other vibrios were investigated by

[124] Heidelberger and Scherp (303); v. Macfadyen et al. (304).
[125] Mudd et al. (305).
[126] Nelson (306), Thompson and Dubos (307).
[127] Heidelberger (303), Sevag and Smolens (309).
[128] v. Zittle (310).
[129] On complexes of nucleoproteins with carbohydrates in streptococci v. Heidelberger et al. (311).
[130] Seibert (312), Coghill et al. (313).
[131] Lancefield (314), Sevag et al. (315) (601), Henriksen and Heidelberger (316), Mudd et al.; Zittle et al. (308), Stamp et al. (317).
[132] Seibert (319); Schaefer (320).

Linton and his colleagues [133] and Bruce White (322), and various fractions were prepared, some characterized by solubility in acid alcohol.[134] Upon applying to vibrio proteins the method of racemization by alkali outlined below (p. 116), these organisms could be classified into two groups which were further subdivided on the basis of diversity of polysaccharides, giving six groups in all (Linton). Two sorts of curves were obtained by plotting optical rotation of the proteins against time of treatment with alkali, and when the racemized proteins were hydrolyzed with acid, differences were observed; for instance, from one type of proteins optically active leucine, and from the second optically inactive leucine, was recovered, while with glutamic acid the result was reversed.

Further studies to be mentioned on the proteins of bacteria were conducted with staphylococci,[135] salmonella bacilli,[136] Bact. coli,[137] C. diphtheriae (328), H. pertussis (329), Cl. parabotulinum (330). The antigens of spores have not been examined chemically (331).

Whereas in the work with toxins culture filtrates are commonly employed, it has been ascertained in several instances, either by producing antitoxins with washed bacteria or by extraction, that these toxic proteins are ingredients of the cells.[138]

Substances which have attracted much attention on account of their diagnostic usefulness are certain proteins passing into culture media, such as tuberculin, mallein, trichophytin.[139] Numerous experiments have been carried out on the chemical nature of tuberculin [140] which induces specific systemic and skin reactions in tuberculous individuals and in animals infected with tubercle bacilli or treated, under certain conditions, with killed bacilli.[141] As stated by Seibert (336), tuberculin after removal of concomitant polysaccharide and nucleic acid is a mixture of proteins in which two practically homo-

---

[133] Linton (321) (review).
[134] v. Landsteiner and Levine (323). Using dilute or acidified dilute alcohol, proteins could be extracted also from Salmonella bacilli and other bacteria [Furth and Landsteiner (324), Bruce White (325), Goadby (326)].
[135] (602) (type-specific protein).
[136] (324), (325), (327).
[137] (327a) (protein containing a high proportion of basic amino acids).
[138] Prigge (332), Eaton (262), Marton et al. (332a).
[139] Similarly, other preparations from fungi and oidia are employed for skin tests in dermatomycoses [Sulzberger (333), Jadassohn et al. (333a); v. (333b)].
[140] The literature is reviewed by Seibert [The chemistry of the proteins of the acid-fast bacilli (312)].
[141] Petroff et al.; Saenz; Freund et al. (334), Flahiff (335).

geneous active fractions, of different electrophoretic mobilities, with the molecular weights of 16000 and 32000 could be identified.[142] The same author has described active proteins of still smaller molecular size, capable of inhibiting the precipitin reactions of tubercle bacillus protein. The substances having higher molecular weight were antigenic, engendering precipitins; those of small molecular size could be rendered antigenic by adsorption to aluminum hydroxide or charcoal.[143]

Concerning the species specificity of bacterial proteins the occurrence of cross reactions between kindred and even distantly related bacteria should be pointed out. Examples are the cross reactions of the proteins of gonococci, meningococci and N. catarrhalis, their reactivity with anti-pneumococcus sera,[144] and overlapping reactions within the salmonella group.[145]

Mention may finally be made of precipitin reactions observed by Lackmann, Mudd et al. (341) of pneumococcus sera with nucleic acids. These reactions did not take place with normal sera and rarely with immune sera other than those to pneumococci. It could not be determined that the globulin in the antiserum which gives the reaction is formed as a response to the injected pneumococcus nucleoprotein. The reactions were inhibited by purines; in the sensitivity to changes in ionic strength they differed from common precipitin reactions.

HORMONES. — Those hormones that are proteins, either simple or, like prolan, which contain active prosthetic groups, would by presumption be endowed with antigenic capacity, as are proteins in general. A contrary opinion was offered on hypothetical grounds (342), and in some instances antibodies observable by reactions in vitro, after injection of hormonal preparations, have probably been mistaken for antibodies to hormones while actually directed towards admixed tissue antigens. Precipitins for thyroglobulin were prepared and thoroughly studied by Stokinger and Heidelberger (343), who established the composition of the precipitates through iodine determinations and characterized the precipitin reactions of thyroglobulin by the unusually low antigen-antibody ratios, owing to the high mo-

---

[142] Seibert, Pedersen and Tiselius; (336). Two fractions of tuberculin, one producing systemic, the second skin reactions, have been described by Maschmann (337); v. Dorset et al. (338).

[143] Seibert; (339).

[144] Boor and Miller (340).

[145] (324), (306).

lecular weight of the antigen. The reports on insulin,[146] including anaphylactic experiments [147] and instances of human allergy, leave not much doubt as regards antigenic capacity, although, fortunately for therapeutic use, of rather low degree. It may be speculated that if large portions of the molecule, responsible for the biological activity, were much alike in the insulins of various species this could tend to reduce species specificity, as reported by Lewis, and at the same time antigenicity. The experiments of Wasserman and co-workers hardly constitute stringent proof of the identity of insulins derived from various species.

After continued administration of various hormones neutralizing agents can appear in the serum for which the term "anti-hormones" has been adopted. This was well shown in the original experiments of Collip and Anderson (348) who in this way induced a resistance in rats to thyrotropic hormone; the sera of these animals and of a treated horse inhibited the action of the hormone in tests on hypophysectomized and normal rats. The anti-hormones developed to thyrotropic and gonadotropic hormones, both high molecular substances, are the best studied instances; in cross tests these anti-hormones were seen to exhibit specificity.

The moot question in this line of research — supposing the mechanism is always the same — is whether the neutralizing substances found in the serum are antibodies in the serological sense or, according to Collip's theory, antagonistic hormones present in small amounts also in normal animals. The subject and the literature have been critically discussed by this author and by Thompson.[148] In favor of the antibody theory is the fact that neutralizing sera have not been reliably demonstrated against low-molecular hormones, e.g. adrenalin or estrin (v. p. 181), and of similar import are observations on the species specificity of anti-hormones including frequent failures to produce anti-hormones with homologous preparations. Thus Zondek and Sulman remark, "antiprolan is exclusively elaborated after a type of prolan foreign to the species has been administered." [149] Divergent results, however, were obtained by some investigators. The most

[146] v. review by Harten and Walzer (344) (gives references on human allergy to insulin, other hormones and enzyme preparations); Wasserman et al. (345); Yasuna; Lowell (345a) (report on neutralizing antibodies). Immune sera produced with conjugates of insulin are mentioned on p. 62.
[147] (346), (347).
[148] (349), (350); v. (351).
[149] v. (352).

definite evidence should come from the chemical identification of anti-hormones, and in studies along this line the antithyrotropic and antigonadotropic principles were recovered, like antibodies, in globulin fractions of the sera (Thompson; Harington et al., [150] Zondek et al.[151]), the latter group of workers noting differences between antiprolan and the common run of antibodies.

The presence of precipitins or complement fixing antibodies along with the neutralizing principle does not necessarily bear directly upon the problem, since in several cases they were found to be not connected with the neutralizing property and to depend possibly upon protein contaminants (355–357).

ENZYMES. — As can be gathered from older reviews,[152] despite repeated assertions of the existence of anti-enzymes,[153] much of the evidence, derived largely from neutralization tests, was regarded as dubious. The suspicion prevailed that immune sera which were induced by and gave precipitin reactions with impure enzyme solutions were reacting not on the enzymes themselves but on accompanying proteins, and that reduction of enzymatic activity was brought about by unspecific adsorption to the precipitates. With the discovery of crystalline enzymes (Sumner, Northrop, Kunitz, etc.) the situation was changed. Not only was the antigenic material thereby made available in higher concentration and much superior in purity, permitting precipitin reactions in the virtual absence of contaminations — previously only neutralization tests had seemed appropriate — but the conclusion, now widely accepted, that many enzymes are proteins made the existence of anti-enzymes seem highly probable.

Indeed, precipitating immune sera have been prepared or anaphylaxis tests performed with the following crystalline enzymes: urease,[154] pepsin, pepsinogen,[155] trypsin, chymotrypsin, chymotrypsinogen (the last three differentiable from each other [156]), catalase,[157]

---

[150] (353).
[151] (354), v. (355).
[152] Wells (358), Pick and Silberstein (359).
[153] v. (360–364).
[154] Kirch and Sumner (365), Pillemer, Ecker et al. (366).
[155] Northrop; Seastone et al. (367). The reactions had to be carried out with alkali-denatured pepsin. The one-sided cross reaction between pepsin and pepsinogen is to be noted.
[156] TenBroeck (368).
[157] Campbell et al. (369), Tria (369a). On anti-hyaluronidase v. (370); on anti-papain (371). Haas (372) reported precipitation and neutralization of papain by immune sera.

and ribonuclease.[158] Species specificity was demonstrated with pepsin, pepsinogen and trypsin. It is perhaps of somewhat lower order than that of serum proteins (367), yet TenBroeck was able in anaphylaxis tests with sensitized uteri to distinguish pig and beef trypsin.

For the demonstration of species differences between gastric enzymes, the action on synthetic peptides as substrates has been utilized by Fruton and Bergmann (373).

With yellow enzyme Várterész and Kestyüs (374) failed to produce any sort of antibodies. This result alone is hardly sufficient for their inference that the protein part of the enzyme is the same in all animals.

Curiously, and in striking distinction to antitoxins, antibodies for enzymes often do not neutralize the biological activity of the antigens. This was again found to be true in a recent study by Adams (374a) on precipitins for tyrosinase from mushrooms. A positive instance is the inhibition of lecithinase activity by antivenin.[159] Antiurease partially inhibits the decomposition of urea by the enzyme, but this is ascribed (Kirk and Sumner) mainly to the decrease in dispersion since the urease precipitates still retain most of the original activity.[160] Whether antisera neutralize pepsin is doubtful, and catalase is practically unimpaired when mixed with anti-catalase which was taken to indicate that the active prosthetic groups of the enzyme — two molecules of hematin per molecule of catalase and another unidentified group — are not directly involved in the antigen-antibody combination. Thus the criticism about the early work was not wholly unjustified as regards neutralizing antibodies, but the conclusion of very weak or wanting antigenicity of enzymes is not warranted in view of the formation of precipitins which, from the accumulated evidence, cannot be attributed to contaminants even though crystallization of an enzyme is not a rigorous proof of purity.

Immune sera made for the thromboplastic lipoprotein from lung did not inactivate its coagulating activity (375a).

---

[158] Smolens and Sevag (368a).

[159] Morgenroth (375); v. (285).

[160] If the common definition of toxins as antigenic poisons is taken literally, urease would be included. The toxicity of urease, however, is caused by the decomposition of urea in the blood and it acts without an incubation period. Chickens which have but little urea in the blood are resistant to urease but succumb immediately if urea is injected at the same time [v. Sumner (376)]. Antiurease protects animals against the toxic action of the enzyme.

VIRUSES.[161] — From the evidence at present available, viruses can be put into two groups according to their chemical composition, those that have been separated in the form of crystalline or paracrystalline nucleoproteins of very large molecular size, as tobacco mosaic and other plant viruses, and composite viruses represented by the elementary bodies of vaccinia which have been found to contain proteins, thymonucleic acid, fat, lipids, riboflavin, biotin and a copper constituent.[162] Consequently, antibodies to the former group should be similar to other antiproteins while the latter, like bacteria or animal cells, may engender immune sera of greater complexity.

The serological reactions of viruses include neutralization tests in vivo, agglutination or precipitation, and complement fixation.[163] In precipitin tests the viruses with elongated, rod-shaped particles, e.g., tobacco mosaic virus, rapidly form voluminous precipitates, the less elongated ones densely packed precipitates, a difference similar to that between the agglutination of motile and non-motile bacteria (380).

The small amounts of virus which may be contained in infected tissues prompted Merrill (387) to estimate the number of particles of various sizes necessary for visible agglutination or precipitation and for the production of antibodies. From experimental data, about 0.001 mg. per cc. would be needed for perceptible aggregation, comprising for example the following number of particles, which clearly increases as the particle size diminishes: red blood cells 4×10^5, B. paratyphos. 4×10^7, vaccinia virus 7.7×10^8, poliomyelitis virus 1.9×10^{12} (calculated value), ovalbumin 1.8×10^{13}.

The given figures indicate that with some of the smallest viruses, such as poliomyelitis virus, it is difficult to produce agglutination or precipitation reactions (378), and the foremost technical task in this sort of work is the concentration of the viruses and purification from host material.

The specificity of the reactions is such that directly or by absorption tests even various strains of the same virus (tobacco mosaic, in-

---

[161] Additional information is presented in the reviews by Rivers (377), Craigie (378), Smith (379), Bawden (380), Bawden and Pirie (381); v. Chester (382). A review by Burnet et al. (383) deals in detail with the reactions of bacteriophages. Of interest are the striking differences found in the antigenic activity of various bacteriophages.

[162] Hoagland, Ward, Smadel and Rivers (384). The presence of several enzymes in the elementary bodies may be due to adsorption from the host tissues; the report on a specific polysaccharide (384a) is unconfirmed.

[163] With tobacco mosaic virus anaphylaxis tests by the Schultz-Dale method were unsuccessful, but typical responses were secured upon intravenous injection [v. Seastone et al. (385)].

fluenza virus, bacteriophages, etc.) can be differentiated, and on the other hand by cross tests unsuspected relationships between viruses have been disclosed. An example of serological relations as revealed by cross absorption is reproduced in the following table from Bawden and Pirie [164] where the letters designate "antigenic fractions."

Tobacco mosaic virus	A, B
Aucuba mosaic virus	A, B, C
Enation mosaic virus	A, C, D

[Concerning the interpretation of such results, the same remarks apply as those made in connection with other cross reactions (p. 114).]

A correlation is indicated between amino acid composition and serological relationship by experiments of Knight and Stanley (388a). Thus, tobacco mosaic and aucuba mosaic viruses are similar in both respects, whereas cucumber viruses (3 and 4) and the rib-grass strain of tobacco mosaic virus differ considerably from these serologically and in chemical composition, particularly in tryptophane and phenylalanine content; the latter virus deviates also in its content of histidine, tyrosine and sulfur. Some chemical differences — viz. in the proportion of arginine — were later found between serologically related strains.

Enzymatic removal of nucleic acid apparently does not change the precipitin reaction of tobacco mosaic virus (386).

In the group of "complex" viruses, vaccinia virus has been most thoroughly examined as regards immunological properties. It has been found to contain a specific protein molecule displaying two serological properties, one easily destroyed by heating, the other more heat resistant; in addition, an antigenic nucleoprotein was detected in the virus. Both these antigens contribute to the agglutinable property of the virus and give precipitin and complement fixation reactions when dissolved. Besides, there is possibly another agglutinogen as well as an antigen eliciting neutralizing antibodies.[165] Soluble specific substances have likewise been detected in other viruses such as those of infectious myxomatosis (which were demonstrable also in the sera of diseased animals),[166] lymphocytic choriomeningitis,[167] and in bacteriophages.[168]

---

[164] (388); v. (380).
[165] Shedlovsky, Smadel, Rivers, Hoagland (389), (390).
[166] Rivers and Ward (391).
[167] Smadel et al. (392).   [168] Burnet (383).

The possibility that a precipitable substance in the serum of monkeys infected with yellow fever virus is derived from the altered tissues of the host is suggested by Hughes (393).

CHEMICALLY MODIFIED PROTEINS

DENATURATION. — The immunological behavior of proteins is affected in different ways by chemical alterations. Apparently the first artificial modification of antigens to have been studied was a process peculiar to proteins, namely denaturation.[169] This, most conveniently carried out by heating of protein solutions, has a definite influence on serological properties.

Although loss of solubility [170] is the most conspicuous effect, denaturation cannot be considered merely as a "physical" modification since chemical differences have been detected between native and denatured protein, in particular liberation of SH and other reducing groups.[171] A final theory of denaturation has not yet been reached; according to prevalent views there occurs breakage of relatively unstable bonds that hold the peptide chains of the protein molecule in a fixed spatial configuration, followed by an irregular rearrangement through the formation of new bonds.[172] The nature of the linkages involved (hydrogen bonds according to Mirsky and Pauling, salt linkages) is still under discussion [v. (401)].

The published results of precipitin and anaphylaxis tests with denatured proteins [173] vary in details, presumably owing to differences in the preparation of the antigens and the individual antisera, yet most investigators agree upon the marked serological difference between native and denatured protein, remarkable since the fundamental "peptide structure" is supposed not to be affected in the process. Antisera to native protein, in general, react much less, if at all, with the denatured products; in the reverse procedure cross reactions have been observed by several investigators.

[169] A presentation of the subject is given by Anson (394).

[170] When solutions of protein are heated above a certain temperature at weakly alkaline reaction, they remain clear but the protein has become insoluble at the isoelectric point and will coagulate upon slight acidification; at a proper pH, therefore, heat coagulation occurs directly, as if in a single step.

[171] v. (395), (531).

[172] Wu (396), Mirsky and Pauling (397); v. (398-400).

[173] A summary is given by Hartley (402); Wu, TenBroeck et al.; Schmidt, and others (403-412); v. (413).

On changes in immunological properties by irradiation studied with a view to the therapeutic use of foreign protein v. Henry (413a).

In experiments of the author (414) immune sera were produced by means of heat-coagulated crystalline ovalbumins or horse serum albumin, which, despite the insolubility of the injection material, succeeded without difficulty even though definite reduction of antigenic activity was noted. These sera were tested against both the native and a solution of denatured protein and compared with antisera for the native protein. The anti-"denatured" immune serum precipitated the denatured and scarcely the native protein, while the anti- "native" immune sera reacted weakly also with denatured ovalbumin,[174] and this was not attributable to reversal or to remnants of denatured albumin in the native antigen used for immunization. Restitution of the original serological properties on reversal of denaturation was described by Spiegel-Adolf.[175]

Horse serum albumin after denaturation by urea and seeming reversion to the undenatured state showed its antigenicity to be still greatly reduced but its specificity to be not different from the native protein (Neurath et al.).

A moderate decrease in species specificity upon denaturation has frequently been observed. An experiment is reproduced in Table 6

TABLE 6. — PRECIPITATION OF VARIOUS ANTIGENS BY AN IMMUNE SERUM FOR HORSE SERUM DENATURED BY HYDROCHLORIC ACID (MAMMALIAN REACTION)

[after Landsteiner and Prásek (407)]

(0.2 cc antigen diluted 1:100, 0.05 cc immune serum, in narrow tubes. Reading after 1 hour at room temperature.)

Sera	Horse	Ox	Man	Chicken	Rabbit*
Unaltered	±	o	o	o	o
Heated	±	o	o	o	o
Treated with hydrochloric acid	++	+	+	o	o

(In this and the following tables increasing degrees of precipitation are expressed by the symbols: o, ± (trace), ±, +, +, +±, ++, etc.)

* The relatively lower capacity of altered protein of an animal to react with immune sera produced in the same species is a general rule.

which shows that an immune serum made against acid albumin from horse serum reacted distinctly with the corresponding preparations from other mammalian sera (bovine and human), though not with

[174] v. (410), (415). In tests made by Bawden and Kleczkowski (528) unheated serum globulin inhibited the precipitation of heated globulin by antiserum made with the latter. The reasons for the difference in precipitin reactions of native and heated proteins are discussed by these workers. It may be of significance in this connection that the difference between unaltered and denatured proteins was not demonstrable in film reactions of proteins (p. 57).

[175] (406); cf. Miller; Hewitt (416), Neurath et al. (417).

chicken acid albumin. So far, a second manifestation of diminished species specificity, namely the production of antibodies following injection of altered protein of the same species, has not been described for acid albumin; but Furth found that when rabbits were injected with heated ox serum, some of the antisera reacted not only with heated sera of other mammals, but weakly with heated rabbit serum also (Table 7); and Uwazumi (412), immunizing rabbits with rabbit serum heated to 120°, obtained antisera which precipitated similarly treated serum proteins of this and various other mammalian species.

TABLE 7. — PRECIPITATION OF VARIOUS NATIVE AND HEATED SERA BY AN IMMUNE SERUM FOR OX SERUM HEATED TO 100°C (MAMMALIAN REACTION) (DILUTION OF THE ANTIGENS 1:1000)

[after Furth (410)]

Sera	Ox	Sheep	Horse	Man	Rabbit	Guinea Pig
Heated	+++	+++	++	++	+	±
Unheated	+++	++	o	o	o	o

In a recent study (414) immune sera for denatured horse globulin gave no exceptional cross reactions, and sera prepared with coagulated ovalbumin, while giving stronger cross reactions with denatured heterologous ovalbumins than those for native ovalbumin, still differentiated the proteins of kindred species (chicken, guinea hen, etc.) and showed sharp specificity upon absorption with heterologous antigens, disproving again the recurring assertions in the literature that denatured proteins are devoid of specific properties [v. (417a), (417b)].

HYDROLYTIC AGENTS. — Treatment with hydrolytic agents, as digestive enzymes, alkalis [176] or acids, apart from primary denaturation effects, decreases or destroys the antigenic activity, alkalis being more effective than acids.[177] As found by Johnson and Wormall, serum albumin loses its reactivity with precipitins within 24 hours if kept at pH 13 and a temperature of 19° C. and, under similar conditions, it loses its immunizing capacity, for reasons not fully understood, although protein of high molecular weight is still present in the solutions.

Dakin explained the loss of antigenicity following treatment with

[176] Dakin (418), TenBroeck (419), Landsteiner and Barron (420), Johnson and Wormall (421).
[177] v. (402).

alkali by the resistance of racemized proteins to proteolytic enzymes so that after injection into animals they are eliminated in the urine unchanged.[178] According to Lin, Wu and Chen (425), however, proteins treated with alkali are not completely resistant to digestion. It might also be held that upon treating proteins with alkali, structures of importance for the antigenic function are destroyed, as it is not known what constitutional features of proteins determine their immunizing activity. The existence of protein derivatives (acylated proteins) resistant to pepsin and trypsin, yet antigenic, does not rule out Dakin's suggestion since these antigens might be attacked by enzymes in the animal body; it is of greater significance, as Hartley (402) pointed out, that antigenic activity is markedly reduced by alkali even before racemization has become complete, and that the antigenic capacity of proteins altered by alkali is restored, to some degree, by nitration or iodination.[179]

PROTEOSES. — The products of hydrolytic cleavage of proteins have been repeatedly examined for immunological activity.[180] The want of methods for separating proteoses as chemical individuals is certainly among the reasons for the vagueness of our knowledge in this field. Amino acids or small peptides, the ultimate products of hydrolysis by enzymes, acids or alkalis, are inactive as antigens [181] and by and large the attempts at producing antibodies for proteoses have given unsatisfactory results.[182] Still, in a number of instances [183] anaphylaxis and formation of antibodies could be induced, for instance with peptic heteroproteoses adsorbed to charcoal. The heteroproteose antisera reacted with the proteose used for immunization, with peptic "metaprotein," and to some extent with the original protein, and cross tests showed species specificity. Anaphylactic sensitization was attained in more recent experiments with proteoses made from various sources.[184] Furthermore, there are reports on sensitization of human beings to proteoses (albumoses) occurring after administration of bacterial culture fluids.[185]

[178] Objections to Dakin's conclusions have been raised by Kober (422); v. (358), (423), (424). [179] (420), (421).
[180] v. Fink (426), Wells (358), Pick and Silberstein (359), Hartley (402).
[181] Inconclusive is the evidence for anaphylactic sensitization by a synthetic peptide [Abderhalden (427)]. The experiments on production of specific enzymes, by injection of peptides (v. 428), are here left out of account.
[182] A report on antigenicity of digested gelatin (429) is open to doubt until it is shown that the antibodies were not directed towards the added enzyme.
[183] (430), (431), (426).
[184] Stull and Hampton (432). [185] (433); v. (434).

Even short peptic digestion has been seen to destroy the capacity to react with precipitins for the unaltered protein at a time when the solution yet contained coagulable material; antisera obtained by injection of peptic metaprotein precipitated also the unaltered protein.[186] In experiments of Holiday (437) precipitability was still apparent after 5 minutes' digestion but was lost in 30 minutes, the solution then partially inhibiting [187] the precipitation of the unaltered serum albumin (v. p. 196).

PLASTEIN. — Loss of specificity was supposed to occur in the formation of so-called plasteins which have been described as proteins arising from proteoses by enzymatic synthesis.[188] Plasteins were reported to be serologically alike when the albumoses used were derived from entirely different sources.[189] This result is difficult to understand because synthesis starting from different albumoses would hardly be expected to yield such similar products. An explanation may exist in the serological work of Flosdorf, Mudd et al. (442) which led them to believe that the antigenicity of plastein is due to enzymes present in the preparations (besides traces of denatured protein) rather than to the synthesis of an antigen. The actual nature of plasteins is still debated. Against the conclusion that plasteins are synthesized proteins, are determinations by Folley in the ultracentrifuge which pointed to a molecular weight of not more than 1000, while experiments by Collier, who criticizes this result, were interpreted as favoring the hypothesis of enzymatic synthesis, although his product was not a typical protein. At any rate, the anaphylactic experiments of Collier showed a very slight degree of antigenicity at best, in practical agreement with Flosdorf.[190]

TREATMENT WITH FORMALDEHYDE, BENZALDEHYDE, NINHYDRIN. — An example of a chemical change with relatively small effect on antigenic properties is the treatment of proteins with formaldehyde, a fact which is made use of in immunization against toxins (v. p. 32) and some viruses. The reaction is more complicated than simply the formation of methylene compounds of the type $CH_2=N-CHR-CO-$, as has been assumed; in any event free amino groups are blocked, chiefly the ϵ-NH_2 in the lysin, in addition to changes in other groups such as indol nuclei and imidazol rings.[191] In part the reaction

---

[186] (435), (436); v. (432); (471) (tryptic digestion).
[187] For these inhibition reactions v. (p. 182).
[188] v. (438–440).
[189] Knaffl-Lenz and Pick (441).
[190] (443–445).   [191] (446–453).

is reversible, which has been shown also by the reactivation of for-molized tobacco mosaic virus.

On testing formolized serum proteins the writer found no pro-nounced change in specificity in the ordinary precipitin reaction,[192] and antisera to the formolized antigen precipitated the native protein. Yet a change was recognizable (455) as rabbit serum after formoliza-tion became antigenic for the rabbit and antibodies were obtained which reacted with the modified protein. Substantially in accord are experiments of Jacobs et al.[193] who believe that the greater deteriora-tion of species specificity observed by Horsfall (458) is due to pro-tracted treatment with concomitant gelatinization. Furthermore, no appreciable serological effect of blocking NH_2 groups by formalde-hyde or removing them by nitrous acid was seen with virus proteins. Potato virus X treated in this manner became inactive, like viruses in general, but precipitated with antisera for the active virus and the denatured product engendered immune sera indistinguishable from the latter.[194]

Rabbit immune sera against horse serum proteins which had been treated with benzaldehyde either showed no characteristic features or precipitated benzaldehyde treated sera (horse, man) and, curiously, unchanged sera of all mammals tested.[195]

By means of ninhydrin, colored derivatives are obtained in which NH_2 groups are no longer demonstrable by formaldehyde titration. The product made from human serum gave cross reactions with similarly modified horse serum.[196]

The reaction of ninhydrin with protein is discussed by Eggerth (464) and compared with the effect of formaldehyde (v. p. 32).

OXIDATION. — Oxidative decomposition of proteins is easily achieved, for instance by means of potassium permanganate.[197] As Obermayer and Pick (465) reported, the so-called oxyprotsulfonic acids formed during the reaction had lost their precipitability with "native" antisera but immune sera prepared with the oxidized prod-uct showed that species specificity was not abolished, in spite of the considerable chemical change.

---

[192] (454), (455).
[193] (456); v. (457).
[194] Bawden (459), Bawden, Pirie and Spooner (460); v. Parker and Rivers (461).
[195] Mutsaars et al. (462).
[196] Dulière (463).
[197] On photo-oxidation v. (464a).

REDUCTION. — The sole chemical effect of reducing agents on proteins so far observed is the conversion of S-S into SH groups which in some cases (insulin, crotoxin) is accompanied by inactivation of hormone or toxin activity. In a serological study Blumenthal (466) found that the precipitability of serum albumin was decreased after reduction with thioglycolic acid, in her opinion owing either to participation of S-S and -SH groups in the serum reactions or to splitting of the molecule; egg albumin which fails to give the cyanide-nitroprusside test showed no serological modification after treatment with cysteine. On account of the irreversibility of the reduction of serum albumin Ecker [198] submits that other changes may be involved. Further instances of immunological differences between oxidized and reduced forms are proteins of the eye lens (124), keratins (116), and urease (366).

NITRATION, IODINATION. — Of greater interest than the chemical changes which mainly cause antigenic deterioration or moderate impairment of species specificity are those that lead to profound transformation of serological properties. Such reactions, discovered in important investigations of Obermayer and Pick (465) were found to result from treating proteins with nitric acid, nitrous acid, and iodine.[199] The effect common to these reactions is the loss of the original specificity, partial or complete, depending upon the intensity of the treatment, and the appearance of a new specificity or, as Pick [200] puts it:

"If, for example, a nitroprotein, the so-called xanthoprotein, is prepared from the protein of rabbit serum by treating it with concentrated nitric acid, it is an easy matter to immunize rabbits with it and to obtain an immune serum that does not differ in its action in any way from an immune precipitin which has been prepared with a xanthoprotein from a different species; both of the immune sera so obtained possess the capacity to react specifically with the xanthoproteins of the entire series of animals and even with plant xanthoproteins, whereas they have lost, more or less, the capacity to precipitate normal serum protein. Similar conditions obtain with iodized and diazotized proteins. In all cases the original species specificity is lost and a new specificity has taken its place . . ." (author's translation).

[198] (467) (Oxydation et réduction dans l'immunologie).

[199] In prior experiments by P. Th. Müller (468) sera to iodized casein reacted with the unchanged protein, and Bauer et al. (469) reported that casein differed from serum proteins in that the species specificity was not destroyed by iodination; v. (470).

[200] (471). Observations — with regard to casein treated with nitrous acid — in disagreement with those of other workers, were described by Lewis (472).

According to Pick, these serological modifications which, incidentally, do not affect merely species specificity, are determined "chiefly
by the character and the position of the substituting groups in the
aromatic nucleus, and the changes in the entire structure of the
molecule attending the process of substitution" (471). In the writings on this subject it is often assumed that the serological behaviour
of the three antigens mentioned depends directly on the nature of the
introduced group itself (NO_2, $N=N$, I). If so, the antibodies obtained with these antigens would be specific for the nitro- or diazo-
group or iodine, and should react with these substituents as such.
This, however, is not the case; in fact, Wormall's [201] investigations
made with the inhibition method (p. 183) led to the result that the
serological determinants in the reactions of iodo- and bromopro-
teins [202] are not iodine or bromine, but parts of the modified protein
molecule, that is to say the tyrosine residues disubstituted by halogen
in 3,5 position (p. 194). In similar manner Mutsaars (480) found
that the characteristic serological specificity of nitroproteins depends
on the nitrotyrosine residues. Concerning nitroproteins [203] it is important to note that, as shown in Table 8, this antigen is scarcely to be

TABLE 8. — PRECIPITATION OF VARIOUS ANTIGENS BY AN IMMUNE SERUM FOR
HORSE-XANTHOPROTEIN (CONCENTRATION OF THE ANTIGEN 0.01%)

Xanthoproteins		Diazoproteins		
Horse	Ox	Horse	Ox	Chicken
++	+++	+++	+++	+++

distinguished serologically from diazoprotein,[204] and Mutsaars (487)
found in absorption experiments that both these antigens react with
the same antibody. The two substituents NO_2 and $N=N$ being quite
different, it follows again that the nature of the substituting group is

[201] (473); v. Jacobs (474). On iodo- and nitroprotein v. (475–476), (492).
   [202] The conflicting results of Bruynoghe et al. (477) are probably attributable
to the drastic treatment with bromine in hot alkaline solution whereby, in addition to bromination, oxidation is apt to occur [v. (478), (479)]. The possibility
of oxidation by iodine and alkali is pointed out by Clutton et al. (479a). Under
the conditions used in bromination chlorine does not substitute (473).
   [203] (481), (475). Nitroproteins prepared with tetra-nitromethane have similar
properties to those made with nitric acid.
   [204] (407), (473). With reference to the formation of diazo compounds by the
action of nitrous acid on proteins and phenols v. Landsteiner (482), Morel and
Sisley (483), Rohrlich (484); as Eagle et al. (485) found, tryptophane may take
part in the reaction. On the participation of tryptophane in the nitration of proteins v. (486). The literature on nitration of proteins is abstracted in (480).

not the salient point. Moreover, bromo- and iodoproteins are only slightly different, and iodized nitroproteins and nitrated iodoproteins are said to be serologically distinct (471) although both contain the iodine and the nitro group.

The intense yellow color of both nitro- and diazoproteins cannot well be attributed only to the presence of a nitro- or diazo-group and seems to indicate that the substituted aromatic rings possess a quinoid structure.[205] Hence, the pronounced serological characteristics of, and the relationship between nitro- and diazoproteins very probably arise from a structural change in the tyrosine residue.[206] In regard to this it is significant that introduction of NO_2 groups or halogen need not alter radically the specificity of aromatic compounds (p. 163) and, as Wormall comments, the differences between halogenated and nitrated proteins need not depend solely on the particular substituent, since these modifications are not simply cases of substituting a different group in the same position.

It is remarkable that in the cases under consideration different proteins react alike, although their original chemical differences cannot have been abolished. Even though characteristic groupings may have been destroyed by the reactions one must, in order to explain this behaviour, assume that it is due to the predominance of structures formed by iodination or nitration which, occurring repeatedly in the molecule, largely mask the other chemical differences.[207] It may well be relevant that iodination greatly increases the acidity of the hydroxyl in tyrosine.[208]

Immune sera for iodoproteins may contain distinct antibodies, reacting with the homologous antigen alone, with all iodoproteins, and with the original protein, their relative proportion (and occurrence) depending on the degree of iodination.

Because of the inhibition reactions referred to above, it is not likely that in addition to the substitution of tyrosine other chemical modifications take a decisive part in the serological change following iodination or nitration. A quantitative study of the course of iodination showed that loss of species specificity became complete just when enough iodine was taken up to substitute all tyrosine groups.[209]

---

[205] v. (488), (489), (473).

[206] The so-called desamidoalbumin (471) obtained from xanthoprotein by reduction has not been studied sufficiently to be discussed; v. (481).

[207] Only slight cross reactions occur between halogenated and nitroproteins (473).

[208] Cohn (490).   [209] Kleczkowski (491); v. Haurowitz et al. (492).

METHYLATION, ACYLATION. — Nearly complete loss of the original specificity is not limited to reactions which involve substitution in the aromatic rings. The other modifications so far discussed, it is true, result in a less marked decrease of specificity and give rise to cross reactions extending only to species of the same zoological class (e.g. mammals) as that which furnished the immunizing antigen. However, the assumption that salt forming groups are of significance in serological reactions led to the detection of alterations which affect profoundly the specificity of proteins. From this point of view esterification of the acid groups in proteins was undertaken and was effected with acid in alcoholic solution, or with diazomethane.[210] With diazomethane, at least on intense treatment, in addition to esterification, hydroxyl-, amino- and imino-groups are methylated. The protein esters and methylated proteins behave in a fashion similar to xanthoproteins and iodoproteins.[211] Their capacity to react with immune sera for the unchanged protein is lost, and immune sera obtained with these derivatives react with other correspondingly treated proteins of various animal species or even plants. The original specificity of the proteins is not entirely destroyed by the alterations, for upon titrating with diminishing quantities of antibodies the reactions are seen to be considerably stronger with the homologous antigen than with preparations made from widely distant proteins; however, this is to some extent also true of iodized protein.[212] The cross reactions between proteins treated with alcoholic acids and with diazomethane are readily comprehensible on account of chemical similarity.

Acylation[213] (v. p. 157), for instance acetylation by means of acetic anhydride, brings about a specificity change which is analogous to and as pronounced as that due to methylation. Of the same order are the changes occasioned by reaction with carbobenzoxychloride $C_6H_5-CH_2-O-COCl$ [Gaunt and Wormall (502)].

Phosphorylated proteins, of interest in view of the existence of natural phosphoproteins (casein, vitellin), were investigated by Heidelberger et al.

[210] (407), (493); v. Edlbacher (494). Changes that occur, in addition to esterification, are considered by Kiesel et al. (495).
[211] On discrepant results obtained in anaphylaxis experiments with esterified glycine v. (496). [212] (473); v. (469).
[213] Landsteiner and Jablons (497); v. (498). It can be concluded from the analysis and the disappearance of the Millon reaction that NH_2 and OH groups take part in the acetylation [v. (499–501)]. On acetylation by ketene v. Herriott and Northrop; Pappenheimer (449).

(499). The immune serum to phosphorylated egg albumin gave cross reactions with denatured, not with native ovalbumin.

TREATMENT WITH DIAZOBENZENE, ISOCYANATES, MUSTARD GAS. — Treatment with diazobenzene was first investigated serologically by Obermayer and Pick, who found this to modify the specificity in such a manner that the antisera produced against the coupled product precipitated this antigen but not the unchanged protein nor proteins of other species likewise coupled with diazobenzene. In later experiments,[214] however, it was seen that sufficient treatment causes a considerable change in species specificity.

Antigens similar in serological behaviour to methyl- or acetyl-derivatives, save for distinct reactivity with antisera to the untreated proteins, are the phenylcarbamido-proteins formed by treatment with phenylisocyanate ($C_6H_5-N=CO$) and p-bromphenyl isocyanate; involved in the reactions are the free ϵ-amino groups of the lysine residues.[215] Less changed in specificity were the protein derivatives made by Berenblum and Wormall (507) with mustard gas ($\beta\beta'$-dichlorodiethyl sulphide) and dichlorodiethyl sulphone; here the combination may take place through amino or sulfhydryl groups.

CONJUGATION WITH SEROLOGICALLY REACTIVE ORGANIC COMPOUNDS. — The observations dealt with in the foregoing concern the specificity of altered protein molecules. A different effect, however, can be demonstrated when proteins are combined with appropriate organic compounds. Such compound antigens can be shown to exhibit a specificity determined by the attached chemical substances and independent of the protein moiety. These conjugated antigens will be discussed in a later chapter.

CROSS REACTIONS. — In order to interpret the overlapping reactions obtained upon testing precipitin sera with the proteins of related animal species an opinion is often held, based upon the assumption of absolute specificity of antibodies, which may be set forth in the following statement translated from Arrhenius: [216]

[214] Landsteiner and Lampl (503).
[215] Hopkins and Wormall (504); (505). Differences between the effects of formaldehyde and isocyanates, and of other substitutions in the amino groups of protein are considered by Marrack (505a) and Kleczkowski (457).
Treatment with methylisourea produced no great change in the immunological properties [Cohn (147)].
It should be mentioned that clupein, which contains no lysin, was found by Gutman (506) to react with phenylisocyanate.
[216] Arrhenius (508); v. Nicolle (509).

"Sheep serum probably contains, in addition to its principal constituent, several other substances which are also present in the sera of goats and cattle and which, upon injection into the veins of rabbits, incite the formation of antibodies against these sera, although in smaller quantity than the precipitin which is produced by the principal substance of the sheep serum."

This hypothesis does not explain the increased range of reactivity of immune sera made with artificially modified, e.g., denatured, proteins and, if applied to a large number of species instead of only a few, as in the example cited, leads to an inadmissible conclusion. One would then have to assume that every normal serum contains numerous proteins identical respectively with those of other species and that their quantities are determined by the degree of zoological relationship. Accordingly, the hemoglobin of an animal should show many different forms of crystals, which is contrary to the fact. The fallacy

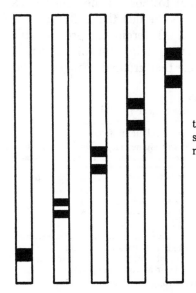

Electrophoretic separation of a mixture of guinea hen and duck ovalbumins; successive schlieren photographs at 30 minute intervals.

of the argument was furthermore evinced by electrophoretic analysis which brought out sharp, discontinuous differences in mobility between cross-precipitating proteins so that mixtures of two such proteins — guinea hen and duck ovalbumin — could be clearly resolved into the components.[217]

In seemingly more acceptable form the idea of like elements shared

[217] Landsteiner, Longsworth and van der Scheer (115).

among different species has been advanced by substituting groupings for whole proteins. For instance, since after exhaustion with duck ovalbumin, antisera for hen ovalbumin still react with the latter,[218] there must be at least two antibodies in play, and the inference might be drawn, accordingly, that hen ovalbumin contains two reacting groups, one peculiar to chicken, the other common to chicken and duck. Here again when this simple reasoning is applied to numerous species one arrives at a more than doubtful picture of protein constitution; for example, of the determinants in hen ovalbumin — at least four from the experiments with heterologous antigens in Tables 9a, 9b — three would be shared with turkey albumin and two be present also in guinea hen albumin, and so forth.[219] A much more likely explanation is provided by the principle established through investigations on azoproteins, namely, that the action of antibodies extends to structures that are chemically similar to those of the homologous antigen. In harmony with this opinion is the fact that following apparent exhaustion of an antiserum with a heterologous protein, still further addition may inhibit and retard the precipitation of the homologous antigen.[220] One may then assume that gradual changes in the

TABLE 9a

[after Landsteiner and van der Scheer (511)]

Hen ovalbumin immune serum absorbed with ovalbumin of	Ovalbumins from				
	Hen 1:2000	Turkey 1:2000	Guinea Hen 1:2000	Duck 1:2000	Goose 1:2000
Turkey	+++±	o	o	o	o
Guinea hen	+++	+±	o	o	o
Duck	+++±	+++	++±	o	o
Goose	+++±	+++	++±	o	o
Unabsorbed	+++±	+++	++±	++	++

Immune serum for hen ovalbumin was exhausted with various albumins. For the tests 0.2 cc of absorbed serum was added to 0.05 cc of antigen dilutions expressed in terms of dry weight. Readings were taken after 1 hour at room temperature.

With the unabsorbed sera all ovalbumins gave strong reactions.

determinants of related proteins rather than the presence or absence of certain groupings — in conjunction with the formation of qualitatively different antibodies in response to single determinants (p. 269) — will account for the complexity of the absorption phenomena.

[218] Hooker and Boyd (510).
[219] Cf. (333a) (Schultz-Dale experiments with trichophytins).
[220] (511), (512).

The multiplicity of antibodies, as well as the variety among antisera made in one species with the same antigen, is clearly seen from the tests presented; it appears that there are antibodies which react only with chicken ovalbumin while others precipitate chicken and turkey

TABLE 9b
(Tests as in table 9a)

Hen ovalbumin immune serum absorbed with ovalbumin of	Ovalbumins from				
	Hen 1:2000	Turkey 1:2000	Guinea hen 1:2000	Duck 1:2000	Goose 1:2000
Turkey	++±	0	0	+	0
Guinea hen	++±	++±	0	++	+±
Duck	+++	+++	+++	0	0
Goose	+++	+++	+++	++	0

albumins, et cetera. Accordingly, it is possible to distinguish the five albumins with selected sera against chicken ovalbumin.

PROTEIN SPECIFICITY AND CHEMICAL CONSTITUTION.[221] — Considering the scanty knowledge of the constitution of individual proteins, it is not surprising that, despite the abundance of observations on natural proteins and those on modified ones, the specificity of immunological protein reactions cannot yet be interpreted concretely in terms of chemical structure. Ehrlich's opinion that serological properties are intimately connected with chemical constitution is now conclusively established, and it is equally certain that the proteins themselves and not contaminating substances, a possibility once considered (514), are responsible for the reactions. The dependence of protein specificity on chemical structure, indeed probable because of the serological relationship of proteins of similar composition, became a certainty as a result of investigations on protein derivatives (Obermayer and Pick). The first point, the significance of the chemical structure rather than of biological origin, was stressed by Wells and Osborne (515) in their extensive work on plant proteins. For example, in the group of alcohol soluble proteins, gliadins from wheat and rye behaved alike, both chemically and in anaphylactic experiments, and hordein of barley and gliadin, differing slightly in composition, were serologically related although readily distinguishable. Or, the legumins of beans, vetches and lentils and the globulins of Cucumis melo and Cucurbita maxima, which could not be distin-

[221] v. (46), (513).

guished chemically with certainty, agreed also in their immune reactions (516).

That differences in particle size, which certainly affect the manifestation of antibody reactions, or in "micellar" structure, influence their specificity is not supported by reliable evidence. On the contrary, there are observations to show that changes in the state of dispersion need not be accompanied by differences in specificity.[222] The presence in proteins, depending on environmental conditions (in particular accompanying proteins and salt content), of smaller units resulting from reversible dissociation of the ordinary molecule, and the formation of aggregates by combination of different proteins was assumed by Sörensen (518), and positive observations of this kind have been made by ultracentrifugation and electrophoresis.[223] With the former technique it was found, for example, that the hemoglobin molecule may split into halves in urea solutions or upon dilution, and under the influence of clupein serum albumin may break up into eighths.[224] Phenomena, however, such as described by Sörensen, are not reflected in the serological behaviour, and one must conclude that the specificity of proteins is an attribute of the units, that is, the smallest components which cannot reversibly be split. Otherwise, by means of rather gentle operations, such as salting out of proteins, it should be possible to produce or destroy serological characteristics, or by mixing different proteins to create new specifically reacting "component systems," which is not the case. A similar view has been taken by Hooker and Boyd (523).

It was claimed by Zoet (524) that on immunizing with a previously heated mixture of horse and pig serum the antisera reacted markedly with this solution and scarcely with a mixture of the separately heated sera, which would indicate that on heating aggregates are formed consisting of both proteins and endowed with a new serological property. In experiments of Nigg (525) the appearance of a new specificity could not be confirmed; however, as was shown by Landsteiner and Prášek (526) and by Bawden and Kleczkowski, loss of precipitability by immune serum may occur when a protein is heated with a large amount of another protein, owing to the formation of mixed aggregates.[225] Formation of complexes upon heating

---

[222] (454), (517).
[223] MacInnes and Longsworth (519). For aggregation of like or even dissimilar hemocyanin units v. Tiselius and Horsfall (520).
[224] v. Svedberg (521), Bernal (522).
[225] (527); antigenicity and the ability to fix complement remained unimpaired (528).

protein mixtures was also demonstrated by ultracentrifugation [226] and salt precipitation.

Immune sera prepared against specific precipitates react with both components (530).

If in trying to form an idea of the chemical basis of protein specificity one adopts the current theory that proteins are substantially made up of amino acids in peptide linkage, then their specificity obviously depends in some way on the assortment and arrangement of these components. Here a new situation arose from the application of physico-chemical methods to the study of proteins (Svedberg's ultracentrifugation, diffusion measurements) which established, for many proteins, a "globular" (ellipsoidal) shape of the molecule and indicated a higher organization than that of a mere peptide chain or a system of several chains.[227] It would then be possible and has in fact been contended that the configuration resulting from the coiling up of the chain is important for the specific reactivity, and it has been further submitted that the considerable change in serologic properties upon denaturation is caused by the breaking up of the weak bonds which hold the chain in its spatial position.[228]

Experiments bearing upon this issue were made with monomolecular layers of proteins; these showed that extended films of a thickness of about 8 Å are capable of absorbing antibodies specifically,[229] as determined by the increase in thickness. Consequently, the globular shape of the protein molecule is not a necessary condition for specific reactivity but, as Pauling remarks,[230] if one assumes unleafing of a layer structure, a two-dimensional configuration of the folded chain could persist in the monolayers.

Furthermore, specific reactions are given by fibrous proteins, keratin and silk which according to present knowledge are essentially long polypeptide chains,[231] and from this and especially in view of specific

---

[226] van der Scheer et al. (529).

[227] Chibnall (531). On the cyclol theory of protein constitution see Wrinch (532); on diketo-piperazines as constituents of proteins see (394).

[228] Wu (396), Mirsky and Pauling (397). If both ideas are correct it requires explanation why denatured proteins still exhibit pronounced, although more or less diminished, species specificity (p. 43). From the symmetry of the molecules, estimated by viscosity measurements, Bull (533) concludes that in denatured globular proteins the peptide chain is still "greatly folded." To be considered is the fact that altered proteins, e.g. azoproteins, in which the original configuration must be affected are nevertheless not without species specific properties.

[229] Rothen and Landsteiner (414).

[230] (534); v. (535).

[231] v. (535).

reactions obtained with hydrolytic products of proteins (p. 196) it may be concluded that polypeptide chains, irrespective of any configuration due to intramolecular folds, are sufficient to provide a specific pattern for the combination with protein antibodies. To what extent in the reactions with native globular proteins the "secondary" configuration plays a part remains to be decided.

One cannot safely offer an opinion concerning the specific groups of proteins ("determinants") as long as it is not known what the maximal size of such a determinant can be. Evidently, determinant groups of small size (known to be sufficient for reactions of conjugated antigens), i.e. amino acids or di- or tripeptides, cannot furnish a sufficient number of combinations [232] to account for the vast variety of protein reactions, and the specificity, in order to afford a pattern characteristic for a single species, must be referable to more complicated structures, probably a number of these existing in the molecule. The thought that the specific effect is brought about by a summation of partial reactions, in other words by the joint action of a number of antibodies severally directed towards small groups, is in disagreement with the low incidence of overlapping reactions between unrelated proteins, which are infrequent even when many antigens are used for immunization at the same time.[233] For another thing, antibodies specifically adjusted to amino acid residues are not demonstrable in immune sera produced against proteins or low molecular peptides.

The existence of multiple binding groups, not improbable a priori on account of the large size of protein molecules, is borne out by quantitative examination of precipitates which established that, for example, ovalbumin (molecular weight about 40,000) can unite with antibody in the ratio 1:5 and therefore has at least five and possibly more binding areas (p. 243), a fact which at the same time fixes an upper limit for the size of the combining sites in a typical protein. For proteins of large molecular size much higher ratios have been found, up to more than 100 in the case of a hemocyanin (molecular weight about 6,000,000).

The intricate problem of identifying the determinants in proteins would be simplified to a certain extent if there were a repetition of similar deter-

---

[232] The number of combinations, allowing for repetition and for differences in the arrangement, is a^n for groups of n amino acids, a representing the number of different amino acids.

[233] v. Hektoen et al. (536), von Gara (537), Roesli (71), Dean et al. (538), Björneboe (539); cf. pp. 103, 104, 140.

minant patterns, i.e. groups of amino acids, in the molecule. Such a condition would not necessarily require, but could be due to, a periodicity in the distribution of the residues.[234] Under the assumption of recurrence of the same amino acids at constant intervals along one chain — which reduces enormously the number of "possible" natural proteins — it is hard to understand the existence of the small serological differences between the proteins of closely related species; however, the difficulty would be moderated if one might suppose independent periodicities in the components of a "system of peptide chains of varied composition" (531).

Evidence relating to the size of the specifically reacting groups in proteins has been acquired through inhibition tests with dialyzable proteoses [235] and with split products of silk which consisted of not more than 8–12 amino acids [Landsteiner (543)], (p. 196). It is probably not too much to say that further work along this line (supplemented by the study of artificial antigens containing known peptides) should ultimately lead to definite information about the specific groupings of proteins. Granted the validity of our experiments, they would seem to support the hypothesis of recurring determinants.

With conjugated antigens it has been shown that peptides (made up of glycine and leucine) are serologically characterized. Whether in proteins special portions of the molecule, because of their chemical nature,[236] preferentially function as determinants is still a matter for speculation; it may be assumed that the groups involved are located . on the surface of the molecule (polar groups will tend to be oriented outward in aqueous solution). A major influence of free acid groups may be surmised from observations on artificial antigens and from the marked effect of esterification on protein specificity; [237] on the other hand, determinants within the chain appeared to play a part in the specificity even of relatively small peptides. The view that aromatic groups have prominent significance was advanced by Obermayer and Pick. Their conclusion that the "species specific structure of the protein molecule is influenced mainly by groupings which are

[234] Bergmann and Niemann (540); Astbury (540a); v. (531).

[235] Landsteiner and van der Scheer, and Chase (541), Holiday (542).

[236] v. Marrack (148). Results obtained with protein derivatives suggest that certain parts of the molecule are of greater influence than others since some chemical modifications of proteins, as treatment with formaldehyde which combines with NH_2 groups, affect the specificity but slightly.

[237] Felix and Reindl (544) assume on the basis of methoxyl determinations on esterified gliadin that the free carboxyl groups do not belong exclusively to the dicarboxylic amino acids. [Cf. (545) and the studies of Chibnall (531) on acid and basic groups in proteins.]

connected with the aromatic nuclei of the protein" (author's translation) was based upon the striking change in immunological properties of proteins by chemical reactions that cause substitution in the aromatic nuclei, and the lack of antigenic activity of gelatin which is deficient in aromatic amino acids. These arguments lost force when it became known, first, that chemical changes of proteins (esterification, coupling with isocyanates) that do not involve substitution in aromatic rings can produce profound serological alteration and, secondly, that substances without aromatic groups, e.g., polysaccharides, react specifically with antibodies as do proteins. Still the possibility that aromatic groups may be of greater importance than others for the specificity of protein reactions cannot be flatly denied. From their observation that azo-serum albumin, in contrast to azo-egg albumin, reacted like unchanged serum albumin with antisera to the latter, Kabat and Heidelberger (546) inferred that tyrosine and perhaps histidine groups have little significance for the specificity of serum albumin but are of importance for the specificity of egg albumin. In experiments of Haurowitz et al., and Kleczkowski,[238] the reactivity of serum globulin with homologous antisera was abolished when enough iodine was introduced to substitute all tyrosine residues; however, the same result can be produced by esterification.

It has already been indicated that in contemplating protein specificity one has to account for both the types of differences within a single species.[239] Considering their common characteristics, proteins of one type must have some structural similarity; for example, all globins possess a peculiarity in composition that is responsible for their basic properties; but of the chemically similar proteins there exists a special variant in each animal species. With regard to the chemical basis for the serological specificity of the sundry types of proteins, the great differences in their content of the various amino acids should first be considered. The constant ratios obtaining between the values of the basic amino acids among hemoglobins and among keratins, and the almost constant cysteine content in several keratins, is here relevant (p. 65). Moreover, some types of proteins are known to contain characteristic groups, an example being the phosphoric acid in casein. More difficult to understand is species specificity since there exist often no conspicuous differences in the chemical properties and constitution in homologous proteins from

[238] (491), (547).
[239] v. Doerr and Berger (548).

various animals, corresponding to the serological distinctions. Any chemical interpretation will have to explain not only the multitude of species specific proteins but also the fact that the serological differences gradually increase with the distance in the zoological scale, and that some similarity exists even in distantly related animals, as revealed by the "mammalian" reaction of chemically altered proteins (p. 43).

There is a further question as to the species specific properties in the several proteins of one kind of animals. These seem to be quite independent, which implies that, for example, the structures which characterize albumins on the one hand and globulins on the other as being of human origin have nothing in common. The possibility that the various proteins within one species are somehow related in structure is not supported by any actual evidence. If they were, one could more easily conceive why usually no antibodies are found after injection of homologous proteins, just as though all of the proteins from the same species were recognized by the cells of the animal. Thus after injection of rabbit hemoglobin into rabbits no antibodies could be demonstrated in the serum. It is not entirely excluded that antibodies are formed but are bound by the homologous substances present in the body. In fact, immunization effects were described in guinea pigs with guinea pig keratin,[240] in rabbits with homologous thyroglobulin,[241] [and perhaps fibrinogen (551)], in goats with goat casein (552), and in chickens with egg albumin [242] (v. pp. 22, 102). Consequently, it would be reasonable to entertain the idea that antibody formation can occur, under physiological conditions, without the aid of foreign material. Several phases of this subject call for further investigation.

ANTIGENICITY OF PROTEINS. — The opinions prevalent not very long ago concerning the nature of antigens are clearly expressed in the following quotation from Doerr (46):

"The typical anaphylactogens are proteins and as such possess in common certain physico-chemical properties, as considerable molecular size, formation of colloidal solutions, digestibility by enzymes, composition of amino acids, the latter mostly being optically active. No one of the characteristics enumerated is sufficient for antigenicity, not even all of them

---

[240] Krusius (549).
[241] Hektoen et al. (550).
[242] Bruynoghe (553); v. (553a). The presence of precipitins for testicle substances in the serum of male birds (pigeon, goose) is reported by Picado (554).

together, as displayed in proteins; . . . However, the view prevails today almost exclusively that the properties named have the character of necessary conditions . . ." (author's translation; "anaphylactogen" is in this connection practically synonymous with "antigen").

The first of these sentences is now obsolete since non-protein antigens even of high activity have been found (p. 225); why proteins have a paramount position as antigens is still not fully accounted for. A considerable molecular size is common to all antigens, but carbohydrates of high molecular weight give rise to antibodies only in some species and may altogether lack antigenic ability; on the other hand, it would seem that antigenic molecules need not be as large as average proteins, for the production of precipitins has been reported with ribonuclease (mol. weight 15000), and with a substance resulting from the action of phenylisocyanate upon clupein whose molecular weight, unless polymerization takes place in the preparation, should be less than or not much above 5000 [Gutman (506)].

Regarding the function of amino acids a great deal of attention has been paid to the lack of antigenic capacity in gelatin [243] which is hypothetically ascribed to deficiency in aromatic acids,[244] especially of tyrosine. By way of proof, aromatic groups were introduced into gelatin, and actually gelatin became antigenic when combined with aromatic diazonium compounds,[245] with O-β-glucosido-N-carbobenzyloxy-tyrosine or O-β-glucosido-tyrosine [Clutton, Harington and Yuill (562), (603)]. These experiments, at first glance suggestive of the necessity of aromatic groups, do not suffice to prove their significance for antigenic behaviour, on several grounds. First, combination with phenylisocyanate did not convert gelatin into an antigen [Hopkins and Wormall (563)]. Secondly, also, insulin which has about 12% tyrosine was rendered antigenic or greatly improved in this respect by condensation with glucosido-tyrosine (562). And thirdly, the aforesaid non-protein antigens are not known to contain aromatic groups, and conversely proteins degraded by alkali but containing tyrosine fail to provoke antibody formation (p. 44). In the case of

[243] Wells (555), Pabis and Ragazzi (556), Starin (557), Hooker and Boyd (558).
[244] In addition to 1.4% phenylalanine, in a recent analysis about 1% tyrosine was found (559).
[245] (560), (558); Nathan and Kallos (561). The immune sera so produced may react less — or not at all — with the gelatin (or insulin) antigens than with antigens made similarly from other proteins. Parenthetically, nitrated gelatin was found not to immunize (480), (513).

phenylureidoclupein, likewise, it may be disputed whether the acquisition of antigenicity is due to introduction of benzene rings as such, considering that histones, basic as are protamines, but containing tyrosine and presumably having higher molecular weight, are reported to lack immunizing properties, for reasons not yet ascertained. Finally, an active bacterial antigen was found — and more of this kind may exist — in which no amino acids at all could be detected (p. 225).

The experience of poor or wanting antigenicity of the animal nucleoproteins [246] which are composed of nucleic acids and histones or protamines is intelligible since, as evidence hitherto existing would indicate, the protein components have no appreciable antigenic capacity. Bacterial and virus nucleoproteins, on the other hand, were found to induce antibody production. In view of positive results with some nucleoproteins of animal origin it would be worth while to resume the subject with the use of well purified and unaltered preparations.

The absence, in proteins, of several non-aromatic amino acids appears to be immaterial. Thus, from published analyses, casein contains little cystine and no glycine, ovalbumin no glycine, zein no lysine, glycine or tryptophane, and hordein only small quantities of diamino acids; yet all these substances have antigenic activity.

Since the chemical basis for antigenicity in general is not well established, one can hardly expect to know the cause of the considerable differences in antigenic power of various proteins; most likely a number of factors are implicated. Some special instances may be mentioned. Although it is in general unsafe to link together the scattered data in the literature obtained with a variety of methods, one can say that insulin (p. 37) and crystalbumin,[247] hemoglobin and denatured proteins are weaker antigens than, for instance, native serum globulin or ovalbumin.[248] That the proteins of the former group contain no carbohydrate or very little may or may not be accidental; it is worth mentioning that artificial glycoside-proteins are quite strongly antigenic [Goebel et al. (568)]. On the other hand, in the case of mucins the high carbohydrate content was taken to be

[246] Wells (564), (358), Boné (565).

[247] Hewitt (566) states: "Although in a given fraction the antigens pseudoglobulin, seroglycoid and crystalbumin are present in the ratio 1:10:70 the corresponding antibodies in a serum prepared by injecting this mixed antigen may be present in the ratios 1:0.6:0.03." Gell and Yuill (567) on the contrary secured good antisera with crystalbumin.

[248] v. Doerr and Berger (158).

responsible for poor antigenicity (Goodner), which is plausible if essential parts of the protein were covered by groupings which have themselves no antigenic activity. Possibly this explanation applies to the difficulty to produce antibodies with conjugated proteins which contain a great number of attached groups or large hydrocarbon residues (p. 160).

The suggestion that antigenicity tends to be greater with proteins of high molecular weight was made because of the ease with which strong antisera to hemocyanins (molecular weight of several millions) were produced (92); and indeed the huge molecules of tobacco mosaic virus are very effective precipitinogens.[249]

In summary, then, there is as yet no consistent theory embracing the experimental facts. For clarification it will be desirable to make comparative experiments with a wide choice of proteins differing in various respects under standardized conditions with regard to routes and modes of injection and quantities, taking into account the great variation in response of individual animals and strains (as well as species differences)[250] and preferably injecting into the same animal mixtures of the antigens to be compared.[251] Better comprehension of antigenicity may, furthermore, come from the studies on protein-free antigens of bacteria, and even a synthetic approach is not inconceivable.

DEMONSTRATION OF SPECIES DIFFERENCES BY CHEMICAL METHODS. — Relatively little work has been done towards developing a comparative chemistry of proteins, a line of research brought to attention by the results of serology. These presented the problem of correlating serological with chemical findings and of discerning the scheme at the bottom of the variation in proteins that reflect so remarkably the zoological relations.

Investigations were undertaken by Dakin (418) on the assumption that the racemization of peptides depends upon keto-enol tautomerism, and in proteins, therefore, only those amino acids would be racemized whose carboxyl groups are in peptide union. Consequently, the cleavage of racemized proteins by hydrolysis should disclose the

---

[249] On immunization with insoluble proteins and proteins adsorbed to particles, v. (p. 43; cf. p. 45); on differences in antigenic activity of tuberculin fractions, depending on molecular size, v. Seibert (569).

[250] It has already been mentioned that for the production of precipitins rabbits have mostly been used; as a sensitive test for antigenicity anaphylactic experiments in guinea pigs are of advantage.

[251] v. (158).

amino acids bearing free carboxyl groups that, presumably not being racemized, would be recovered among the cleavage products in optically active form. According to the communications of Dudley and Woodman, and of Dakin and Dale,[252] sheep and cow casein, and the ovalbumins of ducks and chickens, actually could be distinguished by this method.

Apparently well defined results, inviting further study, were obtained by Obermayer and Willheim (573). These authors found that the ratio of total nitrogen to amino acid nitrogen in certain protein fractions of horse and cattle serum differs considerably from the corresponding values obtained with chicken or goose serum.

The most promising approach is analysis for individual amino acids in corresponding proteins of different species; as yet there are not enough reliable data to make generalizations possible. Analyses of mammalian hemoglobins,[253] presented in Table 10, indicate a constant ratio of the basic amino acids but show wide differences in cysteine and non-cysteine sulfur [Bergmann and Niemann (578)], except in kindred species [Birkofer et al. (579)].

TABLE 10. — MOLECULAR RATIO OF CONSTITUENTS IN VARIOUS HEMOGLOBINS
[after Bergmann et al. (578)]

	Hemoglobin			
	Horse	Sheep	Cattle	Dog
Iron	4	4	4	4
Cysteine	2	3	3	6
Non-cysteine sulfur	6	12	..	6
Arginine	12	12	12	12
Histidine	32	32	32	32
Lysine	36	36	36	36

Figures for tyrosine, tryptophane and phenylalanine are given by Block (580); electrophoretic measurements are reported in (115).

In lens proteins in the reduced form the percentage values found for cysteine were: sheep 5.2, swine 5.0, chicken 4.3, fish 8.5.[254] The cysteine content was found by these authors (582) to be approximately the same (about 13%) in wool, feathers and hair keratin, and determinations for the basic amino acids gave an almost constant

[252] (570), (571); v. (572).
[253] Species differences in hemoglobins were shown with regard to oxygen affinity and denaturation [Anson et al. (574), Haurowitz (575)]. On hemocyanins (and hemoglobins) v. Roche et al. (576); on differences in caseins v. (577).
[254] Ecker and Pillemer (581).

ratio (1:4:12) of histidine:lysine:arginine in these substances, and in nails and horn (580).

Figures for some amino acids in crystalline seed globulins of Cucurbita species indicating variation in accord with botanical relationship are recorded by Vickery et al. (583).

Amino acid analyses on the erythrocyte stromata of several mammalian bloods failed to reveal considerable differences. Determination of eight amino acids in whole muscle tissue of species belonging to five classes gave in most cases similar values (584).

In a number of papers on chemical species differences in serum proteins, globins, casein, etc., the requirement that only single, pure proteins be subjected to analysis has not been fulfilled.[255]

APPLICATIONS OF SEROLOGICAL PROTEIN REACTIONS. — In conclusion, attention may be called to the applications of the serological protein reactions which are sensitive and specific enough to detect a particular protein in quantities less than one hundredth of a milligram per cc (586). These reactions have proved to be dependable as routine tests for the differentiation of human and animal blood in forensic cases and for the examination of foodstuffs (meat, milk and egg products, adulteration and contamination of flour, et cetera) (587).

Of greater importance, from the point of view of biochemical research, is the use of serum reactions for the identification of proteins and, since serologically reactive impurities are easily detected,[256] as an aid in the purification of protein fractions.[257] While immunological reactions give no information on physico-chemical properties or chemical composition, no other single method is so efficient in qualitatively characterizing an individual protein and an increasingly wider use of these tests can be foreseen. The manner in which immunological reactions are employed for preparative protein chemistry is illustrated by the work of Osborne and Wells (210) on plant proteins and recent studies on serum proteins.[258] For instance, Kekwick obtained an albumin fraction which in the ultracentrifuge and also electrophoretically appeared homogeneous yet by precipitin tests was shown to be a mixture. Other examples are the distinction by pre-

---

[255] v. (p. 119); Groh and Faltin (585).

[256] Thus Goldsworthy and Rudd (588) were able to demonstrate globulin in repeatedly recrystallized serum albumin.

[257] A thought provoking discussion of the criteria of purity applied to substances of high molecular weight is presented by Pirie (589).

[258] (164), (168), (169).

cipitin reactions of blood and muscle hemoglobin,[259] demonstrations of thyroglobulin in blood and lymph, and of a phosphoprotein, reacting like ovovitellin, in the serum of laying hens (possibly in small quantity in serum of roosters) (591). In disease, some abnormal proteins have been detected (p. 26), and Bence Jones proteins were demonstrated in urine and serum of myeloma patients.[260] As already remarked, in the field of viruses precipitin and neutralization tests are employed in many ways, e.g. for tracing contamination by viruses or normal plant proteins.

A method for quantitative determination of proteins by means of precipitin reactions has been developed by Goettsch and Kendall (593) and used for examining the composition of blood serum in pathological conditions (Kendall (164)).

Estimation of the plasma volume by intravenous injection of precipitating immune sera has been proposed by Culbertson (594).

Immune sera were further utilized in an investigation of the origin of serum proteins. Chicken fibroblasts were grown for as many as 35 weekly passages in rabbit plasma, and the tissue fluids were tested with precipitins for chicken serum. The results indicated that connective tissue can produce proteins which are identical with or closely related to serum proteins. They showed in addition that the species specific properties are retained unchanged when cells are grown for quite a time in foreign plasma.[261]

BIBLIOGRAPHY

(1) Tchistovitch: AP *13*, 406, 1899; Bordet: AP *13*, 225, 1899. — (1a) Nuttall: Blood Immunity and Blood Relationship, Cambridge, 1904. — (1b) Wolfe: JI *44*, 135, 1942. — (1c) Cohen et al: JI *46*, 59, 1943. — (1d) Youmans et al: JI *46*, 217, 1943. — (1e) Colwell et al: JID *68*, 226, 1941 (B). — (1f) Dingle et al: JI *34*, 357, 1938. — (2) Erhardt: Erg. u. Fortschr. d. Zool. 7, 279, 1929. — (3) Boyden: PZ *15*, 109, 1942. — (4) Wolfe: BB *76*, 108, 1939. — (4a) Wolfe: Z *27*, 17, 1942. — (4b) DeFalco: BB *83*, 205, 1942. — (5) Kellaway: AJE *8*, 123, 1931. — (6) Boyden et al: Am. Mus. Novitates 1933, #606. — (7) Streng: APS, Suppl. *37*, 498, 1938. — (8) von Dungern: Die Antikörper, Fischer, Jena, 1903. — (9) Boyden: CIY *35*, 82, 1936; *38*, 219, 1939. — (10) Erhardt: ZI *60*, 156, 1929. — (11) Makino: ZI *81*, 316, 1934. — (12) Kuramoto: AMO *4*, 249, 1934. — (13) Eisenbrandt: AJH *27*, 117, 1938 (B). — (14) Wilhelmi: BB *82*, 179, 1942. — (15) Uhlenhuth et al: HPM *3*, 365, 1930. — (16) Cromwell: JID *37*,

---

[259] Hektoen et al. (590).
[260] Hektoen (592).
[261] Landsteiner and Parker (595); v. Kimura; Fischer (596).

321, 1925.— **(17)** Satoh: ZI *79*, 117, 1933 (B).— **(18)** Culbertson: JI *23*, 439, 1932 (B).— **(19)** Welsh et al: ZI *9*, 517, 1911.— **(20)** Jones et al: JI *25*, 381, 1933.— **(21)** Leers: CB *54*, 462, 1910.— **(22)** Boyden et al: JI *17*, 29, 1929 (B); Pr *27*, 421, 1930.— **(23)** Manwaring et al: Pr *27*, 14, 1929.— **(24)** Kolmer: Infection, Immunity, etc., Saunders, Philadelphia, 1925.— **(24a)** Hanks: JI *28*, 95, 1935.— **(25)** Hoen et al: ZI *58*, 143, 1928.— **(26)** Collier et al: CB *86*, 505, 1921.— **(27)** Wolfe et al: PZ *11*, 63, 1938.— **(28)** Lowell: JI *46*, 177, 1943 (B).— **(29)** Roberts et al: Pr *47*, 11, 1941.— **(30)** Goodner: S *94*, 241, 1941.— **(31)** Dean et al: JP *29*, 473, 1926; *31*, 89, 1928; SystB *6*, 1931.— **(32)** Taylor et al: JH *34*, 118, 1934 (B).— **(33)** Opie: JI *8*, 24, 1923.— **(34)** Goldsworthy et al: JP *40*, 169, 1935.— **(35)** Taylor et al: JH *35*, 169, 1935.— **(35a)** Boyd: JEM *74*, 369, 1941.— **(36)** Ramon et al: SB *135*, 148, 673, 1941; AP *37*, 1009, 1923.— **(36a)** Felton: Nat. Inst. Health Bull. No. 169, 1937.— **(37)** Wu et al: Pr *25*, 853, 1928; *26*, 737, 1929.— **(38)** Heidelberger et al: JEM *50*, 809, 1929; BaR *3*, 49, 1939; S *97*, 405, 1943 (B).— **(39)** Nungester et al: Pr *48*, 425, 1941.— **(39a)** Paić: AP *62*, 534, 1939.— **(39b)** DeFalco: Pr *46*, 500, 1941.— **(39c)** Pope et al: Br *19*, 397, 1938.— **(40)** Sachs et al: AIF *3*, 85, 71, 1907 (B).— **(41)** Boyden: BB *50*, 73, 1926.— **(42)** Wolfe: PZ *6*, 55, 1933.— **(43)** Cole: ArP *26*, 96, 1938.— **(44)** Landsteiner et al: JEM *71*, 445, 1940.— **(45)** Dean: SystB *6*, 424, 438, 1931.— **(46)** Doerr: HPM *1*, 759, 1929.— **(47)** Dakin et al: BJ *13*, 248, 1919.— **(48)** Wells et al: JID *12*, 341, 1913.— **(49)** Nicolas: SB *109*, 1249, 1932.— **(50)** Hooker et al: JI *30*, 41, 1936.— **(51)** Wolfe: JI *29*, 1, 1935.— **(52)** Hallmann: CB *130*, 234, 1933.— **(53)** Saeki: BPh *66*, 134, 1932.— **(54)** Moritz: Planta *15*, 647, 1932.— **(55)** Sasaki: J. Dep. Agric. Kyushu Imp. Univ. *2*, 117, 1928; Annotationes Zool. Japon *12*, 433, 1930; ZZ *29*, 284, 1934.— **(56)** von Dungern: CB *34*, 355, 1903.— **(57)** Cumley et al: G *27*, 177, 1942.— **(58)** Fujiwara: DZG *1*, 754, 1922.— **(59)** Beger: CB *91*, 519, 1924.— **(60)** Furth: AH *92*, 158, 1923.— **(61)** Nishegorodzeff: ZI *66*, 276, 1930.— **(61a)** Ishihara: ZI *92*, 121, 1938.— **(61b)** Mollison: Scientia *69*, 154, 1941.— **(62)** Uhlenhuth: DM 1905, p. 1673.— **(63)** Landsteiner et al: JI *22*, 397, 1932.— **(64)** Holzer: ZI *84*, 170, 1935.— **(65)** Manteufel et al: ZI *33*, 348, 1921; von Gara: ZI *79*, 171, 1933.— **(66)** Levine et al: PZ *12*, 400, 1939.— **(67)** Heidelberger et al: JEM *62*, 697, 1935.— **(68)** Landsteiner et al: JEM *40*, 91, 1924 (B).— **(69)** Hicks et al: G *16*, 397, 1931.— **(70)** Moody: JI *39*, 113, 1940.— **(71)** Roesli: CB *112*, 151, 1929 (B).— **(72)** Friedberger et al: ZI *36*, 233, 1923 (B).— **(73)** Myers: CB *28*, 237, 1900.— **(74)** Müller: MM 1902, p. 1330.— **(75)** Morgenroth: MM 1902, p. 1033 (B).— **(76)** Dölter: ZI *43*, 95, 1925.— **(77)** Witebsky et al: ZI *52*, 359, 1927.— **(78)** Bauer: MM 1911, p. 71.— **(79)** Ouchi: ZI *53*, 462, 1927.— **(80)** Schiff: K 1924, p. 679.— **(81)** Zuckerman et al: JEB *12*, 222, 1935.— **(82)** Simon: JEM *75*, 315, 1942. — **(83)** Heidelberger et al: JEM *38*, 561, 1923 (B).— **(84)** Fujiwara: DZG *11*, 253, 1928 (B).— **(85)** Hektoen et al: JID *49*, 29, 1931 (B).— **(86)** Giovanardi: AH *117*, 74, 1936 (B).— **(87)** Gay et al: JEM *17*, 535, 1913.— **(88)** Ottensooser et al: BZ *193*, 426, 1928.— **(89)** Johnson et al: JID *57*, 70, 1935 (B).— **(90)** Browning et al: JP *14*, 174, 1909; JI *5*, 417, 1920.— **(91)** Went et al: DZG *30*, 64, 1938 (B).— **(92)** Hooker et al: JI *30*, 33, 1936; BB *73*, 181, 1937.— **(93)** Reichert et al: Carnegie Inst. Publ. #116, 1909.— **(94)** Robson: The Species Problem, pp. 16, 67, Oliver and Boyd, Edinburgh, 1928.— **(95)** Barcroft: The Respiratory Function of the Blood, *2*, p. 40, Cambridge, 1928.— **(96)** Anson et al: RSB *97*, 61, 1924.— **(97)** (Roche: SB *110*, 1084, 1932; *113*, 317, 1933; BCB *16*, 757, 1934.— **(98)** Svedberg et al: The Ultracentrifuge, Oxford, 1940.— **(99)** Schönberger: BZ *267*, 57, 1933 (B).— **(100)** Adair et al: CR *198*, 1456, 1934.— **(101)** Landsteiner et al: JGP *6*, 131, 1923.— **(102)** Rodenbeck:

CB *123*, 460, 1932. — **(103)** Moribe: BPh *64*, 385, 1932. — **(104)** Demanez: AI *8*, 255, 1933. — **(105)** Saeki: BPh *68*, 388, 1932. — **(106)** Kenton; JI *25*, 461, 1933 (B). — **(107)** Kyes et al: JI *20*, 85, 1931. — **(108)** Komatu: BPh *98*, 156, 1937. — **(109)** Lewis et al: JID *40*, 316, 1927. — **(110)** Welsh et al: JH *10*, 177, 1910. — **(111)** Sievers: APS *16*, 44, 1939. — **(112)** Bauer: BK 1910, p. 830. — **(113)** Demanez: AI *8*, 233, 1933. — **(114)** Witebsky: Nw 1929, p. 771; HP *13*, 503, 1929. — **(115)** Landsteiner et al: S *88*, 83, 1938. — **(115a)** Pedersen: KZ *63*, 268, 1933. — **(115b)** Forster: JEM *47*, 903, 1928 (B). — **(116)** Pillemer et al: JEM *69*, 191, 1939. — **(117)** Uhlenhuth et al: CB, Ref. *47*, Beil., 68, 1910. — **(118)** Uhlenhuth: Festschr. Robert Koch, Fischer, Jena 1903. — **(119)** Hektoen et al: JID *34*, 433, 1924. — **(120)** Krusius: ZI *5*, 699, 1910, Z. Augenheilk. *24*, 257, 1910. — **(120a)** Swift et al: JEM *63*, 703, 1936. — **(121)** Tutui: BPh *76*, 551, 1934. — **(122)** Hoffmann: ZI *71*, 171, 1931 (B). — **(123)** Markin et al: JID *65*, 156, 1939. — **(124)** Ecker et al: JEM *71*, 585, 1940 (B). — **(125)** Gotoh: AMO *3*, 172, 222, 1932. — **(126)** Garcia: BPh *96*, 148, 1937. — **(127)** Kimura: Acta Schol. Med. Kioto *8*, 309, 1926. — **(128)** Witebsky: ZI *58*, 297, 1928; HP *13*, 502, 1929. — **(129)** Arnold: ZI *82*, 154, 1934. — **(130)** Witebsky et al: ZI *58*, 271, 1928. — **(131)** Hektoen et al: JID *40*, 641, 1927. — **(132)** Adant et al: SB *117*, 230, 232, 1934. — **(133)** Snapper: NT *79*, 2007, 1935, WK 1935, p. 1199. — **(134)** Stokinger et al: JEM *66*, 251, 1937 (B). — **(135)** Chester: Q *12*, 19, 1937. — **(136)** Moritz: Beitr. Biol. d. Pflanzen *22*, 51, 1934. — **(137)** Wendelstadt et al: ZI *8*, 43, 1910. — **(138)** Thöni et al: ZI *23*, 83, 1914. — **(139)** Galli-Valerio et al: ZI *15*, 229, 1912, *17*, 180, 1913. — **(140)** Gilg et al: BDB *45*, 315, 1927. — **(141)** Eisler: ZI *53*, 136, 1927. — **(142)** Becker: Bot. Arch. *34*, 267, 1932. — **(143)** Moritz: "Der Züchter" *6*, 217, 1934. — **(144)** Chester: AJB *24*, 451, 1937. — **(145)** Lewis et al: JBC *66*, 37, 1925. — **(146)** Abderhalden: Abwehrfermente, ed. 6, Steinkopff, Dresden, 1941. — **(147)** Cohn: ChR *28*, 395, 1941 (B). — **(148)** Marrack: CAA. — **(149)** Went et al: ZI *82*, 392, 474, 1934. — **(150)** Tiselius: ARB *8*, 155, 1939. — **(151)** Kleczkowski: BZ *299*, 311, 1938. — **(152)** Blix et al: JBC *137*, 485, 1941. — **(153)** Bierry et al: SB *127*, 483, 1938, *128*, 700, 1938. — **(154)** Doladilhe et al: SB *131*, 752, 1939. — **(155)** Leblanc: Cellule *18*, 335, 1901. — **(156)** Michaelis: DM 1904, p. 1240. — **(157)** Dale et al: BJ *10*, 408, 1916. — **(158)** Doerr et al: ZH *96*, 191, 258, 1922; K 1922, p. 949. — **(159)** Hektoen et al: JID *35*, 295, 1924. — **(160)** Kimura: ZI *56*, 330, 1928. — **(161)** Györffy: ZI *71*, 428, 1931. — **(162)** Asaba: AMO *3*, 314, 1932. — **(163)** Hooker et al: JBC *100*, 187, 1933 (B). — **(164)** Kendall: CSH *6*, 376, 1938, JCI 16, 921, 1937 (B). — **(165)** Harris et al: JGP *19*, 383, 1935. — **(166)** Hewitt: BJ *21*, 216, 1927, *32*, 26, 1938. — **(167)** Chick: BJ *8*, 404, 1914. — **(168)** Kekwick et al: BJ *32*, 552, 560, 1938. — **(169)** Hewitt: BJ *31*, 1047, 1937, *32*, 1540, 1938 (B). — **(170)** Haurowitz: CA *36*, 2609, 1942. — **(171)** Hektoen et al: JID *42*, 1, 1928. — **(172)** Jukes et al: JEM *56*, 469, 1932. — **(173)** Roepke et al: JI *30*, 109, 1936. — **(174)** Wells et al: JID *29*, 200, 1921. — **(175)** Longsworth et al: JAC *62*, 2580, 1940. — **(176)** Pedersen: BJ *30*, 948, 961, 1936. — **(177)** Woodman: BJ *15*, 187, 1921. — **(177a)** Pappenheimer: JEM *71*, 263, 1940. — **(178)** Linderström-Lang: ZPC *176*, 76, 1928, LC *17*, No. 9, 1929. — **(179)** Carpenter: JAC *57*, 129, 1935 (B). — **(180)** Mellander: BZ *300*, 240, 1939. — **(181)** Carpenter et al: JID *47*, 435, 1930. — **(182)** Demanez: SB *112*, 1561, 1933. — **(182a)** Haurowitz: ZPC *232*, 125, 1935. — **(182b)** Tadokoro et al: JB *14*, 145, 1931. — **(182c)** Rasmussen: ZPC *224*, 97, 1934. — **(182d)** Herriott et al: JGP *23*, 439, 1940. — **(183)** Sörensen: KZ *53*, 102, 1930. — **(183a)** Reiner: JBC *95*, 345, 1932. — **(184)** Marrack et al.: Br *19*, 171, 1938; IC, p. 87. — **(185)** Landsteiner: JEM *75*, 269, 1942. — **(185a)** Fell: CA *36*, 3193, 1942. — **(186)** Loiseleur et al: SB *103*, 776, 1930. — **(186a)**

Michaelis et al: JBC *147*, 91, 1943. — **(187)** Raubitschek: Verh. d. d. path. Ges. *21*, 273, 1910. — **(188)** Hektoen et al: BJ *34*, 487, 1940 (B). — **(189)** Bayne-Jones et al: BJH *33*, 37, 119, 1922. — **(190)** Everett et al: BJH *34*, 385, 1923 (B). — **(191)** Welker et al: JID *53*, 165, 1933. — **(192)** Packalén: APS *17*, 263, 1940. — **(193)** MacLeod et al: JEM *73*, 191, 1941 (B). — **(193a)** Perlman et al: JEM *77*, 97, 1943. — **(194)** Kamekura: ZI *42*, 439, 1925 (B). — **(195)** van der Ende: JE *1*, 356, 1939 (B). — **(196)** Eichbaum et al: ZI *86*, 284, 1935. — **(197)** Witebsky et al: ZI *80*, 108, 1933. — **(198)** Campbell: JID *59*, 266, 1936, AJH *23*, 104, 1936. — **(199)** Taliaferro: Ph *20*, 469, 1940. — **(200)** Hektoen: JID *39*, 342, 1926. — **(201)** Gaase: MM 1941, p. 473. — **(202)** Eisenbrandt: Pr *35*, 322, 1936. — **(203)** Dulaney et al: Pr *48*, 620, 1941. — **(204)** Taliaferro: The Immunology of Parasitic Infections, Century Co., New York, 1929. — **(205)** Culbertson: Immunity against Animal Parasites, Columbia University Press, New York, 1941. — **(206)** Leipert et al: ZPC *267*, 251, 1941 (B). — **(207)** Block: JBC *133*, 71, 1940. — **(208)** Abderhalden et al: ZPC *240*, 237, 1936. — **(209)** Birkhofer et al: ZPC *265*, 94, 1940. — **(209a)** Bálint et al: BZ *308*, 83, 1941. — **(210)** Wells et al: JID *8*, 66, 1911, *12*, 341, 1913, *13*, 103, 1913, *14*, 364, 377, 1914, *17*, 259, 1915, *19*, 183, 1916, *40*, 326, 1927, JBC *66*, 37, 1925. — **(211)** Jones et al: JBC *56*, 79, 1923. — **(212)** Wodehouse: AJB *4*, 417, 1917. — **(212a)** Coulson et al: JI *41*, 375, 1941 (B). — **(212b)** Spies et al: JA *14*, 7, 1942 (B). — **(212c)** Boatner et al: JD *5*, 7, 1942. — **(213)** Newell: JA *13*, 177, 1942. — **(214)** Harley: Br *18*, 469, 1937 (B). — **(214a)** Caulfeild et al: JA *7*, 1, 1935. — **(215)** Stull et al: JA *12*, 117, 606, 1941 (B), *13*, 537, 1942, *10*, 417, 1939. — **(216)** Benjamins et al: JA *6*, 335, 1935. — **(217)** Sanigar: J. Franklin Inst. *230*, 781, 1940. — **(218)** Roth et al: JA *13*, 283, 1942. — **(218a)** Simon: JEM *77*, 185, 1943. — **(219)** Harsh et al: JA *14*, 121, 1943 (B). — **(220)** Abramson et al: JPC *46*, 192, 1942. — **(221)** Rockwell: JI *43*, 259, 1942. — **(222)** Winkenwerder et al: S *90*, 356, 1939. — **(222a)** Sherman: JA *14*, 1, 1942. — **(223)** Elliott: JID *15*, 501, 1914. — **(224)** Goodner: JID *37*, 285, 1925. — **(225)** von Godin: ZI *96*, 320, 1939. — **(226)** Uhlenhuth et al: ZI *96*, 183, 1939. — **(227)** Lustig et al: BZ *289*, 365, 1937. — **(228)** Bierry: CR *192*, 1284, 1931; SB *110*, 889, 1932, *115*, 1168, 1934 (B), *116*, 702, 1934 (B), *130*, 411, 1939. — **(229)** Levene et al: JBC *84*, 49, 63, 1929, *140*, 279, 1941. — **(230)** Hewitt: BJ *32*, 1554, 1938. — **(231)** Neuberger: BJ *32*, 1435, 1938. — **(232)** Sörensen: BZ *269*, 271, 1934. — **(233)** Masamune et al: JB *24*, 219, 1936. — **(234)** Rimington: BJ *25*, 1062, 1931. — **(235)** Ozaki: JB *24*, 73, 1936. — **(236)** Hewitt: BJ *33*, 1496, 1939. — **(237)** Stacey et al: JCS 1940, p. 184. — **(238)** Levene: JAC *39*, 828, 1917. — **(239)** Blix et al: ZPC *234*, III, 1935. — **(240)** Sevag et al: ZI *83*, 464, 1934; BZ *273*, 419, 1934. — **(241)** Ferry et al: JBC *105*, XXVII, 1934. — **(242)** Coghill et al: JI *35*, 477, 1938. — **(243)** Neuberger: BJ *34*, 109, 1940. — **(244)** Gell et al: BJ *32*, 560, 1938. — **(245)** Osborne et al: AJP *14*, 259, 1905. — **(246)** Karrer et al: ZPC *135*, 129, 1924; H *8*, 384, 1925. — **(247)** Kellaway: ARB *8*, 541, 1939. — **(248)** Kraus and Werner: Giftschlangen, etc., Fischer, Jena, 1931 (B). — **(249)** Slotta et al: BC *71*, 1076, 1082, 1938. — **(250)** Gralén et al: BJ *32*, 1375, 1938. — **(251)** van Heyningen: BJ *35*, 1257, 1941 (B). — **(251a)** Herbert and Todd: BJ *35*, 1124, 1941. — **(252)** Chain: Quart. J. Exp. Physiol. 27, 49, 1937. — **(253)** Ghosh et al: N *143*, 380, 1939. — **(254)** Micheel et al: ZPC *249*, 157, 1937 (B); BC *71*, 72. — **(255)** Gessner: Handb. d. exp. Pharmak., Ergänzungsw. *6*, 1, 1938 (B). — **(256)** Ghosh et al: C 1937, I, 3970. — **(256a)** Delezenne: AP *33*, 68, 1919. — **(256b)** Morgenroth: BK *42*, 1550, 1905; Arb. path. Inst. Berlin 1906, p. 437. — **(256c)** Landsteiner et al: ZI *1*, 439, 1909 (B). — **(257)** Barron et al: JBC *137*, 267, 1941. — **(257a)** Hottle et al: JEM *74*, 545, 1941 (B). — **(258)** Tetsch et al: BZ *290*, 394, 1937. — **(259)** Reinert: SMW 1937, p. 515. — **(260)**

Micheel et al: ZPC *265*, 266, 1940. — **(261)** Klobusitzky: AEP *179*, 204, 1935; ZI *92*, 418, 1938. — **(262)** Eaton: BaR *2*, 3, 1938 (B). — **(262a)** Wadsworth: SM. — **(262b)** Ando et al: JI *29*, 439, 1935, *31*, 355, 1936; v. SB *123*, 69, 648, 768, 1936. — **(263)** Mueller: JBa *36*, 499, 1938 (B); JI *40*, 21, 1941. — **(264)** Eaton: JBa *31*, 347, 367, 1936. — **(265)** Pappenheimer: JBC *120*, 543, 1937, *125*, 201, 1938. — **(266)** Wadsworth et al: JID *62*, 129, 1938. — **(267)** Linggood: Br *22*, 255, 1941. — **(268)** Petermann et al: JPC *45*, 1, 1941. — **(269)** Maschmann: ZPC *201*, 219, 1931. — **(270)** Eaton et al: JBa *36*, 423, 1938 (B). — **(271)** Prigge: CB *144*, 12, 1939 (B). — **(272)** Koerber et al: JI *40*, 459, 1941. — **(273)** Pope et al: Br *20*, 297, 502, 1939. — **(274)** Theorell et al: ZI *91*, 62, 1937. — **(274a)** Schmidt et al: APS Suppl. *16*, 407, 1933. — **(274b)** Reiner: JI *22*, 439, 1932. — **(275)** ARB *1*, 659, 1932, *2*, 505, 1933, *8*, 580, 1939; SystB *6*, 122, 1931. — **(275a)** Nelis: AP *52*, 597, 1934. — **(275b)** Parfentjev et al: JI *40*, 189, 1941. — **(275c)** Hallauer: ZH *105*, 138, 1925 (B). — **(276)** Eaton: JI *33*, 419, 1937. — **(277)** Hewitt: BJ *24*, 983, 1930. — **(277a)** Haas: ZI *99*, 121, 1940. — **(278)** Velluz: SB *127*, 35, 1938. — **(278a)** Goldie and Sandor: SB *129*, 454, 1938, *125*, 863, 1937. — **(279)** Velluz: SB *128*, 11, 1938. — **(280)** Todd: JP *39*, 299, 1934; Br *22*, 172, 1941 (B), *23*, 136, 1942. — **(281)** Hooker: NEM *215*, 68, 1936. — **(282)** Hazen: JID *60*, 260, 1937 (B). — **(283)** Clauberg: K 1939, p. 1490. — **(284)** O'Meara: JP *51*, 317, 1940. — **(284a)** Ehrlich: MM 1903, p. 1428. — **(285)** Macfarlane et al: BJ *35*, 884, 1941. — **(286)** Prigge: ZI *81*, 185, 1934. — **(287)** Forssman: APS *17*, 232, 1940. — **(287a)** Morgan et al: JP *43*, 385, 1936. — **(288)** Ramon: AP *37*, 1001, 1923 (B). — **(289)** Otto et al: AIF 1935, H.31. — **(290)** Grasset: Quart. Bull. Health Org. League of Nations *5*, 367, 1936 (B); Bull. Soc. Path. Exot. *29*, 210, 1936; Trans. Roy. Soc. Trop. Med. *31*, 445, 1938. — **(291)** Noguchi: Snake Venoms, Carnegie Inst., Washington, 1909. — **(292)** Arthus: Arch. Int. Physiol. *11*, 265, 1911–12. — **(293)** Lamb: Scient. Memoirs, Med. and San. Dept. India No. 5, 7, 10, 1903–04. — **(294)** Githens: JI *42*, 149, 1941 (B). — **(294a)** Amaral: NK p. 1066. — **(295)** Klobusitzky: EH *24*, 226, 1941. — **(296)** Neisser et al: ZH *36*, 299, 1901. — **(297)** Todd: Trans. Path. Soc. London *53*, 196, 1902. — **(297a)** Pribram: HPM *2*, 575, 1929. — **(297b)** Pearce et al: AJM *128*, 669, 1904. — **(297c)** Fukuhara: ZI *2*, 305, 1909. — **(297d)** Weld et al: Pr *49*, 370, 1942. — **(298)** Buchanan et al: Physiol. and Biochem. of Bact., Williams and Wilkins, Baltimore, *1*, 1928. — **(299)** Nicolle et al: AP *23*, 547, 1909. — **(300)** Tamura: ZPC *80*, 289, 1914. — **(301)** Buchner: MM 1897, p. 209. — **(302)** Boivin et al: Arch. Roumaines Pathol. *7*, 95, 1934. — **(303)** Heidelberger et al: JI *37*, 563, 1939 (B). — **(304)** Macfadyen et al: CB *34*, 618, 1903. — **(305)** Mudd et al: JI *30*, 405, 1940 (B). — **(306)** Nelson: JID *38*, 371, 1926; *40*, 412, 1927. — **(307)** Thompson et al: JBC *125*, 65, 1938 (B). — **(308)** Zittle et al: JI *43*, 31, 47, 61, 1942 (B). — **(309)** Sevag et al: JBC *140*, 833, 1941. — **(310)** Zittle: ANY *43*, 47, 1942, JI *45*, 29, 1942 (B). — **(311)** Heidelberger et al: JI *30*, 267, 1936; RI *4*, 293, 1938. — **(312)** Siebert: BaR *5*, 60, 1941. — **(313)** Coghill et al: JBC *81*, 115, 1929 (B). — **(314)** Lancefield: HL 1941, p. 251 (B). — **(315)** Sevag et al: JBC *124*, 425, 1938. — **(316)** Henriksen et al: JI *42*, 187, 1941 (B). — **(317)** Stamp et al: L *232*, 257, 1937. — **(318)** Menzel et al: JBC *124*, 301, 89, 1938 (B). — **(319)** Seibert: ART *21*, 370, 1930. — **(320)** Schaefer et al: AP *64*, 517, 1940, *53*, 72, 1934. — **(321)** Linton: BaR *4*, 261, 1940. — **(322)** White: IP *51*, 447, 449, 1940 (B). — **(323)** Landsteiner et al: JEM *46*, 213, 1927. — **(324)** Furth et al: JEM *47*, 171, 1928. — **(325)** White: IP *35*, 77, 1932, *36*, 65, 1933 (B). — **(326)** Goadby: RS B, *102*, 137, 1927. — **(327)** Happold: JP *31*, 237, 1928. — **(327a)** Eckstein et al: JBC *91*, 395, 1931. — **(328)** Wong: Pr *45*, 850, 1940. — **(329)** Flosdorf et al: JI *39*, 475, 1940. — **(330)**

Gunnison: JI *26*, 17, 1934. — **(331)** Howie et al: JP *50*, 235, 1940. — **(332)** Prigge: ZI 77, 421, 1932. — **(332a)** Marton et al: JI *45*, 63, 1942. — **(333)** Sulzberger: JI *23*, 73, 1932; Dermatologic Allergy, Ch. C. Thomas, Springfield, Ill., 1940. — **(333a)** Jadassohn et al: JI *32*, 203, 1937. — **(333b)** Duncan: Br *13*, 489, 1932. — **(334)** Freund et al: JI *38*, 67, 1940 (B). — **(335)** Flahiff: AJH *30*, Sec. B 65, 1939. — **(336)** Seibert et al: JAC *65*, 272, 1943 (B); JI *28*, 425, 1938. — **(337)** Maschmann: DM *63*, 778, 1937. — **(338)** Dorset et al: J. Am. Vet. Med. Ass. *71*, 487, 1927. — **(339)** McCarter et al: JI *43*, 85, 1942. — **(340)** Boor et al: JEM *59*, 63, 1934. — **(341)** Lackmann et al: JI *40*, 1, 1941. — **(342)** Harington: JCS 1940, p. 119. — **(343)** Stokinger et al: JEM *66*, 251, 1937. — **(344)** Harten et al: JA *12*, 72, 1940. — **(345)** Wasserman et al: E *31*, 115, 1942. — **(345a)** Lowell: Pr *50*, 167, 1942 (B). — **(346)** Lewis: JAM *108*, 1336, 1937 (B). — **(347)** Bernstein et al: JLC *23*, 938, 1938. — **(348)** Collip et al: L 1934, I, 76, 784. — **(349)** Collip et al: BiR *15*, 1, 1940. — **(350)** Thompson: Ph *21*, 588, 1941. — **(351)** Selye: JI *41*, 259, 1941. — **(352)** Jailer et al: Pr *45*, 506, 1940. — **(353)** Harington et al: BJ *31*, 2049, 1937. — **(354)** Zondek et al: BJ *32*, 1891, 1938 (B); v. JEM *65*, 1, 1937. — **(355)** Zondek et al: Pr *37*, 343, 1937, *39*, 283, 1938; Arch. Int. Pharmac. *61*, 319, 1939. — **(356)** Brandt et al: ZI *88*, 79, 1936. — **(357)** van den Ende: JE *2*, 403, 1941. — **(358)** Wells: CAI. — **(359)** Pick et al: HPM *2*, 317, 1929. — **(360)** Sachs: Fortschr. d. Med. *20*, 425, 1902. — **(361)** Schütze: ZH *48*, 457, 1904. — **(362)** Achalme: AP *15*, 737, 1901. — **(363)** Gessard: AP *15*, 593, 1901. — **(364)** Moll: Hofmeister's Beitr. *2*, 344, 1902. — **(365)** Kirk et al: JI *26*, 495, 1934. — **(366)** Pillemer et al: JBC *123*, 365, 1938. — **(367)** Seastone et al: JGP *20*, 797, 1937 (B), *21*, 575, 1938. — **(368)** TenBroeck: JBC *106*, 729, 1934. — **(368a)** Smolens et al: JGP *26*, 11, 1942. — **(369)** Campbell et al: JBC *129*, 385, 1939 (B). — **(369a)** Tria: JBC *129*, 377, 1939. — **(370)** Hobby et al: JEM 73, 109, 1941. — **(371)** Walter et al: BJ *26*, 1750, 1932. — **(372)** Haas: BZ *305*, 280, 1940. — **(373)** Fruton et al: JBC *136*, 559, 1940. — **(374)** Várterész et al: ZI *99*, 211, 1941. — **(374a)** Adams: JEM *76*, 175, 1942. — **(375)** Morgenroth et al: BZ *25*, 100, 1910. — **(375a)** Cohen et al: JBC *136*, 243, 1930. — **(376)** Sumner: EE *6*, 201, 1937. — **(377)** Rivers: Lane Med. Lectures, Stanford Uni. Press, Publ. *4*, 46, 1939. — **(378)** Craigie: Handb. d. Virusforsch., Springer, Wien, *2*, 1106, 1939. — **(379)** Smith: ibid., p. 1319. — **(380)** Bawden: Plant Viruses and Virus Diseases, Chronica Botanica Co., Leiden, 1939. — **(381)** Bawden et al: Tab. Biol., *16*, 355, 1938. — **(382)** Chester: Phytopathol. *27*, 903, 1937 (B). — **(383)** Burnet et al: AJE *15*, 227, 1937; v. Br *14*, 302, 1933; Monographs Hall Inst., No. 1, MacMillan, Melbourne, 1941. — **(384)** Hoagland et al: JEM *74*, 133, 1941 (B), *76*, 163, 1942. — **(384a)** Ch'en: Pr *32*, 491, 1934. — **(385)** Seastone et al: JI *33*, 407, 1937. — **(386)** Schramm: BC *74*, 532, 1941. — **(387)** Merrill: JI *30*, 169, 1936. — **(388)** Bawden et al: Br *18*, 275, 1937. — **(388a)** Knight et al: JBC *144*, 411, 1942; JAC *64*, 2734, 1942. — **(389)** Shedlovsky et al: JEM *75*, 165, 1942, *77*, 155, 165, 1943; S *95*, 107, 1942. — **(390)** Smadel et al: ArP *34*, 275, 1942. — **(391)** Rivers et al: JEM *69*, 31, 1939 (B). — **(392)** Smadel et al: JEM *72*, 389, 1940. — **(393)** Hughes: JI *25*, 275, 1933. — **(394)** Anson: in Schmidt, Chemistry of Amino Acids and Proteins. Thomas, Springfield, 1938. — **(395)** Holden et al: AJE *6*, 79, 1929. — **(396)** Wu: CJP *5*, 321, 1931. — **(397)** Mirsky et al: PNA *22*, 439, 1936. — **(398)** Cohn et al: JBC *100*, XXVIII, 1933. — **(399)** Wrinch: N *138*, 241, 1936. — **(400)** Harris et al: JBC *132*, 477, 1940. — **(401)** Bull: AE *1*, 39, 1941. — **(402)** Hartley: SystB *6*, 224, 1931. — **(403)** Pick et al: WK 1903, p. 659. — **(404)** Schmidt: JI *6*, 281, 1921 (B). — **(405)** Wu et al: CJP *1*, 277, 1927 (B). — **(406)** Spiegel-Adolf: BZ *170*, 126, 1926. — **(407)** Landsteiner et al: ZI *20*, 211, 1913. — **(408)** Sharma: ZI 77, 79, 1932. — **(409)** Schwab: ZI *92*, 289,

1938.— **(410)** Furth: JI *10*, 777, 1925 (B).— **(411)** Pels Leusden et al: ZI *78*, 393, 1933.— **(412)** Uwazumi: AMO *4*, 53, 1934.— **(413)** Rodenbeck: CB *123*, 460, 1932.— **(413a)** Henry: JEM *76*, 451, 1942 (B).— **(414)** Rothen et al: JEM *76*, 437, 1942; S *90*, 65, 1939.— **(415)** MacPherson et al: Pr *43*, 646, 1940.— **(416)** Hewitt: BJ *28*, 575, 1934 (B).— **(417)** Neurath et al: S *96*, 116, 1942; JPC *46*, 203, 1942 (B).— **(417a)** Fujiwara: DZG *11*, 384, 1928.— **(417b)** Dujarric et al: AP *65*, 63, 1940.— **(418)** Dakin et al: JBC *13*, 357, 1912; *15*, 263, 1913.— **(419)** TenBroeck: JBC *17*, 369, 1914.— **(420)** Landsteiner et al: ZI *26*, 142, 1917.— **(421)** Johnson et al: BJ *26*, 1202, 1932.— **(422)** Kober: JBC *22*, 433, 1915.— **(423)** Groh et al: ZPC *198*, 267, 1931.— **(424)** Csonka et al: JBC *93*, 677, 1931.— **(425)** Lin et al: CJP *2*, 131, 1928. — **(426)** Fink: JID *25*, 97, 1919 (B).— **(427)** Abderhalden et al: ZPC *81*, 315, 1912, *109*, 289, 1920.— **(428)** Abderhalden: ZPC *270*, 9, 1941 (B).— **(429)** Mori: JB *29*, 1, 185, 1939.— **(430)** Landsteiner et al: Pr *25*, 666, 1928, *30*, 1413, 1933.— **(431)** Bailey et al: JEM *73*, 617, 1941.— **(432)** Stull et al: JI *41*, 143, 1941 (B).— **(433)** Cooke et al: JAM *114*, 1854, 1940.— **(434)** Bliss: JI *34*, 337, 1938.— **(435)** Michaelis: DM 1904, p. 1240; Z. klin. Med. *56*, 409, 1905.— **(436)** Landsteiner et al: Pr *28*, 983, 1931.— **(437)** Holiday: RS *A 170*, 79, 1939. — **(438)** Wasteneys et al: Ph *10*, 110, 1930.— **(439)** Blagowestschenski et al: BZ *270*, 66, 1934 (B).— **(440)** Cuthbertson et al: BJ *25*, 2004, 1931.— **(441)** Knaffl-Lenz et al: AEP *71*, 407, 1913.— **(442)** Flosdorf et al: JI *32*, 441, 1937 (B).— **(443)** Collier: Canad. J. Res., Sect. B., *18*, 272, 305, 1940; v. JBC *12*, 233, 1912.— **(444)** Flosdorf: S *93*, 157, 1941.— **(445)** Chen: CJP *15*, 159, 1940. — **(446)** van Slyke et al: JBC *16*, 539, 1913/14.— **(447)** Wadsworth et al: JBC *116*, 423, 1936.— **(448)** Eaton: JI *33*, 419, 1937.— **(449)** Pappenheimer: JBC *125*, 201, 1938.— **(450)** Ross et al: JGP *22*, 165, 1938 (B); Pr *38*, 260, 1938.— **(450a)** Eagle: JEM *67*, 495, 1938 (B).— **(450b)** Follensby and Hooker: JI *31*, 141, 1936.— **(451)** Clarke: Gilman, Organic Chem. *2*, 875, 1938.— **(452)** Gustavson: CA *35*, 1422, 1941.— **(453)** Felix et al: ZPC *205*, 11, 1932.— **(454)** Landsteiner et al: ZI *26*, 133, 1917.— **(455)** Landsteiner et al: ZI *20*, 618, 1914. — **(456)** Jacobs et al: JI *36*, 531, 1939.— **(457)** Kleczkowski: Br *21*, 1, 1940.— **(458)** Horsfall: JI *27*, 553, 1934.— **(459)** Bawden: Br *16*, 435, 1935.— **(460)** Bawden et al: Br *17*, 204, 1936.— **(461)** Parker et al: JEM *62*, 65, 1935.— **(462)** Mutsaars et al: SB *132*, 469, 1939.— **(463)** Dulière: SB *127*, 1122, 1938. — **(464)** Eggerth: JI *42*, 199, 1941.— **(464a)** Smetana et al: JEM *73*, 223, 1941.— **(465)** Obermayer et al: WK 1906, p. 327 (B), 1903, p. 659, 1904, p. 265. — **(466)** Blumenthal: JBC *113*, 433, 1936.— **(467)** Ecker: RI *4*, 543, 1938.— **(468)** Müller: CB *32*, 521, 1902.— **(469)** Bauer et al: Verh. Ges. dtsch. Naturf. *2*, 389, 1912.— **(470)** Demanez: AI *8*, 233, 1933.— **(471)** Pick: HPM 2 ed., *1*, 685, 1912, 3 ed., *2*, 317, 1929.— **(472)** Lewis: JID *55*, 203, 1934.— **(473)** Wormall: JEM *51*, 295, 1930.— **(474)** Jacobs: JI *23*, 361, 375, 1932.— **(475)** Bauer et al: BZ *211*, 163, 1929.— **(476)** Strauss: AIF *21*, 197, 1928.— **(477)** Bruynoghe et al: Bull. Acad. Med. Belg. ser. V, *10*, 298, 1930.— **(478)** Goldschmidt et al: BC *58*, 1346, 1925; AC *456*, 1, 1927.— **(479)** Finkelstein: JI *25*, 179, 1933.— **(479a)** Clutton et al: BJ *32*, 1119, 1938.— **(480)** Mutsaars: AP *62*, 81, 1939.— **(481)** Ottensooser et al: BZ *193*, 426, 1928.— **(482)** Landsteiner: Zbl. Physiol. *9*, 433, 1895.— **(483)** Morel et al: Bull. Soc. Chim. Paris ser. 4, *41*, 1217, 1927, *43*, 881, 1928.— **(484)** Rohrlich: Dissertat. Univ. Berlin 1931.— **(485)** Eagle: Pr *34*, 39, 1936.— **(486)** Lieben: BZ *145*, 535, 1924.— **(487)** Mutsaars: SB *132*, 467, 1939.— **(488)** Armstrong: Proc. Chem. Soc. *8*, 103, 1892.— **(489)** Hantzsch: Diazoverbindungen, Samml. chem. Vortr., Enke, Stuttgart, 1902; BC *39*, 1084, 1906.— **(490)** Cohn: Erg. d. Physiol. *33*, 847, 1931.— **(491)** Kleczkowski: Br *21*, 98, 1940.— **(492)** Haurowitz et al: ZI *95*,

478, 1939. — **(493)** Landsteiner et al: ZI *26*, 122, 1917; BZ *58*, 362, 1913, *67*, 334, 1914, *61*, 458, 1914. — **(494)** Edlbacher: ZPC *112*, 80, 1921. — **(495)** Kiesel et al: ZPC *213*, 89, 1932; v. N *150*, 57, 1942. — **(496)** Leontjew et al: BZ *270*, 116, 1934. — **(497)** Landsteiner et al: ZI *21*, 193, 1914. — **(498)** Landsteiner et al: BZ *74*, 388, 1916. — **(499)** Heidelberger et al: JAC *63*, 498, 1941. — **(500)** Goldschmidt et al: ZPC *165*, 279, 1927. — **(501)** Kiesel: BPh *105*, 15, 1938 (B). — **(502)** Gaunt et al: BJ *33*, 908, 1939. — **(503)** Landsteiner et al: ZI *26*, 293, 1917. — **(504)** Hopkins et al: BJ *27*, 740, 1933, v. *30*, 1915, 1936. — **(505)** Creech et al: AJC *30*, 555, 1937, *35*, 203, 1933. — **(505a)** Marrack: ARB *11*, 644, 1942. — **(506)** Gutman: RI *4*, 111, 1938 (B); v. JAM *113*, 198, 1939. — **(507)** Berenblum et al: BJ *33*, 75, 1939. — **(508)** Arrhenius: Immunochemie, Akademische Verlagsges. Leipzig, 1907, p. 195. — **(509)** Nicolle: Les Antigenes et les Anticorps, Paris, Masson 1920. — **(510)** Hooker et al: JI *26*, 469, 1934, *30*, 41, 1936. — **(511)** Landsteiner et al: JEM *71*, 445, 1940. — **(512)** Cumley et al: PNA *27*, 565, 1941. — **(513)** Boyd: FI. — **(514)** Obermayer et al: Wiener Klin. Rundsch. 1902, No. 15. — **(515)** Wells et al: JID *8*, *12*, *13*, *14*, *17*, *19*. — **(516)** Jones et al: JBC *56*, 79, 1923. — **(517)** Thomsen: Antigens, etc., Copenhagen 1931, p. 43. — **(518)** Sörensen: KZ *53*, 102, 1930. — **(519)** MacInnes et al: S *93*, 438, 1941. — **(520)** Tiselius et al: JEM *69*, 83, 1939. — **(521)** Svedberg: RS *B 127*, 5, 1939; N *139*, 1051, 1937. — **(522)** Bernal: N *143*, 663, 1939. — **(523)** Hooker et al: JBC *100*, 187, 1933 (B). — **(524)** Zoet: Pr *32*, 1469, 1935. — **(525)** Nigg: JI *33*, 229, 1937. — **(526)** Landsteiner et al: ZI *10*, 68, 1911. — **(527)** Bawden et al: Br *22*, 208, 188, 1941. — **(528)** Bawden et al: Br *23*, 169, 1942. — **(529)** van der Scheer et al: JI *40*, 39, 1941. — **(530)** Fujiwara: Mitt. Med. Fak. Kyushu *5*, 325, 1920. — **(531)** Chibnall: RS *B*, *131*, 136, 1942. — **(532)** Wrinch: RS *A*, *161*, 505, 1937. — **(533)** Bull: AE *1*, 36, 1941. — **(534)** Pauling: JAC *62*, 2643, 1940. — **(535)** Astbury et al: N *147*, 696, 1941 — **(536)** Hektoen et al: JID *48*, 588, 1931, *57*, 337, 61, 1935. — **(537)** von Gara: ZI *71*, 1, 1931. — **(538)** Dean et al: JH *35*, 69, 1935. — **(539)** Björneboe: ZI *99*, 245, 1941. — **(540)** Bergmann et al: JBC *122*, 577, 1938 (B). — **(540a)** Astbury: AE *3*, 63, 1943. — **(541)** Landsteiner et al: Pr *28*, 983, 1931, *30*, 1413, 1933. — **(542)** Holiday: RS *B*, *127*, 40, 1939. — **(543)** Landsteiner: JEM *75*, 269, 1942. — **(544)** Felix et al: ZPC *205*, 11, 1932. — **(545)** Simms: JGP *11*, 629, 1928. — **(546)** Kabat et al: JEM *66*, 229, 1937; JAC *60*, 242, 1938. — **(547)** Haurowitz et al: JI *40*, 391, 1941. — **(548)** Doerr et al: ZH *96*, 258, 1922. — **(549)** Krusius: AA *67*, Ergzgsh., 47, 1910. — **(550)** Hektoen et al: PNA *11*, 481, 1925. — **(551)** Hektoen et al: JID *40*, 706, 1927. — **(552)** Lewis: JID *55*, 168, 1934. — **(553)** Bruynoghe: SB *118*, 1260, 1935. — **(553a)** Senges: ZI *92*, 431, 1938. — **(554)** Picado: SB *127*, 1096, 1098, 1938. — **(555)** Wells: JID *5*, 449, 1908. — **(556)** Pabis et al: ZI, Ref., *9*, 411, 1915. — **(557)** Starin: JID *23*, 139, 1918. — **(558)** Hooker et al: JI *24*, 141, 1933 (B). — **(559)** Dirscherl: EH *22*, 368, 1939. — **(560)** Adant: AI *6*, 29, 1930. — **(561)** Kallos et al: EH *19*, 200, 1937. — **(562)** Clutton et al: BJ *32*, 1111, 1938; JCS 1940, p. 119. — **(563)** Hopkins et al: BJ *27*, 1706, 1933. — **(564)** Wells: JBC *28*, 11, 1916. — **(565)** Boné: AI *13*, 177, 1938. — **(566)** Hewitt: BJ *31*, 1047, 1937. — **(567)** Gell et al: BJ *32*, 560, 1938. — **(568)** Goebel et al: JEM *60*, 599, 1934. — **(569)** Seibert: JI *28*, 425, 1935. — **(570)** Dudley et al: BJ *9*, 97, 1915, *15*, 187, 1921. — **(571)** Dakin et al: BJ *13*, 248, 1919. — **(572)** Dale et al: BJ *10*, 408, 1916. — **(573)** Obermayer et al: BZ *50*, 369, 1913. — **(574)** Anson et al: RS *B*, *97*, 61, 1924. — **(575)** Haurowitz: ZPC *183*, 78, 1929. — **(576)** Roche et al: BCB *16*, 757, 769, 1934; SB *115*, 1645, 1934, *118*, 174, 1935 (B). — **(577)** Kovacs: BZ *306*, 74, 1940. — **(578)** Bergmann et al: JBC *118*, 301, 1937 (B). — **(579)** Birkofer et al: ZPC *265*, 94, 1940. — **(580)** Block: CSH *6*, 79, 1938. — **(581)** Ecker et al:

JEM *71*, 585, 1940. — **(582)** Pillemer et al: JEM *69*, 191, 1939. — **(583)** Vickery et al: JBC *140*, 613, 1941. — **(584)** Beach et al: JBC *128*, 339, 1939, *148*, 431, 1943. — **(585)** Groh et al: ZPC *199*, 13, 1931. — **(586)** Heidelberger: JEM *73*, 695, 1941. — **(587)** Manteufel: Handb. biol. Arbtsmeth. *4*, Teil 8, 1809, 1928. — **(588)** Goldsworthy et al: JP *40*, 169, 1935. — **(589)** Pirie: BiR *15*, 377, 1940. — **(590)** Hektoen et al: JID *42*, 31, 1928. — **(591)** Roepke et al: JI *30*, 109, 1936. — **(592)** Hektoen et al: JAM *81*, 86, 1923, *84*, 114, 1925; PNA *11*, 481, 1925; JID *34*, 440, 1924. — **(593)** Goettsch et al: JBC *109*, 221, 1935. — **(594)** Culbertson: Pr *30*, 102, 1932, AJP *107*, 120, 1934. — **(595)** Landsteiner et al: JEM *71*, 231, 1940. — **(596)** Fischer: Pflüger's Arch. Phys. *223*, 163, 1930. — **(597)** Boyd et al: JI *33*, 111, 1937. — **(598)** Reiner et al: JBC *146*, 583, 1942. — **(599)** Moriyama: Jap. JEM *12*, 437, 1934. — **(600)** Bachrach et al: AH *70*, 1, 1909. — **(601)** Sevag et al: JBC *134*, 523, 1940. — **(602)** Verwey: JEM *71*, 635, 1940. — **(603)** Harington et al: IC, p. 822.

III

CELL ANTIGENS

AMONG antibodies against cells, the agglutinins and lysins for bacteria and erythrocytes have been most extensively investigated. In the fields of bacteriology and medicine bacterial agglutinins have found wide application; they aid in the identification of bacteria when known immune sera are used and in the diagnosis of infectious diseases where the patient's serum is allowed to act upon known bacteria, as in the Gruber-Widal test for typhoid fever. The introduction of these methods started a new chapter in bacteriology when the discovery was made, first in the group of vibrios,[1] that microbes indistinguishable morphologically and in culture can be subdivided by immunological tests. Hemolysins and hemagglutinins were made use of in numerous investigations on general serological problems inasmuch as lysis of red blood corpuscles is readily observed, owing to the pigment contained in these cells and to the ease of preparing suspensions that are suitable for agglutination reactions.

It was formerly taken for granted, because of the belief in the sole importance of proteins for immune reactions, that the antibodies to bacteria and blood corpuscles bear an exclusive relation to the proteins in the cells. This view became doubtful when it was noticed that hemolysis by normal sera may be inhibited by alcohol and ether

[1] Pfeiffer; Kolle and Gotschlich (1).

extracts of blood corpuscles,[2] and that hemolysins may be formed upon injection of such extracts.[3] These facts and studies on other animal and bacterial antigens suggested that lipids take part in serological reactions and in the production of antibodies. Zinsser and Parker (5) then discovered that in bacteria there are serologically reactive substances other than proteins ("residue antigens"), which Heidelberger and Avery, working with pneumococci, recognized as polysaccharides, and it was soon found that, in addition to proteins, high molecular carbohydrates are responsible for many bacterial agglutinin and precipitin reactions. When the usual methods of immunization were employed, however, it appeared that the specific protein-free substances of animal or bacterial origin, although active in vitro, often induced no or only slight antibody response. For substances of this sort, in contradistinction to typical antigens which possess both properties, the term hapten has been proposed by the author (6).

Apart from the chemical data, certain findings, commonly not fully appreciated, indicated that many of the antigens found in animal cells belong to a separate class with respect to specificity. The distinctive characteristics to be discussed below are the striking serological differences between the cells of individuals of the same species, the frequent occurrence of so-called heterogenetic antigens, i.e., similarly reacting substances present in unrelated kinds of organisms, and the fact that blood cells of closely related species exhibit sometimes much greater differences than the respective serum proteins. These features, along with observations on the chemical nature of the antigens bear out in the writer's opinion the conclusion that there exist two systems of species specificity in the animal kingdom, the specificity of proteins and that of cell haptens (7). The proteins, it would seem, undergo gradual variation in the course of evolution, while haptens are subject to sudden changes not linked by intermediary stages.

A peculiarity of cell reactions is that, unlike precipitins for proteins, agglutinins and lysins are often demonstrable in normal sera, a contrast which probably does not depend solely on the respective techniques. The regular antibody response to cell antigens and the high activity of many cell antigens — as little as 0.05 mm³ of red cells, or 10⁵ bacterial cells, were seen to evoke antibodies [4] — might also seem distinctive, but it should

---

[2] Landsteiner and von Eisler (2), Misawa (3).
[3] Bang and Forssman (4).   [4] (8), (8a).

be noted that the anaphylactic state can be induced by minute quantities of proteins and that precipitating immune sera have been obtained upon injection of very little protein.[5]

RED CELLS AND TISSUES;[6] SPECIES DIFFERENTIATION. — Chemical analysis of the antigens in erythrocytes has lagged behind the extensive work put into the serology of these cells. The specific substances are contained in the blood stromata, and it might be assumed a priori that the stroma proteins participate in the reactions, yet it is difficult to assess their role since soluble serologically reactive proteins have not been separated and because it is uncertain to what extent the serological deterioration of stromata upon extraction, as with alcohol, is due to removal of active substances or to changes in the protein (10); a closer investigation of this point would be needed.[7, 8] With specific non-protein substances of the red cells some definite chemical results have been obtained, which will be taken up later (p. 231 ff).

Standard in the serological examination of free cells are agglutination and lysis tests, and precipitation and complement fixation are employed in testing soluble constituents of tissues (and cells) which cannot readily be made into homogeneous suspensions. For motile cells, spermatozoa, protista, and ciliated epithelia,[9] immobilization can serve as a criterion. Measurement of surface potentials was applied by Mudd (22) for spermatozoa.

Titration of agglutinating and lytic sera,[10] satisfactory for common purposes, is carried out by determining the highest dilution of immune serum at which a visible reaction or a reaction of certain intensity still occurs; colorimetric measurements may be used for determining the degree of hemolysis. Titration of lysins requires the addition and standardization of complement. An absolute quantitative estimation of agglutinins, introduced by Heidelberger and Kabat, consists of nitrogen analyses on the bacterial sediment after exhaustion of antisera by bacteria (24).[11]

[5] Hektoen and Cole (9).

[6] Detailed reviews on agglutination and hemolysis are given by Sachs (9a), and Schiff (9b).

[7] Material soluble in organic solvents is about one fifth of the dry weight. On stroma proteins v. Jorpes (11), Boehm (12). In the experiments reported on the formation of agglutinins and lysins by injecting "globulins" of blood cells (13), (14), errors may have been caused by the presence of stromata in the solutions. Whether weak inhibition effects of peptone preparations on normal hetero-agglutinins (14a) and plant agglutinins are due to proteoses is not fully assured.

[8] The method of modification by chemical treatment has not contributed significant information on the nature of erythrocyte antigens (15), (16).

[9] (17–19) (serological groups in paramecia); (20) (differentiation of related algae); (21) (trypanosomes).

[10] For a compilation of technical procedures v. (22a), (23), (23a).

[11] On the method of optimal proportions v. (24), (25).

Immune sera prepared with blood corpuscles of one animal species will, like precipitins, give reactions with the bloods of related species, too. In Table 11 such results are presented from tests with agglutinating immune sera for human and monkey blood. The figures indicate the titers obtained by determining the highest active dilutions of the sera.

TABLE 11
[after Landsteiner and Miller (26)]

Blood Corpuscles	Immune Serum for		
	Human Blood		Rhesus Blood
	# 1	# 2	
Man	1500	750	125
Chimpanzee	750	750	125
Orang-Utan	500	...	250
Macacus Rhesus	125	...	750

Blood cells of closely allied species can be distinguished either by the agglutinin titer or, more regularly, by absorption experiments.[12] Thus, if to an antiserum for erythrocytes is added heterologous blood of another species, likewise acted upon by the immune serum, the cross reacting agglutinins are absorbed, and after centrifugation there remain in the supernatant fluid agglutinins for the homologous blood.

The following test was carried out with immune sera for horse blood which also agglutinated donkey blood to high titer. The solutions obtained after absorption with blood corpuscles of the donkey no longer acted on these cells but agglutinated horse blood in high dilutions, and the outcome was analogous in the inverse experiment — immune serum for donkey blood, absorption with horse corpuscles (Table 12). These results acquire *additional* significance when contrasted with the difficulty in differentiating clearly horse and donkey serum proteins by precipitin reactions.

In the same manner, marked differences were shown to exist between the blood cells of man and chimpanzee (26), sheep and goat, fox and dog (29), and species of mice and doves (30), (31).

It is worthy of remark that differences between closely related species and individuals of the same species are often demonstrable by means of natural agglutinins, upon direct tests or with absorbed sera (32). For instance, some human sera give strong agglutination with chimpanzee blood (irrespective of the blood groups) although the

[12] Ehrlich and Morgenroth (27), Landsteiner et al. (28); (26).

serological differentiation of the serum proteins is in this case a somewhat delicate matter. On the other hand, the occurrence or absence of agglutination or hemolysis, when normal serum and blood of various species are allowed to interact, cannot be used for estimating

TABLE 12
[after Landsteiner et al. (28)]

Immune Serum for	Treatment	Test Bloods		
		Horse	Donkey	Mule
Horse Blood	Not absorbed	8000	4000	6000
	abs. with Horse	<40	<40	<40
	" " Donkey	2400	<40	1600
	" " Mule	200	<40	<40
Donkey Blood	Not absorbed	8000	8000	8000
	abs. with Horse	40	3200	1600
	" " Donkey	40	<40	40
	" " Mule	<40	160	40
Mule Blood[+]	Not absorbed	3200	1600	3200
	abs. with Horse	<40	80	40
	" " Donkey	800	<40	400
	" " Mule	<40	<40	<40

The sera were absorbed with erythrocytes as indicated.   [+] vide p. (117).

zoological relationships because of a randomness in the reactions of normal hemagglutinins (p. 129), not to mention that, owing to the presence of isoantibodies, a serum may agglutinate blood of its own and not that of a widely distant species; also, immune hemagglutinins may give misleading results on account of heterogenetic reactions.

On antibodies agglutinating red cells infected with plasmodia v. Eaton (32a).

Serological inquiry has extended to almost all component parts of the animal body.[18] The antisera against tissues display to various degrees organ specificity and species specificity. Among results on species specificity those of Henle on spermatozoa (44) and of Grund

[18] Reviews (and bibliography): (33–37); leucocytes (pus, caseous tissue): (38–42); spermatozoa: (36), (43–44); v. (44a); eggs: (36), (44b), (44c), (351); epithelia, also ciliated: (45–46); brain: (47–53); brain tumors: (54); liver: (55–56); kidney: (57–58); stomach, intestines: (59–61); testicle: (62–63); lung: (64); hypophysis: (65); epiphysis: (66); suprarenal gland: (67); placenta: (68–70); embryonic tissues: (71–72); milk: (73–74); saliva: (75); organ-specific anaphylactic antibodies produced with bacterial cultures grown on broth made from organs: (76); various organs: (77–80); autolyzed liver (80c).

with organ juices (77) may be cited. In Table 13 some tests demonstrating organ specificity are presented. The term means that a serum reacts with homologous organs of unrelated species or differentiates one organ from others within a species.

TABLE 13. — COMPLEMENT FIXATION TESTS WITH ANTISERA PRODUCED BY INJECTING RABBITS WITH SMALL QUANTITIES OF CELLULAR MATERIAL FROM CATTLE

[after Landsteiner et al. (81)]

| Antigens | Immune Sera | | | |
	Trachea	Thymus	Kidney	Sperm
Trachea	80	<10	<10	0
Thymus	20	40	0	<10
Kidney	<10	0	80	<10
Sperm	0	0	0	40

Cross reactions were observed with various organs not tabulated and some sera contained lysin for ox blood.

The figures give the limiting dilution of the antisera still causing complete fixation of complement; o signifies practically no inhibition in 1:10 dilution.

The occurrence of overlapping reactions between various kinds of cells from one species is definitely shown by the presence of hemolysins (and hemagglutinins) in antisera produced to tracheal epithelia (45), or spermatozoa (80a) — free from admixture with blood; these antibodies are bound by the cells which served as antigens. Hemagglutinins and hemolysins are developed also upon injection of serum;[14] further, hemolysin formation can be stimulated by urine (80b), and milk elicits antibodies for epithelial cells and erythrocytes (45).

The chemistry of the substances associated with the immune reactions of organs is still in a preliminary stage. Formerly the reactions were commonly attributed to proteins, without attempts being made to separate well defined fractions, and up to now only some well characterized organ proteins, already noted, as thyroglobulin, insulin, keratin, mucins, hemoglobins, ferritin, have been the subject of serological examination. A brief account is given by Witebsky (33) on sharply distinctive reactions of globulins from suprarenal gland and pancreas. Organ nucleoproteins asserted by some workers to be responsible for specific reactions were found by others to be without pronounced antigenic or reactive properties.[15]

[14] Ref. chapter II (73–79). The reverse, production of serum precipitins by injecting washed red cells, has been observed, presumably owing to adsorbed proteins.
[15] v. (p. 63); (82), (83).

Of interest chemically and promising as material for further research are the serologically active substances extractable by alcohol, with the Forssman hapten as paradigm; another well known instance is the substance, practically ubiquitous as regards organs, which underlies the Wassermann reaction (p. 230). Upon injection of brain, Witebsky obtained two sorts of immune sera.[16] Some were species specific alone, presumably directed toward proteins, while others reacted with emulsions of alcoholic extracts of brain (and testicle)[17] of various animals, a behaviour recalling that of anti-lens sera. A second brain-specific hapten, soluble in hot alcohol and present in the protagon fraction, was described by Schwab.[18] Other organ-specific antibodies, reacting with alcoholic extracts, were obtained by injection of liver, lung, leucocytes, and so forth, often together with Wassermann antibodies.[19]

A new method for the examination of tissues consists of the separation of particles by differential centrifugation. Submicroscopic particles thrown down at 20,000 r.p.m., chiefly consisting of ribonucleoprotein and phospholipids (Claude),[20] exhibit species and, to some extent, organ specificity;[21] the particles were found to contain "Wassermann hapten" and those from mice and chickens to carry Forssman antigen. Cross reactions between organs, more pronounced in complement fixation than in agglutination, were largely eliminated by absorption. Sera for brain and liver particles gave overlapping reactions between unrelated species; the brain antigen was exceptional in that it withstood heating to 100° and was devoid of species specificity. Brain particle antisera reacted with alcoholic brain extracts.

Serological distinction of morphologically different parts of a cell was demonstrated by Henle et al. with heads and tails of spermatozoa, separated by sonic vibration. Antigens showing characteristic agglutination patterns were detected, one in the tails and two in the heads, in addition to a heat stable antigen shared by both (87).

Experiments of Levit et al. (88) purporting to show that the Y (sex) chromosome in Drosophila can be serologically recognized have not been repeated by others. Immunological tests on Drosophila species agreed fairly

---

[16] Brandt, Guth and Müller (47); (48).
[17] (62), (63); v. (51).
[18] (84); v. (84a) (extraction with phenol).
[19] v. (48).
[20] (84b); v. (84c).
[21] Furth and Kabat (85), Henle et al. (86).

well with taxonomic expectations despite the complex nature of the antigenic substrates (whole flies).[22]

Numerous experiments, often unsuccessful or dubious as to the specificity of the effects, have been undertaken with the aim of demonstrating changes in organs following the injection of organ antibodies (cytotoxic effects),[23] a subject which is of theoretical and practical importance, to mention only the possibilities of studying organ functions, inhibition of pathological growth, or sterilization with sera against spermatozoa (43). Definite results in inducing nephritis have been obtained by Masugi (92) by means of anti-kidney sera.[24] Evidence has been adduced to prove that the nephritic changes are not caused by a direct toxic effect of antibodies [Kay (95)] but the original claim is supported by other investigators [Swift and Smadel (96)]. With lymphocytes and neutrophilic leucocytes a decrease in number was seen following injection of the corresponding antisera [Cruickshank (97)]. Anti-platelet sera engender, as shown by Ledingham et al., hemorrhages simulating purpura hemorrhagica.[25]

Substances which, among other effects, agglutinate sperm and eggs and probably perform a part in fertilization exist in the germ cells of echinoderms (also molluscs, annelids). Although having specific properties and apparently of protein nature, these substances are hardly, to judge by the thermostability of some, closely related to serum agglutinins.[26]

That much attention has been paid to the serology of tumors with the aim of developing a diagnostic method for malignancy or even to effect immunization against malignant growth is easily understood. Reactions (complement fixation) between tumor extracts and the serum of patients suffering from malignant disease have repeatedly been observed; [27] as it has turned out, these take place with other sera too (pregnancy, syphilis, etc.) and the method is not serviceable for diagnostic purposes. Immune sera prepared against human or

[22] Cumley (89).
[23] (34–36), (80). For demonstrating cytotoxic effects tissue cultures have also been employed (90), (91), (72).
[24] Tikamitu (92a) reported that glomerulonephritis can be produced also by anti-lung immune sera; v. Seegall et al. (92b). Cirrhotic changes of the liver have been produced by long continued administration of liver antisera (93); v. (94).
[25] v. (98), (99).
[26] Lillie; Tyler (100); von Dungern (44b); (101) (displacement by anti-fertilizin of complement bound to fertilizin).
[27] Hirszfeld and Halber, Ewing and others.

animal tumors [28] have in general exhibited imperfect specificity and have given overlapping reactions with normal tissues, although according to some reports antisera that react selectively with the tumor used as antigen and related ones can be obtained, also in animals of the species from which the growth is derived.

Definite results as regards specificity were secured by Kidd [29] who, using complement fixation tests with the serum of rabbits bearing transplantable tumors, was able to differentiate sharply from each other and from normal tissues the rabbit tumor described by Brown and Pearce and the carcinoma (V 2) developed after infection of rabbits with papilloma virus; the antibodies against carcinoma V 2 were different from those in the same serum which gave complement fixation with and neutralized the papilloma virus. The complement fixing antigens in the tumors are thrown down within an hour by centrifuging at 20,000 r.p.m. like the particulate antigens of normal tissue previously noted.[30] It cannot yet be said whether these immune bodies, which develop much more frequently in one strain of rabbits than in others, are really directed against substances characteristic of malignant growths, or against the special types of cells which constitute the tumors, or else whether they might perhaps represent isoantibodies developed to cells of the genetically different animals in which the tumors arose.

Obviously, it would be of prime interest if antibodies, such as those occurring in the serum of animals with transplanted tumors, could be found in the case of autochthonous ones.

Antibodies neutralizing the causative filterable agents of fowl tumors — mostly believed to be exogenous in origin — and giving reactions in vitro are produced upon immunization of rabbits and may be present in the serum of chickens bearing the tumor.[31] Sera neutralizing the agent of filterable sarcomas have also been obtained from birds bearing non-filterable tumors, spontaneous or induced by carcinogenic hydrocarbons; possibly these tumors are intrinsically related to the filterable tumors.

The action of anti-tumor sera on cultures of normal and neoplastic cells was studied by Lumsden (116) and Phelps (117). The existence of a

---

[28] (102–107), (107a).
[29] (108); v. Dmochowski (109), Cheever (110).
[30] In sera of tumor-bearing mice complement fixing antibodies reacting with alcoholic tumor extracts were described by Hoyle (111).
[31] v. Barret (112); (113), (114); Gye; Andrewes; Dmochowski and Knox (115).

special tumor protein present in the serum of patients has been claimed (118).

DIFFERENCES IN CELLS OF INDIVIDUALS OF THE SAME SPECIES. — The principal facts concerning this subject are that normal sera may agglutinate or hemolyze the erythrocytes of other individuals of the same species, and that on injection of red cells antibodies may be formed which by agglutination or hemolysis differentiate the blood corpuscles of various individuals within a species.[32]

One would have supposed that reactions distinguishing individuals, if they occurred at all, would be much weaker, indeed of a different order of magnitude than species specific reactions. That on the contrary they are in many cases very marked is significant, and there is another characteristic, namely the existence of well-defined and clearly separated types. From these reactions it follows that the cells of various individuals of a species contain different substances which, however, for the time being, are distinguishable only by serological methods.

A striking example of individual blood differences is furnished by the four human blood groups.[33] Their distinctive properties arise from the distribution of two agglutinogens in the erythrocytes and two isoagglutinins in the sera. The agglutinogens, designated as A and B, can be lacking or one or both be present in a given individual; the serum contains those isoagglutinins which react upon the agglutinable substances not present in the cells. Hence there results the scheme in Table 14.

Agglutinins for group O cells exist in exceptional human and some animal sera. Group A (and similarly AB) can be further divided into subgroups (A_1 and A_2) (von Dungern and Hirszfeld), distinguishable by sera of group B (or O), previously absorbed with A_2 cells. After such absorptions an agglutinin fraction (a_1) is left in the supernatant which reacts almost selectively with A_1 blood. In rare instances the existence of still other inheritable varieties (A_3, A_4), of very weak reactivity with a agglutinins, has been demonstrated.

Phenomena encountered as a cause of error in making blood grouping tests when the materials have become infected are the appearance of agglutinable properties in erythrocytes and the development of hemagglutinins in serum, produced by the enzymatic action of certain bacteria.[34]

[32] Landsteiner (119), Ehrlich and Morgenroth (120).
[33] Landsteiner (121). Complete reviews are given by Wiener (122), Snyder (123), Schiff and Boyd (124), Steffan (125).   [34] v. (125a).

The blood groups, best named after the agglutinable substances [35] in the cells, are sharply differentiated, and this is emphasized by the fact that the group properties A and B are inherited as dominant genetic units.[36] According to Bernstein's theory (127), abundantly corroborated, the heredity of the four groups is determined by three allelic genes O, A, B, as presented in the diagram on page 86.

TABLE 14

Groups	Agglutinins in the serum	Red blood corpuscles of groups			
		O	A	B	AB
O	α and β	o	+	+	+
A	β	o	o	+	+
B	α	o	+	o	+
AB	..	o	o	o	o

(Agglutination is indicated by the sign +, negative reaction by o.)

In the scheme of the blood groups in Table 14 no account is taken of the normal autoagglutinins that act on the red cells of the same individual and other individuals of the species, a phenomenon frequently seen in man and animals (130). Save in certain pathological cases,[37] this autoagglutination is perceptible only at low temperature [38] and therefore does not interfere with the blood grouping tests. In the susceptibility to changes in temperature autoagglutinins are similar to the autohemolysins of paroxysmal hemoglobinuria where lysis takes place after chilling of the patient's blood and subsequent warming to body temperature (Donath and Landsteiner) [39] (p. 102).

The application of isoagglutination in blood transfusion is based on the fact that when the serum of the recipient contains antibodies acting on the donor's cells — these critical combinations are evident from a glance at the table — the latter are liable to be destroyed with resulting serious sequelae. Because of the relatively small quantity of introduced serum, as compared with the blood and tissues of the recipient, the donor's antibodies are usually not harmful and blood of group O can as a rule be used for recipients of any group.

[35] The expressions agglutinogen (antigen) and agglutinable substance are used in this review regardless of whether or not the components in question can be isolated from the whole antigen complex.

[36] von Dungern and Hirschfeld (126). On the inheritance of blood properties in animals v. Todd; Schermer (128), Kaempfer (129), Wiener (122).

[37] v. (131).

[38] (130); v. (132) (serum proteins precipitating in the cold); see p. 250.

[39] (133); Mackenzie (133a) (review).

GENOTYPES AND PHENOTYPES OF CHILDREN FROM MATINGS BETWEEN PARENTS OF THE SIX GENOTYPES *

Genotypes of Parents	→ OO	AA	AO	BB	BO	AB
↓ OO	OO O	AO	$\frac{1}{2}$AO; $\frac{1}{2}$OO	BO	$\frac{1}{2}$BO; $\frac{1}{2}$OO	$\frac{1}{2}$AO; $\frac{1}{2}$BO
AA	A	AA A	$\frac{1}{2}$AA; $\frac{1}{2}$AO	AB	$\frac{1}{2}$AB; $\frac{1}{2}$AO	$\frac{1}{2}$AA; $\frac{1}{2}$AB
AO	$\frac{1}{2}$A; $\frac{1}{2}$O	A	$\frac{1}{4}$AA; $\frac{1}{2}$AO $\frac{1}{4}$OO $\frac{3}{4}$A $\frac{1}{4}$O	$\frac{1}{2}$AB; $\frac{1}{2}$BO	$\frac{1}{4}$AB; $\frac{1}{4}$AO $\frac{1}{4}$BO; $\frac{1}{4}$OO	$\frac{1}{4}$AA; $\frac{1}{4}$AB $\frac{1}{4}$AO; $\frac{1}{4}$BO
BB	B	AB	$\frac{1}{2}$AB; $\frac{1}{2}$B	BB B	$\frac{1}{2}$BB; $\frac{1}{2}$BO	$\frac{1}{2}$AB; $\frac{1}{2}$BB
BO	$\frac{1}{2}$B; $\frac{1}{2}$O	$\frac{1}{2}$AB; $\frac{1}{2}$A	$\frac{1}{4}$AB; $\frac{1}{4}$A $\frac{1}{4}$B; $\frac{1}{4}$O	B	$\frac{1}{4}$BB; $\frac{1}{2}$BO $\frac{1}{4}$OO $\frac{3}{4}$B $\frac{1}{4}$O	$\frac{1}{4}$AB; $\frac{1}{4}$BB $\frac{1}{4}$AO; $\frac{1}{4}$BO
AB	$\frac{1}{2}$A; $\frac{1}{2}$B	$\frac{1}{2}$A; $\frac{1}{2}$AB	$\frac{1}{4}$AB; $\frac{1}{2}$A $\frac{1}{4}$B	$\frac{1}{2}$AB; $\frac{1}{2}$B	$\frac{1}{4}$AB; $\frac{1}{2}$B $\frac{1}{4}$A	$\frac{1}{4}$AA; $\frac{1}{2}$AB $\frac{1}{4}$BB $\frac{1}{4}$AA $\frac{1}{2}$AB; $\frac{1}{4}$B

* The genotypes of the children are given in the upper right portion of the rectangle, the phenotypes in the lower left portion. The expected frequencies of the genotypes (and phenotypes) of the children are indicated by fractions.

Besides the agglutinogens in the blood cells, related water-soluble group-specific substances are present in tissues, body fluids and secretions (p. 232), except that in a minority of individuals the secretions, as saliva, gastric juice, and the corresponding glands are practically devoid of the substances, a condition which is determined by heredity.

Reactions for blood groups on stains (blood, saliva, semen) are used in forensic medicine. These are done by demonstrating either the specific substances A and B in absorption or inhibition tests, or the isoagglutinins which however do not last for so long a time in the dried state.

The blood of anthropoid apes contains isoagglutinogens and iso-agglutinins indistinguishable from those present in human blood.[40] Chimpanzees belong largely to group A, a few to O, and in orang-utans and gibbons groups A, B or AB have been recognized. Blood of lower animals often contains agglutinins reactive with human A or B cells, and agglutinable substances, mostly characteristic for the species, acted upon by human α or β isoagglutinins although different from the human isoagglutinogens.[41] Individual properties demonstrable by means of natural isoagglutinins exist in various vertebrates, in some instances allowing a classification into groups (not coinciding with that of human blood). But the conditions are often more complicated than with human blood because of the greater variety of reactions and the irregular occurrence of isoagglutinins, and for the latter reason the most instructive results have been obtained with immune sera.[42] While at first the use of isoantibodies seemed necessary, it was soon found that normal and immune sera from a foreign species[43] can be successfully employed; here the agglutinins and lysins active for all the erythrocytes of the species examined must first be removed by absorption with suitable blood cells.

The experiments made on cattle and chickens[44] yielded the remarkable result that the blood of almost every single individual can be identified, and this seemed to constitute a distinction between animal and human blood. As a matter of fact, by using various methods numerous individual blood differences can also be demonstrated

[40] Landsteiner and Miller (26), (32), von Dungern and Hirschfeld (134), Troisier (135). The blood of gorillas behaves differently: Candela, Wiener and Goss (136).
[41] On the relationship between the blood properties of various animal species and of man, v. (134), (137–140).
[42] References may be found in (122), (125), (141–144).
[43] (145), (134), (146–149).
[44] Todd et al. (150), (147); v. (120).

in man. In particular, on immunizing rabbits with human blood, it was possible to obtain, in addition to antibodies for the agglutinogens A and B, agglutinins for three other "factors," denoted as M, N and P [Landsteiner and Levine (151)] which are found with equal frequency in all four groups.[45] The agglutinogens M and N provide a second instance of simple Mendelian inheritance in man. Their heredity is determined by a pair of allelomorphic genes which give rise to the three phenotypes M, N and MN. The genes are almost certainly located in a different pair of chromosomes from that governing the blood groups.

An additional property of human blood (Rh) was recently detected by means of immune sera prepared with the blood of rhesus monkeys.[46] The factor is inherited as a dominant Mendelian character. It is of clinical importance as it leads at times to the formation of isoantibodies upon transfusion [Wiener and Peters (154)] or in mothers lacking the agglutinogen and bearing a child possessing it. Consequently transfusion accidents may happen; moreover, the antibodies have been found to be the chief cause of erythroblastosis fetalis, a blood disease in new-born children [Levine et al. (155)]. In this connection it should be mentioned that the agglutinogens A and B likewise can stimulate antibodies in individuals not having these antigens, following transfusion or during pregnancy.[47] On the other hand, it very rarely happens that after blood transfusions antibodies against M or P are formed, and no certain cases of human anti-N are known.

With the aid of the individuality reactions exclusions can be made in about one third of all cases of falsely alleged paternity.[48] To quote the chief instances, the dominant factors A, B, M, N, Rh, do not appear in children unless present in the blood of at least one parent, and M or N parents cannot have offspring of type N or M respectively. Also, according to Bernstein's theory, a child of a group O parent cannot be AB, and vice-versa. In the study of human and animal heredity the individual blood characters have acquired importance because of their easy and unambig-

[45] Natural isoagglutinins for M have been found only very exceptionally; P agglutinins, mostly weak, are not quite so rare. On the detection of P with normal animal sera and the inheritance of this property v. Dahr (152). The production of P immune antibodies is not readily attained with present methods.

[46] Landsteiner and Wiener (153).

[47] v. (122). On isoimmunization in pregnancy v. Hirszfeld; Jonsson (156); Hookei (157).

[48] Less well-known is the application, based upon the identity of serological blood properties in identical twins, as criterion for zygosity.

uous determination, constancy under varied environmental conditions and the simple genetic mechanism.

The examination of families has shown that the known blood properties are not sex linked and have rendered it reasonably certain that A-B-O and M-N are inherited independently of one another; the same probably holds for the factors Rh and P. Consequently, several pairs of chromosomes in man appear to be marked by the serological blood characters so far recognized.

Since, with the exception that bloods lacking both M and N do not exist, each of the factors named may be present or absent in any one blood, there result 72 different types of human blood cells, if one counts the two subgroups of groups A and AB. And this does not include rare varieties of the properties A and N, subtypes of Rh, and reactions with abnormal human isoagglutinins (122) which cannot be easily reproduced because of difficulty in procuring the necessary reagents.[49] The actual number of individual differences already established is, therefore, certainly much greater. Moreover, not every difference need be reflected in the antibodies formed on immunization and therefore there is no assurance that all of them will soon, if ever, be detected.

For the investigation of chicken blood Todd used immune isoagglutinins. When a number of such immune sera were pooled to increase the variety of antibodies, and absorbed with a single blood, as a rule the exhausted sera acted on the blood corpuscles of every chicken except the one used for absorption. Analogous results had been obtained by Todd and White with isolysins and cattle blood. Certainly these reactions, apparently exhibiting complete individual specificity, are not fundamentally different from those seen with human blood and do not bespeak the existence of special properties possessed by single individuals, an assumption inadmissible for genetic reasons; indeed it was possible to demonstrate serologically defined factors in the blood of chickens,[50] and in cattle blood many — not less than thirty — factors were identified and their heredity investigated.[51] One need not then assume the existence of a huge number of individual cell substances in one species (v. Loeb) since even a moderate number of characters would furnish enough combinations (2^n for n independent factors) to explain the results of Todd's absorp-

---

[49] Andresen (158); (122).
[50] v. (149); Wiener (159), Little (160), Hofferber and Winter (161).
[51] Ferguson, Stormont and Irwin (162).

tion experiments. His results on heredity of individual blood differences in chickens were discussed by Wiener (159) from this point of view. By selective inbreeding, lines of chickens with like blood properties could be established (Todd).

A similar conception may explain transplantation specificity, which manifests itself in that normal tissues or even spontaneous tumors take rarely, if at all, unless derived from the same or a genetically closely related individual.[52] The supposition that the specificity of transplantation and of serum reactions rests upon a similar chemical basis is plausible because of the analogy between the two phenomena, and is substantiated by the demonstration of individual serological differences in tissues and cells. Repeated attempts to improve the outcome of transplantation (skin grafts) in man by selecting donors of the same blood group or to establish any correlation between the success of transplantation of normal tissues and serological blood properties of host and donor have failed to give decisive results.[53] This does not settle the main issue since it is possible that significant blood properties have escaped detection or, perhaps more probable, the tissues contain individually specific substances other than those of the blood cells. And it is not difficult to understand that differences too small to be of significance in blood transfusion may prevent the permanent survival of tissue grafts. At any rate, Gorer observed in experiments with a mouse sarcoma that the latter could be transferred only to mice whose blood had a certain agglutinogen detectable by rabbit immune sera. The same antigen was contained also in the sarcoma and consequently these mice failed to produce antibodies which might be injurious to the tumor cells; like observations were made with mouse leukemia [Gorer (166)], and with rat sarcoma [Lumsden (166a)]. The interference of antibodies with transplantation is further indicated by experiments of Claude on the protection of mice against leukemia with particulate material from leukemic tissues [(166b); v. (166c)].

The conclusion which seems to follow from transplantation experiments, that individual differences are much less pronounced in invertebrates and in the lower vertebrates than in higher species, has not yet been sufficiently tested with immunological methods. It is known, however, that in lower animals even transplants of tissues from other species may grow, an ob-

---

[52] See the comprehensive experimental investigations by L. Loeb (163), Kozelka (164).

[53] (164); Haddow (165).

servation which indicates a dissimilarity between these and higher species with respect to tissue reactions.[54]

RACIAL DIFFERENCES. — The discovery of the blood groups suggested at once the investigation of serological differences among human races. As L. and H. Hirschfeld (169) demonstrated, the relative frequency of the four blood groups varies in different human races, and is, to a certain degree, characteristic. The most conspicuous findings among those reported by these authors and in numerous subsequent papers, are the predominance of agglutinogen A over B in Europeans and Australian aborigines, the inversion of this ratio in a number of Asiatic peoples,[55] and the distribution of the blood groups in the majority of full-blooded American Indians [56] who belong almost exclusively to group O and, as was found later, have an unusually high frequency of M and Rh. Similar differences in the distribution of blood properties have been obtained in investigations on races of animals.[57] The peculiarities observed, then, do not serve to demonstrate the existence of constant differences by which the race of an individual could be ascertained but have only statistical significance. In this regard, the serological qualities are comparable to other more or less distinguishing but not constant attributes of a race, like the color of the eyes or hair, shape of skull, or body height.

BACTERIAL ANTIGENS AND TYPES. — In bacteria various kinds of specific substances [58] — proteins, complex antigens, carbohydrates — have been determined serologically, and there are antigens and haptens peculiar to separate parts of the bacterial cell.[59] Certain specific polysaccharides are localized in the capsules of pneumococci and capsulated bacilli, and differences have been demonstrated in the reactions of bacterial bodies and flagella. This was first discovered in an investigation by Smith and Reagh [60] on a flagellated and a nonmotile strain of hog-cholera bacilli. Immunization with the latter

[54] On individual differences in protozoa v. (167), on self-sterility (168).

[55] e.g., about 39% A, 8% B, 50% O, 3% AB in Englishmen, 19% A, 41% B, 31% O, 9% AB in Hindus (from two representative reports).

[56] (170–173). Matson (174) found in two tribes of American Indians a frequency of group A as high as 76% and 83%. A rare blood property was detectable in 7% of American negroes and in less than 1% of white individuals (175).

[57] (176–177), (161).

[58] These are dealt with in other sections (pp. 33, 210).

[59] Reference should be made to the interesting discussion by Topley and Wilson of the localization of antigens on or beneath the bacterial surface and its bearing on the immunological manifestations (178).

[60] (179), Beyer and Reagh (180); v. Weil and Felix (181).

variant gave rise to low titred immune sera which agglutinated both strains and brought about compact, granular agglutination; sera made with the motile strain caused the formation of large, fluffy aggregates and contained two agglutinins separable by absorption, one identical with the antibody for the non-motile strain, the other agglutinating only the flagellated bacilli. Similar observations were subsequently made with many other motile strains. The usual distinction based upon these findings of one somatic (O) antigen and one flagellar (H) antigen, sharply separable and distinguished by the thermolability of the latter, appears to be oversimplified in view of later studies including the examination of separated flagella and "stripped" cells.[61]

Both the somatic and flagellar antigens are liable to variation. Frequently, typical "smooth" (S) forms of bacteria produce rough (R) variants, mostly avirulent and showing granular growth, and this goes along with a change in the reactions of the somatic antigens [62] (p. 215).

Other variations involve the flagellar antigens. As Andrewes (186) observed, when a bacterial culture cross-reacts with a second motile strain of the same organism, one may be able to isolate on plates colonies which possess only an agglutinogen of their own, and others having instead an antigen shared by both strains and responsible for the cross reaction of the original cultures, and in the further cultivation of the two variants of such "diphasic" strains conversion of the specific phase into group phase, or vice versa, often takes place. In addition to this, a variation of H antigens was recognized in the Salmonella group which consists of the interconversion of two specific phases (α, β phase).[63]

Flagellar and somatic antigens appear to be concerned in the agglutination by acid, and the pH optimum at which bacteria flocculate is in some degree characteristic, so that typhoid and paratyphoid bacilli, for instance, can often be distinguished by acid agglutination.[64] Overlapping of the agglutination zones and variation within a species, however, prevent a practical application of the method.[65]

Apart from the variations described and still others not here touched upon, considerable complication in serological practice arises from

[61] Craigie (182), White (183).
[62] Arkwright (184). Antigenically different, so-called ρ forms were described by White (185), (183).
[63] Kauffmann et al. (187).
[64] Michaelis; v. (188), (189).   [65] (190), (191).

the existence of serological types which would seem to be similar in biological significance to the serological differences within animal species, even granting that the species concept is less definite in bacteria than in sexually reproducing organisms.[66] While bacteria which are distinguishable on morphological and biochemical grounds can be differentiated readily by serological means, as can animal cells of various species, the converse does not always hold; frequently bacteria which are otherwise alike or very similar in many respects and classed in one species exhibit antigenic differences, and here again their sharpness is striking. Such types have been found in many microorganisms (cocci, bacteria, clostridia, etc.),[67] and in some bacteria a large number of varieties were discovered. The immunological properties may be the sole criterion for distinguishing the types or be associated with differences in cultural and pathogenic behaviour.

Positive proof of the existence of pneumococcus types, which have assumed prominence in the study of bacterial antigens and in serotherapy, was given by Neufeld and Haendel (200) who found that immune sera which protect mice against certain strains of pneumococci are ineffective for other strains, and that these strains can be differentiated as well by agglutination reactions. A classification which subdivided pneumococci into three main types and a group comprising all other strains, resulted from the careful investigations of Cole and his coworkers.[68] Many of the types in the last named group have since been serologically defined (Cooper), and in a recent paper about fifty types were listed.[69]

The chemical basis for the type differences between strains of pneumococci is known since Heidelberger, Avery and Goebel (203) in fundamental studies showed that in the capsules of the three classic types there exist an equal number of carbohydrates of rather high molecular weight that are sharply differentiated by precipitin reactions (p. 210). On the other hand, all strains, as far as could be ascertained by serological tests, contain identical or very similar proteins. Polysaccharides are now recognized as carriers of type specificity in many microbes (B. Friedländer, staphylococci, meningococci, gonococci) but proteins, too, have been found to underlie certain type distinctions, as in fusobacteria [v. (204)], staphylococci (205) and,

---

[66] Dobzhanski (192).
[67] (193–197). For yeasts v. (198), (199).
[68] v. Avery, Chickering, Cole and Dochez (201).
[69] (194), (202).

in addition to type specific polysaccharides, in streptococci [Lancefield (206)].

An instructive example, examined in great detail, of the complicated relationships among closely allied bacteria is the group of Salmonella bacilli in which, by means of suitably absorbed immune sera, it has been possible to identify many types. Some of these, together with their serological unit factors, are listed in the following scheme taken from a more extensive table of Kauffmann and Bruce White; in a recent monograph by Kauffmann (207) more than 100 varieties are enumerated.[70] The factors, occurring in various combinations, correspond in part to the somatic antigens which owe their specificity to polysaccharides [71] or else to flagellar antigens whose chemical nature is yet to be fully determined (182).

TABLE 15
[after Bergey's Manual, 5th ed. (210)]

	O-antigen	H-antigen		
		Specific phase		Non-specific phase
		α	β	
S.paratyphi-A	I, II	a	..	..
S.typhi-murium (Aertrycke)	IV, V, XII	i	..	1, 2, 3
S.paratyphi-B	IV, V, XII	b	..	1, 2
S.brandenburg	IV, XII	lv	en	..
S.derby	IV, XII	fg	..	..
S.reading	IV, XII	eh	..	1, 4, 5
S.cholerae-suis (Suipestifer)	VI, VII	c	..	1, 3, 4, 5
S.oslo	VI, VII	a	enx	..
S.newport	VI, VIII	eh	..	1, 2, 3
Eberthella typhosa	IX, XII	d	j	..
S.enteritidis	IX, XII	gom	..	..

From the table one can gather that, for instance, immune sera for Bact. paratyphosum B agglutinate all strains having factors IV or V but that after exhaustion with types Reading or Derby, reactions will occur with those bacteria possessing factor V. This absorption technique, to which reference has been made on previous occasions, plays a prominent part in bacterial diagnosis (Castellani's experiment) (211).

In view of the great diversity in the polysaccharides within groups of closely related bacteria, and of their variation as seen in the inter-

[70] For a critical discussion v. Topley and Wilson (178).
[71] White (208), Furth and Landsteiner (209).

change of smooth and rough forms, one might suppose that poly-saccharides are not so intrinsic a species character of bacteria as are the somatic proteins. Such a conclusion follows in any event from the existence of substances other than typical proteins (p. 231) which are strikingly different in individuals of the same animal species and therefore, although conditioned by heredity, appear to be of less consequence than proteins in determining structure and function. An interesting parallel among others is the great variety of flower pigments, their genetically determined diversity within the species, and the occurrence of the same or similar pigments in different species of plants.[72]

Variability of bacterial enzymes is often encountered yet certain enzymatic activities are characteristic for a species, and fermentation tests are commonly used in the diagnosis of bacteria.

HETEROGENETIC REACTIONS. — The occurrence of serologically related substances — some recognized as non-protein haptens — in animals that are widely separated in the zoological system became manifest from an experiment by Forssman [73] which showed that upon injection of guinea pig organs into rabbits, lysins of high titer were formed for sheep blood. The so-called Forssman substances, characterized by capacity for giving rise to sheep hemolysins, were then found in numerous animals, in the organs or erythrocytes, or in both (sometimes in the serum), and even in bacteria. The presence or absence of these antigens in the cells of an animal is, as a rule, an attribute of the particular species. Schiff and Adelsberger,[74] however, observed that a substance related to the Forssman antigen exists in human blood only in individuals having the agglutinogen A since some of the Forssman immune sera react with cells of groups A or AB, but not of O or B and, conversely, immune sera specific for A blood hemolyze sheep cells. This fact established a connection between the substances underlying heterogenetic and individual reactions of animal cells. An individual variation supposed to have arisen as a genetic mutation, namely, the absence of the Forssman antigen from the blood corpuscles of one sheep, has been described by Mutermilch (221).

[72] v. Lawrence and Price (212); Haldane (213).

[73] (214), Myers (215); (216), (217); v. review by Buchbinder (218). The expressions "heterogenetic" or "heterophile" have often been used to designate Forssman antigens and antibodies. Since a number of similar instances have been detected it is now preferable to apply this terminology generally to reactions of antibodies with antigens not related in their origin.

[74] (219), v. (220).

The distribution of the antigen among animals is usually regarded as solely random;[75] it is present in guinea pigs but absent in other rodents (rabbits, rats) and occurs in distant species (horse, cat, chicken, etc.). Nevertheless, there undoubtedly exist regularities, because in the author's investigations[76] Forssman antigens were found in the organs of all the Felidae examined (tiger, puma, lion, ocelot, cat), in Procyonidae (Cercoleptes caudivolvulus, Procyon lotor) and Canidae (American grey wolf, fox, Lycaon pictus, dog); yet with the same technique[77] they could not be demonstrated in primates (chimpanzee, siamang gibbon, Papio hamadryas, Macacus rhesus, Presbytis maurus, Cercocebus fulginosus and albigens, Pygathrix cristata, two species of Cebus, Hapala jacchus), except for species Nyctipithecus trivirgatus belonging to a special genus or subgroup, the single species of lemurs examined, and one dubious result. Similar correlations have been met with in the examination of bird eggs by Sievers (223), and by Streng with fishes (fish muscle) (224). The presence of the Forssman antigen is therefore peculiar not only to species, but also to certain genera and families.

This result has general bearing inasmuch as other properties defined by heterophile serum reactions exhibit regularities in their distribution. The β-agglutinin of normal human sera of individuals of groups O and A, which acts on blood cells of groups B and AB and, as von Dungern and Hirszfeld demonstrated, also on many animal bloods (rabbit, cattle, etc.), was found to react with the blood of 12 species of American monkeys (Platyrrhinae) representing 7 genera, and 6 species of lemurs, while the property defined by the agglutinin was not detected in 18 species of Old World monkeys (Cercopithecidae) belonging to 4 genera.[78] Thus it seems that the substances sensitive to the β-agglutinins run in entire families or genera of animals. This is illustrated in the following diagram by Keith upon which some of the specific factors (A, B, O, M, N) in the blood of primates have been entered.

Another heterogenetic system was found by Buchbinder (226) who injected rabbits with bacteria of the hemorrhagic septicemia group and obtained a hemolysin that dissolves the blood cells of

---

[75] v. (222).

[76] Carried out in collaboration with Dr. H. Fox.

[77] The experiments were performed with extracts which had been prepared by heating the organ material with alcohol. The sera used did not react with human blood A.

[78] Landsteiner and Miller (225).

many species of birds but not, on the contrary, the erythrocytes of mammals.

Evidently, then, the occurrence of immunologically similar substances in the cells of unrelated species is not limited to Forssman antigens. The wide range of the phenomenon is further indicated by group reactions, which were discovered through employment of immune sera and emulsions of alcoholic blood extracts (p. 100), be-

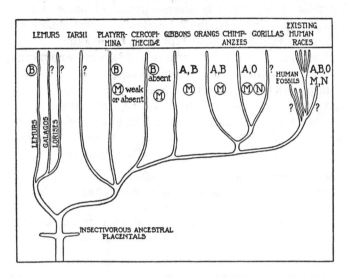

tween the species Macacus rhesus-pig, horse-rat, man-dog-pig, cat-horse, human blood A-cattle, rabbit and a type of pig blood.[79] Other relationships between antigens have been detected by absorption experiments with hemagglutinins of normal sera (p. 128) and a pertinent case is the occurrence of "A substances" in horse saliva and commercial pepsin preparations (p. 232).

With the findings on blood corpuscles may be ranked heterogenetic cross reactions of microbes, mostly due to polysaccharides,[80] which were found to occur not only between Streptococcus viridans and a type of pneumococci (232) or strains of dysentery and coli bacilli, but between such different forms as: yeast and gram-negative bacteria; [81]

---

[79] (227–230), (220).
[80] Heterogenetic reactions between pneumococci and gonococci or meningococci described by Boor and Miller are attributable, according to these authors, partly to carbohydrates, partly to nucleoproteins (231).
[81] (233–235).

Friedländer bacilli B and Pneumococcus II;[82] certain pneumococcus types and various organisms (yeast, a strain of Bact. coli, leuconostoc, Bact. lepisepticum, types of salmonella, influenza bacilli[83]); Shiga bacilli and enterococci (240); Bact. proteus X19 and rickettsiae. The similarity of the polysaccharides contained in the last named microorganisms appears to be responsible for the Weil-Felix reaction in typhus fever and Rocky Mountain spotted fever (pp. 221, 223).

That relationships exist between antigens of bacteria and blood corpuscles, as evidenced by the formation of sheep hemolysins upon injection of certain bacteria, is still more remarkable. Thus Forssman antigens, first detected in strains of dysentery and paratyphoid bacilli,[84] B. lepisepticus, and pneumococci,[85] were shown to be contained in a variety of bacteria — Neisseria catarrhalis, B. welchii, B. megatherium, B. mucosus, etc.[86] In the same category are agglutinogens of human (and rabbit) blood which react with horse sera to Pneumococcus XIV,[87] and anti-dysentery goat sera, the latter sometimes agglutinating preferentially red cells of group O.[88]

The possibility that bacteria acquire reactive substances (agar, group A substance) from culture media must not be overlooked.[89] This possibility, while probably an occasional cause of mistakes, cannot account for the differences in the heterogenetic antigens of various microbes, and moreover such antigens have been demonstrated in bacteria grown on synthetic media (253).

As explanation for the frequent heterogenetic reactions of cell antigens the suggestion may be offered that these reactions depend upon the special chemical nature of the haptens. Whereas proteins are built up from many different amino acids, the carbohydrates which determine the specificity of numerous bacterial antigens have a high

---

[82] (236–237).
[83] (238–239b).
[84] Jijama; Fujita; K. Meyer; bibliography in (241); Shorb and Bailey (242).
[85] (243), v. (244), (245).
[86] Shorb and Bailey; Rockwell and van Kirk (246). Forssman antigen was reported to be present also in plants (247), (246). Its occurrence in ragweed pollen is disclaimed by Arbesman and Witebsky (248).
[87] (249), (250).
[88] Eisler; Schmidt-Schleicher (251); (252).
[89] (pp. 217, 233). Schiff (253); Bliss; Goebel (254).
Some of the cross reactions of pneumococci have been called in question by Neter on that account (239b).
It is stated by Holtman (255) that heterogenetic properties acquired from media or animal hosts are retained through a number of sub-cultures.

molecular weight, yet, as far as is known, each of them contains but a few different "building stones." Consequently, the chance occurrence of similar specific groupings giving rise to overlapping reactions is much more likely in polysaccharides than in proteins where specificity, one may assume, is determined by more complicated structures. Similar reasoning may apply in the case of animal haptens.

Heterogenetic reactions do not signify that the same substance is present in the serologically similar materials. It may suffice to mention differences in the group substances A and B (256a, b) and in the paradigmatic Forssman antigens. In absorption and immunization experiments, and also by comparing various immune sera,[90] it has been demonstrated that the Forssman reactions, which possibly depend upon the presence of relatively small characteristic groupings, are due to similarity rather than to identity of the respective substances. Differences between the antigens in sheep blood and human blood A were observed by Schiff and Adelsberger (p. 115), and the distinction between the antigens in chicken and sheep blood appears from the dissimilar properties of immune sera prepared with these two kinds of blood (217). Another instance, investigated recently, is the peculiar behaviour of the heterophile antigens in mouse tissues (256c).

TABLE 16

Immune sera prepared by injection of	Emulsion of alcoholic extract of				
	Human blood (group A)	Horse kidney	Sheep blood	Chicken blood	Dog blood
Human blood (group A) ...	+++	o	+±	o	±
Horse kidney	o	++±	+++	++	..
Sheep blood 1	+++	+++	+++	++	..
Sheep blood 2	o	++±	+++	+±	+±
Chicken blood 1	o	±	+++	++±	+±
Chicken blood 2	o	±	+++	+±	±̲

As evidence, the following experiment (227) may be cited (Table 16) in which emulsions of alcoholic extracts were tested for flocculation with a number of immune sera, all of which have in common the capacity for hemolyzing sheep red cells and flocculating alcoholic extracts of sheep blood, yet differ markedly in their action upon various materials containing Forssman substances.

A more comprehensive table including bacterial Forssman antigens

[90] (222); cf. (256).

and lysins is given by Shorb and Bailey. The bacterial lysins are bound in all cases by sheep blood and the homologous strain, but in their absorption by other bacteria containing Forssman antigens they are unlike so that most of them, if not all, must be different; [91] only some of the strains are capable of absorbing the bulk of the lysins produced by injection of animal material, as horse kidney or sheep blood.

Of clinical significance as an aid in the diagnosis of infectious mononucleosis are agglutinins (and lysins), frequently of high titer, for sheep and beef blood which occur in the patients' sera.[92] These antibodies have been ascribed to the antigenic effect of substances, present in the causative agent of the disease and similar to substances contained in erythrocytes, but the suggestion has been made (without experimental evidence) that the agglutination may be caused by the virus itself (p. 5).

The fact alone that beef cells are affected separates the mononucleosis antibodies from typical Forssman antibodies,[93] and while the reactive substances in beef and sheep cells resist boiling, as do the typical Forssman substances, they are not, like the latter, dissolved by alcohol (258). Further, the antibodies are not absorbed by tissues containing the common Forssman antigen (guinea pig kidney).

Agglutinins of heterogenetic origin are found in the serum of patients after therapeutic administration of horse or rabbit serum, as in cases of serum sickness.[94] The agglutinins act upon various erythrocytes — horse, rabbit, sheep, ox, etc. — an indication of the presence of similar antigens in the sera of the horse and rabbit and in various erythrocytes (259). Apparently, the sheep agglutinins found in normal human sera, cases of mononucleosis, and serum sickness are all different.

In connection with these findings of medical interest, the occurrence of Forssman antibodies in a human case of Streptococcus viridans infection (263), and in the serum of rabbits infected with trichinae will deserve notice (264).

ANTIGENIC ACTIVITY OF NON-PROTEIN CELL CONSTITUENTS.[95] — The antigenic activity of alcoholic extracts of animal cells and bacteria have been referred to briefly, and this is to be supplemented by citing the observations of K. Meyer [96] on reactions and immunization

[91] Jungeblut and Ross; K. Meyer; Shorb and Bailey (242).
[92] Paul and Bunnell (257).
[93] Stuart et al. (258), Schiff (259), Bailey and Raffel (260), Beer (261).
[94] Hanganutziu, Deicher, Davidsohn, Bailey and Raffel, Stuart et al. (262), Schiff (259).
[95] Cf. chapter VI.
[96] (265); (266) (serology of echinococcus infection); (267), (268).

effects of alcoholic extracts of tapeworms and tubercle bacilli lipids (269). In view of the strong reactions between antisera and some of the substances soluble in organic solvents one would have expected that with these preparations immune sera could be produced without difficulty. Hence it was puzzling that such lipid fractions were found in several instances to incite much weaker immunization effects [97] than the original antigens even though in others the results were plain. In this way, uncertainty prevailed with regard to the antigens of blood corpuscles. Bang and Forssman (4) obtained hemolysins by inject- ing ethereal extracts of red blood corpuscles and therefore concluded that the antigens were lipoid in nature. The hemolytic action of the antibodies obtained by blood injections, however, was not neutralized by the extracts; and the immunizing effect, weaker than that with intact blood corpuscles, was not reproducible in the hands of several workers (272). Consequently, it was still possible to attribute the formation of antibodies to traces of proteins in the extracts.

New investigations were suggested by the preparation of artificial complex antigens from proteins and non-antigenic compounds (p. 156). It seemed reasonable to assume that the structure of the cell antigens under discussion was similar in principle. The Forssman antigen, already dealt with, proved to be suitable for testing this hypothesis. From organs, as horse kidney, a specifically reacting material can be extracted with alcohol [98] and finally obtained free of protein, but despite its high affinity for Forssman antibodies the alcoholic extract alone has only slight immunizing power.[99] The ap- parent contradiction was explained when, on the basis of the pre- sumed analogy to conjugated antigens, it was found possible to re- store the original immunizing capacity by adding antigenic protein to the specific substance extracted with alcohol.[100] When mixtures of such extracts and serum from a foreign species were injected into rabbits, active sheep hemolysin was formed quite regularly, in addi- tion to antibodies against the injected protein.

This method, designated by Sachs as "Kombinationsimmuni- sierung," was found to be rather widely applicable and has been used for extracts of organs such as brain, testicle, etc. (p. 111). The im- munizing effect of alcoholic blood extracts is likewise considerably

[97] v. Sachs and Klopstock (270), Ninni (271).
[98] Doerr and Pick (273), Georgi (274).
[99] Sordelli et al. (275), Landsteiner (276), Taniguchi (277).
[100] Landsteiner (276), (278). The effect is influenced by the protein used, and the manner in which the injection mixture is prepared (271), (279).

increased by the addition of protein.[101] Horse hemolysins so prepared were neutralized by the blood extracts, and the immune sera gave specific flocculation reactions with emulsions of the extracts. In pursuing this investigation it was found that ordinary species specific immune sera, prepared with unaltered blood cells, not infrequently give complement fixation with and flocculate emulsions of alcoholic extracts of blood.[102]

Among the results obtained with the "combination method" the work of Sachs, Klopstock and Weil (287) on the Wassermann reaction in syphilis is to be noted. This test is based upon the fact that sera of syphilitic patients give complement fixation (and flocculation) reactions with extracts of organs containing syphilis spirochaetes as well as with alcoholic extracts of practically any normal organ [103] (p. 230). Immune sera of like properties, Sachs and coworkers showed, can be obtained in rabbits by injections of alcoholic extracts of organs of rabbits (or other animals) when mixed with heterologous serum.[104] These authors maintained, therefore, that the antibodies effective in the Wassermann reaction are produced in consequence of an "autoimmunization" with components of pathologically altered tissues aided, perhaps, by proteins of the syphilis spirochaete (Weil and Braun). The specificity of the reaction for syphilis (and for leprosy and framboesia) is more easily understood if one assumes that the Wassermann reagins are antibodies engendered by the spirochaete and give overlapping reactions with organ extracts owing to the presence of chemically similar substances, probably lipids, in both spirochaetes and animal tissues; this view is advocated by Eagle on account of absorption experiments with culture spirochaetes.[105] In the case of the autohemolysin which is responsible for the blood destruction in paroxysmal hemoglobinuria,[106] a disease almost in-

---

[101] (280–282); cf. (283). On antisera for the group specific substances of human blood v. (284), (229).

[102] Landsteiner et al. (227), Bordet and Renaux (285), Krah and Witebsky (286).

[103] On Wassermann reactions after injections of milk, v. (288).

[104] Several similar experiments in human beings were not successful; Hedén (289) had positive results [v. (299)].

[105] v. Eagle and Hogan (290); Gaehtgens; Beck (290a); Sachs et al. (291) (reactions of antisera for spirochaetes with brain extracts); v. (291a). Wassermann-positive sera could be produced by injection of spirochaetes, and trypanosomes (292–294).

[106] Donath and Landsteiner (295). By injecting extracts of rabbit blood mixed with protein, Oe. Fischer (296) was able to stimulate the formation of autohemolysins in rabbits; v. (297–298).

variably syphilitic in origin, the first hypothesis is more probable because of the specificity of the lysin for human blood. The opinion sometimes offered that it is merely the presence of unstable globulins which causes the Wassermann reaction would seem to be contradicted by the analogy in the methods for producing Wassermann reagins and Forssman antibodies by means of alcoholic organ extracts.[107]

The possibility of immunizing animals with substances derived from tissues of their own species (p. 61) was further demonstrated by Lewis (300) who obtained brain-specific antibodies upon injecting rabbits with alcoholic extracts of rabbit brain along with foreign proteins, and by Schwentker and Rivers (301) directly with autolyzed brain or brain emulsions from rabbits infected intracerebrally with vaccine virus.[108] Bearing upon this subject are observations by Halpern (304), and immunization of animals to lens [109] and spermatozoa [110] of their own species.

The combination method gave certain immunization results which were not obtainable otherwise. Apparently antibodies with special properties are sometimes formed when the inciting substances are separated from the natural antigenic complex.[111] For instance, immune sera produced with alcoholic extracts of horse blood had a higher ratio of lysins to agglutinins than common antisera for horse blood (280), and sera made in a similar way with extracts of human blood of groups O and B neither agglutinated nor hemolyzed appreciably the intact blood cells but reacted with blood extracts in complement fixation tests.[112]

In general, it has been found that immune sera do not always convey a true picture of the structure of cell antigens. Certain antibodies are more readily formed than others; thus species specific hemagglutinins or lysins are regularly produced by immunizing with foreign blood, while it is often not easy to prepare immune sera which define individual properties, and some of these entail greater difficulties than others.[118] Besides, specific substances have been seen to interfere with each other in their antigenic action, e.g., the highly active Forssman substance may inhibit the formation of antibodies against other specific substances. This interference, which German writers

---

[107] The various hypotheses on the Wassermann reaction and other aspects of this subject are reviewed in (299).

[108] Injection of rabbits with autolyzed rabbit kidney gave no definite evidence of antibody formation (302) but positive results were obtained with mixtures of kidney emulsions and bacterial toxins (303).

[109] Krusius (305), Hektoen and Schulhof (306).

[110] Metalnikoff (307), Henle (44).

[111] v. (299), (87).

[112] (284), (308); v. (256).　　　　　　　　[118] v. Irwin et al. (309).

have termed "Konkurrenz der Antigene" (competition of antigens),[114] does by no means occur regularly when several antigens are injected at the same time,[115] as may be seen from the use of multiple antigens in the preparation of therapeutic sera. However, the phenomenon was demonstrated beyond dispute by Doerr and Berger (325) who showed that a large quantity of euglobulin suppressed anaphylactic sensitization to a small amount, simultaneously injected, of a less active antigen, i.e. serum albumin; in the converse experiment the response to euglobulin was not hindered even by a hundredfold dose of albumin. The properties of immune sera depend, moreover, upon the species and the individual constitution of the experimental animals, and the response of the same animal can be quite different with various antigens.[116] Experiments by Witebsky [117] and others are instructive: It was found that no potent agglutinins for human A cells were formed when group A blood was injected into rabbits whose tissues and serum contained substances related to the agglutinogen A, and whose serum consequently had not much or no natural anti-A agglutinin. As Wheeler, Sawin and Stuart established, the occurrence of A substance in the tissues of rabbits is inherited as a dominant Mendelian factor, and in an inbred strain lacking this character sera with uniformly good titers were secured upon injecting with human A erythrocytes.[118] From studies of the same authors the capacity of rabbits to produce antibodies for the M-antigen in human erythrocytes appears to be inheritable; here, however, the lack of this ability could not be traced to the presence of M substances in the animals.

The fact that the antibodies produced in guinea pigs with human blood A, in contrast to anti-A rabbit sera, do not react with the Forssman substance in animal organs is explained by the presence of Forssman antigen in the animals immunized.[119] As was demonstrated in experiments previously quoted, however, the rule that antibodies are not formed against substances present in the animal used for immunization is not entirely valid (pp. 61, 102); some apparent exceptions are explicable by minor specificity differences between the antigens (e.g. group A substances) in the immunized animal and the material injected.

[114] (310), (37), (282), (311–313), (256); (314) (mutual influence of azoproteins).

Its counterpart is the modification of reactions in vitro by other haptens present (315–319), (73); (320), (321) (intensification of complement fixation and flocculation by sterols and lecithin).

[115] (322–324).

[116] v., for instance, Gibson and Banzhaf (326) (immunization of horses with various toxins).

[117] Witebsky (327), Wheeler, Sawin and Stuart (328), Andersen (329); v. (330). As has been pointed out, only a few of the immune sera produced by Forssman antigen react with A-blood.

[118] (331), (328). On variations in antibody response to agglutinogens in dove erythrocytes v. Irwin et al. (309); on protein antibodies v. (331a).

[119] (332); v. (333), (222).

The author's assumption that immunization with hapten-protein mixtures is due to a loose combination with antigenic protein was based on the observation that protein from the same species, which has no immunizing effect, could not be substituted for that of a foreign species.[120] Furthermore, if the two substances were injected separately instead of mixed, no significant immunizing effect was obtained and, on the other hand, protein-lipid precipitates proved particularly suitable for immunizing.[121] Also, Doerr and Girard (338) found that protein which has lost its antigenic power by treatment with alkali does not produce the effect. Objections to the above interpretation were raised by Sachs.[122] In his opinion the antigenic property of the proteins added is not the significant thing, the foreign protein merely serving to carry the specific substances into the cells in which the formation of antibodies takes place. Gonzalez, Armangué, and Romero (340) in fact were able to prepare potent Forssman antisera in another way than by the addition of foreign proteins, namely with the use of adsorbents like kaolin or charcoal. Analogous results with bacterial haptens were reported by Zozaya.[123] Whilst the latter experiments could not be reproduced with purified polysaccharides, the author and Jacobs, and Plaut and Rudy, using Forssman's hapten, and extracts of brain, had no difficulty in repeating the experiments of Gonzalez and Armangué.[124] It was found, however, that after a certain degree of purification the Forssman substance no longer could be activated in immunization experiments by adsorption on to kaolin but could still be rendered antigenic by adding serum. Plaut and Rudy had similar results with cholesterol [125] (p. 111), and Armangué and Gonzalez failed to prepare immune sera in rabbits by alcoholic extracts of rabbit organs adsorbed to kaolin. Hence it would appear that the effects of proteins and various adsorbents are not equivalent. The latter effect may be attributable to enhancement of the weak immunizing capacity of alcoholic extracts of Forssman antigen and similar substances,[126] comparable to the

---

[120] v. Doerr and Hallauer (334), Sachs (335); (336).
[121] Cesari (336), Eagle (337).
[122] (339), (256).
[123] (341), (342).
[124] (343), (344); v. (345), (345a).
[125] Mutsaars reported positive results; tests for specificity are not mentioned (346).
[126] (272), (280); v. (347), (348), (283), (349), (314), (350), (351). The output of sheep hemolysins was markedly increased when Forssman hapten was injected in mixture with arsphenamine and other chemotherapeutic drugs (352).

observations of Ramon and Glenny on the increased yield of antitoxin when toxins are injected along with alum, tapioca or tannin.[127] Similar experiments have been made with proteoses (p. 45), by Seibert with tuberculin (357), by Caulfeild with pollen extracts (358), with serum proteins and hemoglobins.[128] The factors which may be operative in these experiments are slow absorption and delayed removal of the injected material, the particulate state and consequent engulfing of the particles by the cells producing antibodies, and the stimulation of cell activity.[129] As regards the immunization with the aid of added proteins, the original explanation that, by combining with haptens, proteins impart their antigenic property to the complex is still the most probable one. Apart from the reasons already stated, there is the fact that with simple chemical substances the production of antibodies could be induced by combination with proteins but not with the use of adsorbents.

The immunization experiments with alcoholic extracts favour the said concept [130] that there exist natural complex antigens which, like those prepared artificially, consist of two parts: the one a protein essential for the formation of antibodies, and the other a substance (hapten) which reacts specifically in vitro but whose capacity to immunize is wanting or inconsiderable as compared with its binding power. Additional evidence is the marked difference between foreign organs and organs derived from the same species in the stimulation of antibodies against haptens. This is well demonstrated by the behaviour of brain substance. Unaltered rabbit brain does not stimulate antibodies in rabbits but foreign brain engenders immune sera which react also with the hapten prepared from rabbit brain, a result reasonably attributable to dissimilarity of the proteins.

In so far as the specific substances in alcoholic extracts of red cells and animal tissues are concerned, these arguments are not invalidated even though some of the substances in question, for instance those extracted from eggs,[131] worms,[132] and several bacteria,[133] especially acid-fast bacilli, were

---

[127] Ramon et al. (353), Glenny et al. (354), Prigge (355), Freund (356).

[128] Hektoen and Welker; Holford et al. (359).

[129] This interpretation will apply to the subsidiary influence of toxins [Swift and Schultz (360)], and of the dead cells or lipids of tubercle bacilli [Sabin and Joyner (361)].

[130] Landsteiner (276), (278), Taniguchi (277).

[131] (351); v. (362) (brain).

[132] (363); cf. (268).

[133] Bibliography in (271), (299), (364).

said to show definite antigenic activity without the addition of proteins. The active principles in alcoholic extracts of tubercle bacilli appear to be lipids (v. p. 225) which are capable of eliciting antibodies different from those obtained with the protein or carbohydrate fractions.[134]

An unexpected development in the subject of bacterial antigens was brought about by the analysis of the specific substances of pneumococci. In investigations of Avery and Heidelberger [135] it was discovered that injection of intact pneumococci leads to the production of type specific antibodies which not only agglutinate the cells but precipitate the polysaccharides present in the S form of the respective types. Cocci disintegrated by autolysis, bile, or freezing and thawing give rise to another sort of antibody; such sera act neither on the capsulated forms nor on their carbohydrates, but agglutinate the R forms which are devoid of capsules and type specific carbohydrates, and precipitate the species protein common to the S and R forms of the various types. (Such antibodies are induced also by the S forms.) In the experiments of the authors cited, the polysaccharides themselves did not give rise to antibodies in rabbits or anaphylactic sensitization in guinea pigs, and later investigators, working with carbohydrates of other bacilli, obtained conformable results. Hence, these carbohydrates seemed to be haptens which possess antigenic activity only when combined with proteins, or perhaps other cell constituents.

The scheme, shown below, illustrating these relations was offered

Pneumococci

Type I $\quad$ [P $\{$ S_I]

" II $\quad$ [P $\{$ S_{II}]

" III $\quad$ [P $\{$ S_{III}]

P represents protein, S the type specific carbohydrate.

tentatively by Avery and Heidelberger (369), but it is now thought [136] that antigenicity of the capsular substance may be mediated by components other than, or possibly in addition to somatic protein. Whatever the ultimate solution of this problem may be, it is of great inter-

---

[134] (365), (366). Increased activity upon addition of protein was found by Ninni (366a).
[135] (367), (368).
[136] Personal communication from Dr. Avery.

est that Goebel and Avery (370) succeeded in converting the polysaccharide of Pneumococcus III into a full antigen by attaching it to serum globulins by means of a method to be described later (p. 176). When injected with this product, rabbits became immune to virulent Pneumococcus III, and the serum of the animals precipitated the unaltered polysaccharide, agglutinated the pneumococci and protected mice against infection with these microorganisms.

Mere admixture with protein does not suffice to confer antigenic properties upon bacterial carbohydrates. Antibody production with typhoid polysaccharide, after precipitation with immune sera, was claimed by Spassky and Danenfeldt (371), but similar experiments of Partridge and Morgan (371a) were unsuccessful.

Polysaccharide containing preparations that elicit immunity responses, especially in mice, have subsequently been recovered from pneumococci by a number of authors,[137] and Francis and Tillett observed that the polysaccharide of Pneumococci I, as originally prepared, which has no demonstrable antigenic activity in rabbits, gives rise to antibody formation when minute amounts are injected intradermally into human beings.[138]

An antigenic substance separated by Enders from Pneumococcus I, the reaction being kept on the acid side during the preparation, was found by Pappenheimer and Enders (379) to be related to the specific polysaccharide, and these authors suggested that the latter substance may represent a hydrolytic product of the former. Actually, Avery and Goebel (375) obtained from Pneumococci I an immunizing substance, identical with that of Enders, which they found to be an acetyl derivative of the known type specific polysaccharide, and their results were confirmed by Enders and Wu.[139] In very small, but not in larger doses,[140] the substance induced active immunity in mice; yet, in contrast to unaltered pneumococci, it failed to excite antibody production in rabbits.[141] On treatment with alkali, labile acetyl

---

[137] Schiemann et al., Felton, Wadsworth and Brown, and others. v. (375); (376), (376a) (reviews); (377), (481), (378).

[138] Francis and Tillett (372), Zozaya and Clark (373), Finland and Dowling (374).

[139] (380). Arguments against the significance of acetyl groups in the production of immunity were raised by Felton and Prescott (381); v. (382).

[140] Cf. analogous observations with Forssman substance (383), (349).

[141] The carbohydrate is also antigenic in other animals (man, horse, dog, cat) belonging to one group of the two characterized by Goodner and Horsfall (384), (p. 145).

groups, probably combined in ester linkage with hydroxyl, were split off and at the same time the immunizing capacity was destroyed whilst the reactivity in vitro with antibodies, though altered to some extent, was retained (p. 213).

The importance of the results obtained with the polysaccharide of Pneumococcus I lies in the proof that a carbohydrate can possess antigenic capacity. The antigen as it exists in the intact cell is more active and is almost certainly of higher complexity. Although effective antigenic preparations have been described, it has so far been impossible to obtain in solution the full antigen which, for example, engenders type-specific antibodies in the rabbit, and its composition is still undetermined. Significant facts were disclosed in studies by Dubos (385) in which enzymes were employed for the immunological analysis of pneumococci. Hydrolysis by polysaccharidase of the free polysaccharide contained in the capsules did not destroy the type specific antigen and this persisted even when the greater part of the cells was digested away by tryptic enzymes. The type antigen, however, is rapidly destroyed by the autolytic enzyme of the cocci which at the same time lose their gram-positive character, and in the process the polysaccharide and a nucleoprotein, perhaps a part of the complete antigen, go into solution.

Material progress in the investigation of bacterial antigens was made by Boivin and his coworkers (386), and by Raistrick and Topley (387), who isolated highly antigenic preparations from gram-negative bacilli,[142] especially of the Salmonella group, containing polysaccharides, lipids and "peptide-like substances" (later recognized as proteins). Significant information on the antigenicity of these substances was furnished in studies by Morgan and Partridge [143] with the complex antigen of Bact. Shigae (p. 223). Whereas the lipoid part of this material proved unnecessary for the immunizing properties, a fatty constituent appeared to be essential for the antigenicity of the Forssman substance in pneumococci, as shown in the work of Goebel and his colleagues [144] (p. 225). Instances of similar purport are presumably antigens which are contained in acid fast bacilli.[145]

Summarizing the studies on bacterial antigens one may say that,

[142] In gram-negative cocci antigens of this sort were detected by Miller and Boor (482).
[143] (388), (371a).
[144] Goebel, Shedlovsky, Lavin and Adams (389).
[145] (p. 225); (365), (366), (390).

at least in certain cases, and depending upon special experimental conditions, polysaccharides or their simple derivatives are to some degree endowed with immunizing properties, and that in bacteria there exist compounds of polysaccharides with other non-protein substances that have full antigenic activity.

Considering the demonstration of distinct immunizing properties in certain non-protein cellular substances, together with the gradual differences in antigenic capacity of the products formed on breaking up an antigenic complex, the term hapten has become less definite and the question arises whether, save for simple synthetic compounds, it is still justified. However, the lack of a sharp borderline, as between antigens and haptens, is shared by not a few of our conventional classifications, and the general concept holds to the extent that serological reactivity and antigenic capacity are entirely separate functions, and that a substance can possess antigenic properties when included in a larger complex, not necessarily containing proteins. Therefore the expression hapten will still be convenient for distinguishing from full antigens specifically reacting parts thereof which are not antigenic or but weakly so in comparison to the complete antigen. To designate as antigens, against common usage (and etymology), substances which react with antisera but do not or are not known to immunize (Wassermann substance, precipitable simple compounds) would seem to complicate the nomenclature without cogent reason.

The chemistry of complex bacterial antigens will be gone into later on; it is left to review experiments on antibody formation to various well-known carbohydrates and lipids.

ANTIGENICITY OF PHOSPHATIDES, STEROLS, STARCH ET CETERA. — That antibodies can be produced with starch, inulin, dextrin and glycogen has been occasionally asserted.[146] With starch this response was found by Fujimura (394) to be due to impurities, and Uhlenhuth and Remy [147] reported negative results with starch (and simple sugars) though some, apparently positive, with glycogen adsorbed to aluminum hydroxide. Campbell (397) obtained precipitins against glycogen preparations from clams following injection of ground tissue but was unable to demonstrate antibodies upon immunization with liver suspensions and testing against liver- or clam-glycogen (v. p. 217). Whether antibody production and sensitization observed with gums can be attributed to the polysaccharides themselves is uncertain.[148] The presence in antibacterial horse immune sera of anti-

[146] (391–393).
[147] (395), (396).
[148] (398), (398a), (395). Immunization results were also claimed for pectin

bodies for agar, observed by Sordelli and Mayer, and Morgan, is probably brought about by combination with components of bacteria (pp. 98, 217).

Attempts to immunize with several glucosides (including saponins and solanin) had negative results;[149] there is however an early report on the production of antihemolytic sera by injection of extracts of Amanita phalloides which was believed ascribable to the hemolytic glucoside itself.[150]

Serum reactions with cholesterol and lecithin were first described by Sachs and Klopstock [151] who obtained complement fixing and flocculating immune sera in rabbits upon injecting these substances together with serum of a foreign species. These statements are the more interesting inasmuch as both lecithin and cholesterol are normal constituents of animal tissues.[152]

When commercial egg lecithin was used in a repetition of these experiments by Levene, van der Scheer and the author (408), positive results were obtained without difficulty, but other (laboratory) preparations, and also brain lecithin in experiments of Plaut and Rudy,[153] had no distinct immunizing effect and failed to give significant flocculation or complement fixation reactions with the immune sera to impure lecithin. Hence the antibodies readily engendered by commercial lecithin are most probably due to a contaminant.[154] On the other hand, Weil and Besser, and H. Maier [155] recorded the formation of antibodies following the injection, in mixtures with serum, of the synthetic distearyl lecithin prepared by Grün and Limpächer (417). When immune sera obtained with commercial and synthetic lecithin

---

(399); strangely, sera to the latter were said to give complement fixation with galacturonic acid.

[149] (400), (396), (401).

[150] Ford and Rockwood (402). Ford's statements on the production of protecting antibodies against toxic extracts of Rhus plants were not confirmed by Adelung (403).

[151] (404); (404a) (anaphylaxis experiments with lipids); v. (405) (immunization with cephalin and cerebrosides); (73). Brain cephalin has been recognized by Folch (405a) as a mixture of phosphatides.

[152] Antibody production following injection of fats was traced to protein impurities [Uhlenhuth et al. (406); v. (407)]. Anaphylactic symptoms upon injection of oleyl-N-methyl taurine in guinea pigs previously treated with this compound were later found to occur also in normal animals [Fierz et al. (407a)].

[153] (409), v. (410), (411).

[154] From a report by Fujimura (412) the contaminant might be ovovitellin; v. (413).

[155] (414), (415); v. (416) (appearance of non-specific reactivity with "lipids" in the serum after injections of synthetic kephalin).

were tested against both substances, H. Maier found overlapping reactions, whereas similar tests carried out by Weil and Besser displayed definite specificity. Again, in later papers Wadsworth and Maltaner [156] state that they did not succeed in calling forth antibodies with purified kephalin, or lecithin from brain or liver. It would seem desirable, in view of the conflicting results, to pursue the investigations on the immunizing capacity of lecithin with chemically well defined preparations. Antisera specific for a lipid assumed to be a polydiaminophosphatide have been described by Tropp and Baserga (420).

The experiments on the production of complement fixing antibodies to cholesterol by injection of the substance along with serum proteins were confirmed [157] and extended to derivatives of cholesterol and sterols including related hormones,[158] and it is stated that the sera are of low titer but can be produced even with fractions of a milligram (426). Weil and Besser [159] obtained antibodies with cholesterol, hydroxycholesterol (Lifschitz) and dihydrocholesterol while no antibody response was incited by cholesterol oxide, or the dibromide or esters of cholesterol. Antisera for cholesterol and dihydrocholesterol differentiated these two compounds and did not act on the other substances mentioned; cross reactions were observed with hydroxycholesterol antiserum and cholesterol or dihydrocholesterol.

Sera to cholesterol and ergosterol were tested by Berger and Scholer (428) against a series of sterols and cholesterol derivatives; the authors report that the sera enabled them serologically to distinguish cholesterol and ergosterol, as well as unaltered and irradiated ergosterol (vitamin D).[160]

Not to be disregarded is the fact that some immune sera give complement fixation reactions with emulsions of lecithin or cholesterol (and organ extracts) although the antigens, various azoproteins,[161] employed for immunization are quite unrelated to these substances; and that sera of normal or infected animals may react non-specifically with divers lipids as lecithin, cholesterol, sitosterol. Moreover, in papers by Wadsworth and Maltaner (430a) differences are pointed out between the mode of action of sera obtained with cholesterol, and ordinary immune sera, which led them to doubt the existence of true

---

[156] (418); v. (419).
[157] (421), (422); v. (423).
[158] Brandt and Goldhammer (424); v. (425).
[159] (414), (427).
[160] v. (429).   [161] (430).

anti-cholesterol antibodies. Against this criticism counter-evidence has been offered (431). A final decision will hinge upon the validity of the tests for specificity of the sera prepared with various sterols, which have so far not been disproved.

The studies discussed have a bearing upon the question of the lipoid nature of alcohol soluble haptens since, granting the accuracy of the data presented, these haptens and chemically known lipids are the only materials that appear to acquire antigenicity by the addition of proteins. It is not evident, however, why substances chemically so dissimilar as sterols and phosphatides should have a special property in common.

THE MOSAIC STRUCTURE OF CELL ANTIGENS. — The observations discussed in preceding sections on species specific, individual and heterophile reactions of a single kind of cells suggest a mosaic structure of cell antigens.[162] The various properties of a cell that are characterized by immunological reactions (irrespective of assumptions concerning their chemical nature) have been designated as serological "factors," an expression similar in import to the receptors of Ehrlich (120) but different in that his concept embodies hypothetical elements. While receptors, according to Ehrlich's views, are the binding groups in any antigen, it may be remarked that the designation has been used mainly in connection with agglutination and lysis, whereas in precipitin reactions with proteins special reacting units are usually spoken of only when one is dealing with mixtures of various proteins. But a descriptive term such as factors (or receptors) cannot be dispensed with in the discussion of cell reactions, as appears from examples already mentioned — Salmonella bacilli, and human blood groups and types. In this respect the characteristics of human and animal blood cells that distinguish individuals are of particular interest; here, the notion of serological units gains additional meaning from their association with genetic factors.

The prevailing views on the complex structure of cell antigens have been aptly expressed by White (433) in the following sentences, referring to bacterial agglutinins: "The agglutinative and agglutinogenic complex of an organism consists of definite qualitatively different chemical substances or components — or in the limiting case a single component. Each antigenic component stimulates in the animal body its own serum counterpart or agglutinin component which is qualitatively different from that of any other antigenic constituent."

[162] v. Durham (432); Nicolle.

Adopting this statement provisionally without qualification, and assuming, from chemical results, that serological factors may be correlated with non-protein haptens, one might give the following scheme, applying to animal cells, as an illustration which is meant to indicate that some of the components may be similar in unrelated species and different in individuals of the same species.

A	A	B	D'
B	B	C'	L
C	C	F	M
D	D	G	N
E	F	H	O
.	.	.	.
.	.	.	.

Columns 1 and 2 represent the determinant components of two individuals of one species, column 3 those of a zoologically related, column 4 those of an unrelated species. The meaning of the scheme is evident if one considers experiments like those of Todd and those of Irwin on multiple individual blood properties and their Mendelian inheritance (pp. 89, 117–118).

In the author's opinion the assumption of different specific structures to explain diversified reactions is not invariably correct (p. 266).[163] As has been pointed out, it is a fact that one antibody can react with substances of related constitution, and different immune sera with the same substance. Consequently, it is not requisite that an antigen which reacts with several antibodies has an equivalent number of specific substances or distinctive, binding groups.[164] Nor is the fractionation of antisera by partial exhaustion with heterologous antigens by itself unfailing proof of this conclusion, and an alternative explanation is to be considered, namely, the formation of divers antibodies in response to a single determinant group. Results indicative of the latter possibility were obtained in experiments with immune sera for azoproteins, where, on account of the chemical constitution of the azocomponent, the formation of several antibodies specific for distinct groups seems inadmissible; and yet the absorption effects, were the antigen chemically unknown, could be adduced as evidence for an antigenic mosaic (p. 270).

[163] Cf. the discussion by Tulloch (433a); (434), (435).
[164] Attention may here be called to the frequent occurrence of very small quantitative and qualitative differences in cell reactions. Examples are to be found in the tables of Kolle and Gotschlich (1) on the agglutination of vibrios, and in isoagglutination reactions (436).

As an example let us take the heterogenetic antigen of sheep blood. The incomplete absorption of sheep hemolysins by human blood A (Schiff and Adelsberger [165]) and of the lysins in group-specific anti-A sera by sheep blood can be explained if one assumes that antibodies of different reactivity are formed through the action of single determinant structures, similar but not identical, in each of the two antigens, less specific antibodies being responsible for the cross reactions. Usually, and perhaps correctly, it is supposed that the two sorts of blood have a substance or group in common,[166] and in addition each has a specific structure of its own. Such an explanation, however, seems inadequate for the sheep hemolysins which are formed by the injection of bacteria. As stated before, the hemolysins produced by various bacteria are serologically different, but all of them are absorbed by sheep blood. Yet one would hardly be warranted to suppose, for this reason, that sheep erythrocytes possess special binding groups [167] corresponding to each of these antibodies; in fact, immune sera obtained with sheep blood only exceptionally react with "Forssman bacteria." It is much more plausible that the antigen present in sheep blood is capable of combining with a number of antibodies all engendered by antigens containing chemically related substances (222). Doubtless, the agglutination of a great many bloods by plant agglutinins calls for this kind of explanation (p. 5).

The presence of several specific substances in an antigenic material can evidently be established on chemical grounds. As such instances may be cited the demonstration of more than one polysaccharide in bacterial antigens, antisera having antibodies for carbohydrates and antibodies for proteins, and the agglutinogens A, B and M, N occurring in various combinations in human blood cells, and of which only the first two pass to some extent into alcoholic solution. Clearly due to different structures within a molecule is the formation upon immunization with azoproteins of species specific antibodies and antibodies against the azocomponent, separable by absorption (p. 159).

A plurality of specific substances or groupings can also be reasonably affirmed when an antiserum becomes completely exhausted by a combination of heterologous antigens, such as a serum produced with human cells of group AB by a mixture of A and B cells, or where

---

[165] (219); v. (437), (438).
[166] v. (256).
[167] Indeed, in this and other instances the number of hypothetical groups can be increased almost at will by including additional cross reacting antigens.

genetic experiments establish sharp segregation of multiple factors in the offspring;[168] the opposite conclusion seems more probable when the properties fail to segregate or when continued absorption with an heterologous antigen markedly and progressively weakens the homologous reaction,[169] and likewise in cases of non-reciprocal cross reactions, for the presence of the same determinant in two antigens should cause reactions in either direction.[170] However, the interpretation of cross reactions and multiple serological characters must remain hypothetical until chemical information is available.

Some additional examples may be offered. On the basis of absorption experiments it has been suggested that differences between the reactions of isoagglutinin β with human blood B and animal bloods [171] may depend upon the occurrence of several structures B_I, B_{II}, B_{III} ... in human blood of which only B_{II}, B_{III} ... are represented in animal erythrocytes; from findings of Thomsen (442) the assumption of qualitative differences in the B-substances of various species seems more probable. Furthermore, the hypothetical partial B agglutinogens never separate in the offspring, and this is true also of the property M even though anti-M immune sera produced in rabbits with human M blood contain agglutinins which react differently with the M factors in the erythrocytes of diverse monkey species.[172]

That the human isoagglutinin α_1 (acting on human blood of subgroup A_1) is bound in small measure by agglutinogen A_2,[173] can be understood by imperfect specificity, provided the two agglutinogens are qualitatively different. The alternative explanation, maintained by Lattes and Thomsen, that the same agglutinogen may be present in A_2 cells as in A_1 cells but in smaller amounts, is probably incorrect [174] although in general the possibility exists that in absorption tests quantitative differences simulate the presence of diverse substances. Another instance, the appreciable absorption of agglutinins for the factor N by human blood which apparently does not contain the agglutinable substance N (151) may be explained by chemical similarity between M and N or on the supposition that the antibodies have

[168] It would be difficult to conceive of the independent inheritance of specific groupings (e.g. of A and B in human AB cells) if they are contained in one molecule, unless one assumes them to be structures of rather large size held together in a complex.

The apparent contrast between the genetic results on animal cells and unsuccessful attempts to separate from bacterial polysaccharides (p. 222) fractions corresponding to the serological factors calls for further investigation.

[169] v. Krumwiede (439).
[170] v. (37).
[171] v. (440), (256a) (134), (441), (225).
[172] Landsteiner and Wiener (443).
[173] (444), (445), (446).
[174] v. (122); Sachs (263).

affinity, not only for the specific structure N, but also for other parts of the antigen.

INDIVIDUAL VARIATIONS AND SPECIES DIFFERENCES. — The existence of serologically demonstrable individual differences in blood, inherited according to the Mendelian laws, prompts the question whether they are related to the variations that underlie the evolution of species. In many cases serological blood differences between individuals and between species will be similar in nature.[175] Indeed, a serological cell property may be the constant attribute of one species while distinguishing individuals in another. Pertinent instances are the Forssman antigen characterizing sheep blood but occurring — as a related substance — in man only in the groups A and AB, and the M agglutinogens in the red cells of all rhesus monkeys but restricted in human beings to the types M and MN.

As noted before, related species are commonly distinguishable by absorbed hemagglutinating sera, but in two species of ducks (Anas boscas and Dafila acuta) whose close relationship is indicated by complete fertility of the hybrids, it was possible, with the aid of rabbit immune sera, to demonstrate an agglutinable factor which occurs more frequently in one species but does not permit the differentiation of every individual.[176] A similar observation (relating to one animal) was made with Cavia rufescens and Cavia porcellus. Impurity of the strains cannot be ruled out entirely; otherwise the findings would indicate identity of serological cell characters in individuals of very closely related species.

A second point concerns the inheritance of species characters which is in accordance with the hereditary transmission of individual blood properties. This was first investigated with horse-donkey hybrids.[177] It turned out that the red cells of mules are readily distinguishable from the cells of horses and donkeys and contain substantially the agglutinable substances common to both parental species as well as those peculiar to one or the other (Table 12).[178] Comprehensive investigations on this topic have been conducted with hybrids of dove and pigeon species by Irwin and his associates (451). The situation

[175] v. (447) (iso- and heteroagglutination of horse blood); Irwin (p. 89).
[176] Unpublished experiments.
[177] Immune sera from mule blood behaved much like anti-horse sera, showing a predominance of the character of the female parent, in keeping with some other observations in genetics (192).
On the inheritance of agglutinogens in guinea pig hybrids v. Landsteiner (448), Holzer (449).
[178] (450).

was found to be the same as with mule blood except for the occasional observation of new "hybrid substances," presumably due to the combined action of genes. Further significant results have been brought to light in the examination of successive back-crosses. With the aid of agglutinins absorbed with the blood of various back-cross birds, segregation of the factors particular to one parent species was clearly established, and in this way, for instance, more than ten serological factors present in Pearlneck doves but not in Ringneck doves could be demonstrated. On comparing four species of pigeons and doves, information presumably of general validity was secured on the sharing of apparently identical agglutinable cell constituents by kindred species. These relationships in their essentials are conveyed by the following diagram.

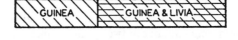

All things taken together, the observations on blood corpuscles conform to the conceptions of modern genetics which consider the process of evolution to have as its basis individual variations resulting from gene mutations and causing, as they accumulate, races and new species to arise in a continuous series (192). Since single genes may affect more than one phenotypical characteristic, a correlation between morphologically recognizable variations and alterations in cell antigens would not seem unlikely.[179] Here it may be worth mentioning that in domestic chickens, of which so many breeds have been produced and maintained, there are many and distinct blood differences, apparently more than in guinea fowl or turkeys (453). Sex differences in the agglutinogens which would reflect the chromosome differences have not been detected.

The parallelism between morphological and biochemical evolution is more difficult to understand in the case of the species differences of proteins. One assumption would be that when mutations occur, the

[179] See Morgan (452).

proteins regularly or frequently also undergo changes. If this is actually the case, one would expect proteins, like cell antigens, to exhibit constitutional differences in individuals of the same species, a hypothetical view that has been put forth repeatedly.

The importance of the question justifies adopting a critical attitude toward some of the relevant experiments.[180] For example, the assertion that the S:Fe ratio of hemoglobin varies within wide limits in horses would require authoritative confirmation. The reports in general are open to the objection that the differences observed may be due to variations in the relative quantities of various protein fractions (p. 66). A priori, the chances of obtaining positive results by chemical investigation with animals of the same kind are rather small since it is usually no easy task to distinguish by chemical analysis proteins of the same type but derived from different species. Serological methods have proved to be most suitable for the latter purpose, and consequently the demonstration of individual differences ought to be accomplished more readily with the aid of serum reactions. The immunization of animals with proteins derived from the same species would seem most promising line to pursue. Uhlenhuth, who used this method, did not succeed in producing isoprecipitins by immunizing rabbits with the serum of other rabbits of the same or a different race, while it was easy, by "cross immunization," to obtain antibodies differentiating the sera of rabbit and hare (p. 18). Likewise, when chickens were transfused with chicken blood no isoprecipitins were formed.[181] Recently, however, individual differences in human sera were observed by Cumley and Irwin by means of precipitin reactions with absorbed rabbit immune sera which may be referable to proteins (466a). In appraising the negative results of isoimmunization, in contrast to the relative ease of isoagglutinin formation in chickens, cattle and rabbits (467), it should be considered that very small variations in the proteins, for example, the substitution of a few amino acids for others, need not be demonstrable by precipitin reactions. The existence of individual protein variations of a special kind, namely

---

[180] v. (454–459).

[181] Experiments by the author and Levine. A few findings from which one may infer, though not with certainty, the existence of individual differences in proteins are the formation of isoprecipitins in one rabbit [(460); v. (461)] and the rare occurrence of anaphylactic reactions following therapeutic blood or plasma transfusions, e.g. a case reported by György and Witebsky (462) in which complement fixing antibodies appeared after a transfusion. As to antibody formation after the injection of sera from animals of different age, v. (463–466).

differences in normal antibodies within one species, may be recalled (see, however, p. 149).

Attempts to demonstrate racial protein differences by serological methods have not yielded uniform results. A report by Glock (468) is unconvincing, but the work of Sasaki (469) and Lühning (470) on protein differences in races of chickens and pigs merits attention; possibly the latter author was dealing with species hybrids. The investigations of Bruck (471) on the differentiation of serum proteins among human races, according to which the white race would hold a special position, could not be confirmed by Marshall and Teague (472), or by FitzGerald (473), and newer experiments suggesting, perhaps, a difference between the sera of white individuals and negroes have not been followed up (474). If the positive results cited are correct, they would signify the existence of constant racial differences, unlike the variations in frequency revealed by isohemagglutination.

Until recently, only scattered data were available concerning the inheritance of proteins.[182] Lately, a systematic attack on the problem was made by Cumley and Irwin (478), the material investigated being the serum of doves and their hybrids. The conclusion of the authors is that the species hybrids possess the proteins of both parent species, and that the proteins are controlled by gene action and consequently segregate in back cross hybrids, parallel to the inheritance of cellular agglutinogens. In view of the fundamental nature of the subject, it will be desirable to establish with full certainty that the observed distinctive reactions are not connected with antigens other than proteins, although the conclusion of the authors would seem the most plausible (v. p. 20).

On account of the meagre experimental evidence one can at present only conjecture the mode of evolution of the proteins in the scale of living beings. In order to explain the change in all proteins of the body, in the genesis of a new species, one either could suppose, in keeping with current theories on evolution, the occurrence of many small variations accompanying single mutations of genes which become detectable by present methods only after accumulation, or one could assume that special mutations cause the transformation. If the latter alternative were correct, then, in so far as proteins are concerned, a line of demarcation between species would be conceivable.

[182] (475) (examination of a chicken-guinea hen hybrid); (476) (anaphylaxis tests with plant hybrids); (477) (measurement of crystals of mule hemoglobin).

The questions whether cell and protein antigens influence each other and whether the effect of genes and the direction of mutations is modified by the nature of the proteins, are still matters for speculation.[183]

BIBLIOGRAPHY

(1) Kolle et al: ZH *44*, 1, 1903. — (2) Landsteiner et al: CB *39*, 309, 1905. — (3) Misawa: JJ Sect. VII *1*, 105, 1932. — (4) Bang et al: BCP *8*, 238, 1906. — (5) Zinsser et al: JEM *37*, 275, 1923. — (6) Landsteiner: BZ *119*, 294, 1921. — (7) Landsteiner: JI *15*, 589, 1928. — (8) Friedberger et al: CB *38*, 544, 1905. — (8a) Topley: JP *33*, 339, 1930. — (9) Hektoen et al: JID *50*, 171, 1932. — (9a) Sachs: HPM *2*, 779, 1929. — (9b) Schiff: OH *3*, Fischer, Jena, 1924. — (10) Landsteiner et al: ZI *13*, 403, 1912. — (11) Jorpes: BJ *26*, 1488, 1932. — (12) Boehm: BZ *282*, 32, 1935. — (13) Bennett et al: JI *4*, 29, 1919. — (14) Schmidt et al: Pr *19*, 345, 1922. — (14a) Eisler: ZI *79*, 293, 1933. — (15) Kosakai: JP *23*, 425, 1920 (B). — (16) Landsteiner et al: ZI *17*, 363, 1913. — (17) Tanzer: JI *42*, 291, 1941 (B). — (18) Sauer: JI *29*, 157, 1935 (B). — (19) Bernheimer et al: JI *41*, 201, 1941. — (20) Beckwith: Pr *30*, 788, 1933. — (21) Taliaferro: JI *35*, 303, 1938 (B). — (22) Mudd et al: JI *17*, 39, 1929. — (22a) Boyd: FI. — (23) Kolmer et al: Approved Laboratory Technic, New York Appleton, 1938. — (23a) Wadsworth: SM. — (24) Heidelberger et al: JEM *60*, 643, 1934. — (25) Miles: Br *14*, 43, 1933. — (26) Landsteiner et al: JEM *42*, 841, 1925. — (27) Ehrlich et al: BK 1899, p. 6; 1901, p. 569. — (28) Landsteiner et al: JI *9*, 213, 1924; JEM *40*, 91, 1924. — (29) von Dungern et al: ZI *4*, 531, 1910. — (30) Irwin: JG *35*, 351, 1938; G *24*, 709, 1939; AN *74*, 222, 1940. — (31) Moody: J. Mammalogy *22*, 40, 1941. — (32) Landsteiner et al: SB *99*, 658, 1928; JI *9*, 221, 1924, *22*, 75, 1932. — (32a) Eaton: JEM *67*, 857, 1938. — (33) Witebsky: Nw 1929, p. 771; HP *13*, 503, 1929. — (34) Sata: BPA *39*, 1, 1906 — (35) Fleischmann et al: Fol. serol. *1*, 173, 1908. — (36) Graetz: EH *6*, 397, 1924. — (37) Doerr: HPM *1*, 759, 1929. — (38) Witebsky et al: ZI *67*, 480, 1930. — (39) Dmochowski: ZI *93*, 311, 1938. — (40) Leschke: ZI *16*, 627, 1913. — (41) Matsuno: Tohoku JEM *19*, 168, 1932. — (42) Atumi: BPh *95*, 244, 1936. — (43) McCartney: AJP *63*, 207, 1923 (B). — (44) Henle: JI *34*, 325, 1938 (B). — (44a) Walsh: JI *10*, 803, 1925. — (44b) von Dungern: Z. allg. Physiol. *1*, 34, 1902. — (44c) Sievers: APS *16*, 44, 1939. — (45) von Dungern: MM 1899, p. 1228; 1900, p. 962; v. BPh *90*, 405, 1936. — (46) Tanaka: Sci. Rep. Inst. Inf. Dis. Tokio *5*, 171, 1926. — (47) Brandt et al: K 1926, p. 655. — (48) Witebsky et al: ZI *58*, 271, 1928. — (49) Heimann et al: ZI *58*, 181, 1928. — (50) Reichner et al: ZI *81*, 410, 1934. — (51) Plaut: ZI *82*, 65, 1934. — (52) Lewis: JI *41*, 397, 1941 (B). — (53) Bailey et al: JI *39*, 543, 1940; JEM *72*, 499, 1940. — (54) Weil et al: JI *30*, 291, 1936. — (55) Weil: ZI *58*, 172, 1928. — (56) Kishioka: BPh *85*, 425, 1935. — (57) Hahn: ZI *86*, 31, 1935. — (58) Ogata: JJ VII, *2*, 65, 1934. — (59) Witebsky et al: ZI *76*, 266, 1932. — (60) Gotoh: BPh *72*, 169, 1933. — (61) Miyagawa: JP *23*, 462, 1920. — (62) Lewis: JI *27*, 473, 1934. — (63) Krüpe: ZI *85*, 487, 1935. — (64) Salfeld et al: JI *31*, 429, 1936. — (65) Witebsky et al: ZI *73*, 415, 1932. — (66) Witebsky et al: ZI *79*, 335, 1933. — (67) Witebsky et al: ZI *78*, 509, 1933. — (68) De Gaetani: ZI

[183] See (479), (480).

77, 43, 1932. — **(69)** Seegal et al: Pr *45*, 248, 1940. — **(70)** Cohen et al: Pr *43*, 22, 249, 1940. — **(71)** Hirszfeld et al: ZI *75*, 193, 1932. — **(71a)** Van der Scheer et al: JI *41*, 391, 1941. — **(72)** Sigurdsson: Pr *50*, 62, 1942. — **(73)** Baeyer: ZI *56*, 241, 1928. — **(74)** Hiro: K 1935, p. 344. — **(75)** Witebsky et al: ZI *80*, 108, 1933. — **(76)** Bailey et al: JEM *73*, 617, 1941. — **(77)** Grund: DA *87*, 148, 1906 (B). — **(78)** Moran: ZI *67*, 115, 1930. — **(79)** Spinka et al: Pr *38*, 447, 1938. — **(80)** Pearce et al: JEM *14*, 44, 1911. — **(80a)** Rosenthal: BZ *42*, 7, 1912. — **(80b)** Schattenfroh: AH *44*, 339, 1902. — **(80c)** Franceschelli: AH *70*, 163, 1909. — **(81)** Landsteiner et al: Pr *25*, 140, 1927. — **(82)** Lake: JID *14*, 385, 1914. — **(83)** Doerr et al: BZ *60*, 257, 1914. — **(84)** Schwab: ZI *87*, 426, 1936 (B). — **(84a)** d'Alessandro et al: C 1941, I, 3383. — **(84b)** Claude: S *91*, 77, 1940; CSH *9*, 263, 1941. — **(84c)** Bensley: S *96*, 389, 1942. — **(85)** Furth et al: JEM *74*, 247, 1941. — **(86)** Henle et al: JEM *74*, 495, 1941, v. 77, 315, 1942. — **(87)** Henle et al: JEM *68*, 335, 1938. — **(88)** Levit et al: N *138*, 78, 1936. — **(89)** Cumley: J. N. Y. Entomolog. Soc. *48*, 265, 1940; J. Exp. Zool. *80*, 299, 1939. — **(90)** Walton: JEM *22*, 194, 1915 (B). — **(91)** Hadda et al: ZI *16*, 524, 1913. — **(92)** Masugi: BPA *92*, 429, 1934. — **(92a)** Tikamitu: CA *36*, 4184, 1942. — **(92b)** Seegall et al: FP *2*, 99, 101, 1943. — **(93)** Joannovics: WK 1909, p. 228. — **(94)** Yokouti: BPh *108*, 151, 1938. — **(95)** Kay: JEM *72*, 559, 1940; JI *42*, 369, 1941. — **(96)** Swift et al: JEM *65*, 557, 1937. — **(97)** Cruickshank: Br *22*, 126, 1941 (B). — **(98)** Gottlieb: JI *4*, 309, 1919. — **(99)** Menne: JID *31*, 455, 1922 (B). — **(100)** Tyler: BB *81*, 190, 364, 1941 (B); Western J. Surg. *50*, 126, 1942. — **(101)** Tyler; PNA *28*, 391, 1942. — **(102)** Dmochowski: AJC *37*, 252, 1939 (B). — **(103)** Witebsky: ZI *78*, 179, 1933. — **(104)** Wilke: ZI *87*, 252, 1936. — **(105)** Trawinski: ZI *90*, 85, 1937 (B). — **(106)** Oswald: ZI *97*, 219, 1939. — **(107)** Bogdanovic: ZI *98*, 283, 1940. — **(107a)** Mann et al: CaR *3*, 193, 196, 1943. — **(108)** Kidd: JEM *71*, 335, 351, 1940; JBa *39*, 349, 1940. — **(109)** Dmochowski: SB *129*, 349, 350, 1938. — **(110)** Cheever: Pr *45*, 517, 1940. — **(111)** Hoyle: AJC *39*, 224, 1940. — **(112)** Barret: CaR *1*, 543, 1941 (B). — **(113)** editorial: Brit. Med. J. *1*, 113, 1942 (B). — **(114)** Graffi: ZK *50*, 501, 1940. — **(115)** Dmochowski et al: Br *20*, 466, 458, 1939 (B). — **(116)** Lumsden: AJC *31*, 430, 1937. — **(117)** Phelps: AJC *31*, 441, 1937. — **(118)** Mann et al: AJC *39*, 360, 1940. — **(119)** Landsteiner: CB *27*, 357, 1900. — **(120)** Ehrlich et al: BK 1900, p. 453. — **(121)** Landsteiner: WK 1901, p. 1132; S *73*, 403, 1931. — **(122)** Wiener: Blood Groups and Transfusion, Thomas, Springfield, 3rd ed., 1943. — **(123)** Snyder: Blood Grouping, etc., Baltimore, Williams and Wilkins, 1929. — **(124)** Schiff and Boyd: Blood Grouping Technic, Interscience Pub., New York, 1942. — **(125)** Steffan: Handb. d. Blutgruppenkunde, Lehmann, München, 1932. — **(125a)** Davidsohn et al: JID *67*, 25, 1940. — **(126)** von Dungern et al: ZI *6*, 284, 1910. — **(127)** Bernstein: Z. indukt. Abstammgslehre *37*, 237, 1925. — **(128)** Schermer et al: A*1* *64*, 518, 1932; ZZ B *24*, 103, 1932. — **(129)** Kaempffer: ZZ B *32*, 169, 1935 (B). — **(130)** Landsteiner: MM 1903, p. 1812. — **(131)** McCombs et al: AIM *59*, 107, 1937; v. N *151*, 419, 1943. — **(132)** Stein et al; N *149*, 528, 1942. — **(133)** Donath et al: MM 1904, p. 1590. — **(133a)** Mackenzie: Medicine *8*, 159, 1929. — **(134)** von Dungern et al: ZI *8*, 526, 1911. — **(135)** Troisier: AP *42*, 363, 1928. — **(136)** Candela et al: Z *25*, 513, 1940. — **(137)** Schiff et al: ZI *40*, 335, 1924. — **(138)** Hirszfeld et al: ZI *59*, 17, 1928 (B). — **(139)** Schermer et al: ZI *68*, 437, 1930. — **(140)** Witebsky et al: K 1927, p. 1095. — **(141)** Schäper: ZZ *20*, 419, 1931. — **(142)** Schermer: ZI *58*, 130, 1928, *80*, 146, 1933; ZR *7*, 33, 1934 (B). — **(143)** Jettmar: ZI *65*, 288, 1930. — **(144)** Castle et al: PNA *19*, 92, 1933. — **(145)** Landsteiner: Wien. Klin. Rdschr. 1902, p. 774. — **(146)** Hooker et al: JI *6*, 419, 1921. — **(147)** Landsteiner et al:

Pr *22*, 100, 1924. — **(148)** Kozelka: JI *24*, 519, 1933. — **(149)** Landsteiner et al: Pr *30*, 209, 1932. — **(150)** Todd et al: JH *10*, 185, 1910; RS *B*, *106*, 20, 1930, *107*, 197, 1930, *117*, 358, 1935. — **(151)** Landsteiner et al: JEM *47*, 757, 1928, *48*, 731, 1928. — **(152)** Dahr: K 1939, p. 806; ZI *97*, 168, 1939. — **(153)** Landsteiner et al: Pr *43*, 223, 1940; JEM *74*, 309, 1941. — **(154)** Wiener et al: AM *13*, 2306, 1940. — **(155)** Levine et al: Am. J. Obst. *42*, 925, 1941. — **(156)** Jonsson: APS *13*, 424, 1936. — **(157)** Hooker: NEM *225*, 871, 1941. — **(158)** Andresen: ZI *85*, 227, 1935. — **(159)** Wiener; JG *29*, 1, 1934. — **(160)** Little: JI *17*, 377, 1929. — **(161)** Hofferber et al: AT *64*, 510, 1932. — **(162)** Ferguson et al: JI *44*, 147, 1942. — **(163)** Loeb: AN *54*, 45, 1920; BB *40*, 143, 1921. — **(164)** Kozelka: PZ *6*, 159, 1933 (B). — **(165)** Haddow: JP *39*, 345, 1934 (B). — **(166)** Gorer: JP *54*, 51, 1942 (B). — **(166a)** Lumsden: AJC *32*, 395, 1938. — **(166b)** Claude: CIY *40*, 248, 1941. — **(166c)** Gross: CaR *3*, 26, 1943. — **(167)** Jensen: Arch. Ges. Physiol. *62*, 172, 1895. — **(168)** Morgan: Exp. Embryology, Columbia University Press, New York, 1927. — **(169)** Hirschfeld: L *2*, 675, 1919; L'Anthrop. *29*, 505, 1918/19. — **(170)** Coca et al: JI *8*, 487, 1923. — **(171)** Snyder: AJA *9*, 233, 1926. — **(172)** Nigg: JI *11*, 319, 1926. — **(173)** Landsteiner et al: JEM *76*, 73, 1942. — **(174)** Matson: AJA *24*, 81, 1938. — **(175)** Landsteiner et al: JI *27*, 469, 1934. — **(176)** Bialbsuknia et al: JI *9*, 593, 1924. — **(177)** Boll. Ist. sieroter milan. *10*, 260, 1931. — **(178)** Topley et al: PB. — **(179)** Smith et al: JM *10*, 89, 1903. — **(180)** Beyer et al: JM *12*, 313, 1904. — **(181)** Weil et al: WK 1917, p. 1509. — **(182)** Craigie: JI *21*, 417, 1931. — **(183)** White: JP *51*, 446, 447, 1940. — **(184)** Arkwright: JP *24*, 36, 1921. — **(185)** White: JP *36*, 65, 1933. — **(186)** Andrewes: JP *25*, 505, 1922, *28*, 345, 1925. — **(187)** Kauffmann et al: ZH *111*, 740, 1930. — **(188)** Beniasch: ZI *12*, 268, 1912. — **(189)** Arkwright: JP *31*, 665, 1928. — **(190)** Gouwens: JID *33*, 113, 1923. — **(191)** Kemper: JID *18*, 209, 1916. — **(192)** Dobzhansky: Genetics, etc., Columbia University Press, New York, 1937. — **(193)** Hooker et al: JI *16*, 291, 1929. — **(194)** Cooper et al: JEM *55*, 531, 1932. — **(195)** Gundel et al: ZH *113*, 498, 1932. — **(196)** Park: J. State Med. *38*, 621, 1930, *39*, 3, 1931. — **(197)** Pfeiffer: CB *121*, 249, 1931. — **(198)** Benham: JID *49*, 183, 1931. — **(199)** Tomcsik: ZI *66*, 8, 1930 (B). — **(200)** Neufeld et al: Arb. Kais. Gesdhamt *34*, 166, 293, 1910. — **(201)** Avery et al: Monogr. Rockefeller Inst. 1917, No. 7. — **(202)** Walter et al: JI *41*, 279, 1941. — **(203)** Heidelberger et al: JEM *38*, 73, 1923, *40*, 301, 1924, *42*, 727, 1925. — **(204)** Weiss et al: JEM *67*, 49, 1938. — **(205)** Verwey: JEM *71*, 635, 1940. — **(206)** Lancefield: HL *36*, 251, 1941. — **(207)** Kauffmann: Die Bakteriologie der Salmonella-Gruppe, Munksgaard, Copenhagen, 1941. — **(208)** White: JP *31*, 423, 1928, *34*, 325, 1931. — **(209)** Furth et al: JEM *49*, 727, 1929. — **(210)** Bergey's Manual, 5th ed., Williams and Wilkins, Baltimore, 1939. — **(211)** Castellani: ZH *40*, 1, 1902. — **(212)** Lawrence et al: BiR *15*, 35, 1940 (B). — **(213)** Haldane: SB *119*, 1481, 1935. — **(214)** Forssman: BZ *37*, 78, 1911. — **(215)** Myers: CB *28*, 237, 1900. — **(216)** Kritschewski et al: ZI *56*, 130, 1928 (B). — **(217)** Hyde: AJH *8*, 205, 1928. — **(218)** Buchbinder: ArP *19*, 841, 1935. — **(219)** Schiff et al: ZI *40*, 335, 1924. — **(220)** Andersen: ZR *4*, 49, 1931. — **(221)** Mutermilch: AP *38*, 1002, 1924. — **(222)** Forssman: HPM *3*, 469, 1928; WK 1929, p. 669. — **(223)** Sievers: APS Suppl. *37*, 458, 1938; APS *16*, 44, 1939. — **(224)** Streng: APS Suppl. *37*, 498, 1938. — **(225)** Landsteiner et al: JEM *42*, 863, 1925. — **(226)** Buchbinder: JI *26*, 215, 1934. — **(227)** Landsteiner et al: JEM *42*, 123, 1925. — **(228)** Kritschewski et al: ZI *56*, 130, 1928. — **(229)** Witebsky et al: ZI *49*, 1, 517, 1926/27, *65*, 473, 1930, *59*, 139, 1928. — **(230)** Witebsky et al: K 1927, p. 1095. — **(231)** Boor et al: JEM *59*, 63, 75, 1934. — **(232)** Eyre et al: JID *58*, 190, 1936. — **(233)** Ballner et al: AH *51*, 245, 1904. — **(234)** Cohn: ZH *104*, 680, 1925. —

(235) Lubowski et al: DA *79*, 396, 1904. — **(236)** Avery et al: JEM *42*, 709, 1925. — **(237)** Beeson et al: JI *38*, 231, 1940. — **(238)** Sugg et al: JEM *53*, 527, 1931. — **(239)** Barnes et al: JEM *62*, 281, 1935. — **(239a)** Sugg et al: JI *43*, 119, 1942. — **(239b)** Neter: Pr *52*, 289, 1943; JI *46*, 239, 1943 (B). — **(240)** Ingalls: JI *33*, 123, 1937. — **(241)** Landsteiner et al: JI *22*, 75, 1932. — **(242)** Shorb et al: AJH *19*, 148, 1934. — **(243)** Bailey et al: AJH *13*, 831, 1931, *17*, 329, 358, 1933. — **(244)** Eisler et al: ZI *76*, 461, 1932. — **(245)** Powell et al: Pr *39*, 248, 1938. — **(246)** Rockwell et al: JID *59*, 171, 1936 (B). — **(247)** Hyde et al: AJH *20*, 465, 1934. — **(248)** Arbesman et al: JA *13*, 85, 1941. — **(249)** Finland et al: S *87*, 417, 1938. — **(249a)** Beeson et al: JEM *70*, 239, 1939. — **(250)** Weil et al: JI *36*, 139, 1939. — **(251)** Schmidt-Schleicher: ZI *97*, 14, 1939 (B). — **(252)** Landsteiner et al: Pr *28*, 309, 1930. — **(253)** Schiff: ZI *82*, 46, 1934. — **(254)** Goebel: JEM *68*, 221, 1938. — **(255)** Holtman: JI *36*, 405, 413, 1939. — **(256)** Sachs: EH *9*, 1, 1928. — **(256a)** Dahr et al: ZI *91*, 211, 470, 1937, *93*, 480, 1938. — **(256b)** Friedenreich et al: JI *37*, 435, 1939. — **(256c)** Brown: JI *46*, 325, 1943. — **(257)** Paul et al: AJM *183*, 90, 1932. — **(258)** Stuart et al: Pr *34*, 209, 212, 530, 1936 (B); AIM *58*, 512, 1936; AJ *26*, 677, 1936. — **(259)** Schiff: JI *33*, 305, 315, 1937 (B). — **(260)** Bailey et al: JCI *14*, 228, 1935. — **(261)** Beer: JCI *15*, 591, 1936. — **(262)** Stuart et al: JID *59*, 65, 1936. — **(263)** Sachs: JP *54*, 105, 1942. — **(264)** Mauss: JI *42*, 71, 1941. — **(265)** Meyer: ZI *19*, 313, 1913 (B), *57*, 42, 1928. — **(266)** Botteri: Z. Ges. Exp. Med. 77, 490, 1931. — **(267)** Sievers: ZI *84*, 208, 1935 (B). — **(268)** Fairley: JP *30*, 97, 1927. — **(269)** Meyer: ZI *15*, 245, 1912. — **(270)** Sachs et al: ZI *55*, 341, 1928. — **(271)** Ninni: AP *52*, 502, 1934. — **(272)** HPM *1*, 1084, 1929 (B). — **(273)** Doerr et al: BZ *50*, 129, 1913, *60*, 257, 1914. — **(274)** Georgi: AIF *9*, 33, 1919. — **(275)** Sordelli et al: Rev. Inst. Bact. Buenos Aires *1*, 229, 1918. — **(276)** Landsteiner: BZ *119*, 294, 1921. — **(277)** Taniguchi: JP *24*, 253, 1921. — **(278)** Landsteiner et al: JEM *38*, 127, 1923. — **(279)** Sachs et al: ZI *82*, 287, 1934. — **(280)** Landsteiner et al: JEM *41*, 427, 1925. — **(281)** Misawa: cited in ZI *76*, 386, 1932. — **(282)** Witebsky: ZI *51*, 161, 1927. — **(283)** Kamada: ZI *71*, 522, 1931. — **(284)** Witebsky: ZI *48*, 369, 1926. — **(285)** Bordet et al: SB *95*, 887, 1926. — **(286)** Krah et al: ZI *65*, 473, 1930. — **(287)** Sachs et al: DM 1925, p. 589, 1017; 1926, p. 650; 1927, p. 394; v. ZI *51*, 73, 1927. — **(288)** Förtig: ZI *52*, 328, 1927. — **(289)** Hedén: ADV *17*, Suppl. 2, p. 1, 1936. — **(290)** Eagle et al: JEM *71*, 215, 1940. — **(290a)** Beck: JH *39*, 298, 1939 (B). **(291)** Sachs et al: APS Suppl. *16*, 388, 1933; ZI *80*, 222, 1933. — **(291a)** Plaut: Z. ges. Neurol. u. Psychiatrie *123*, 365, 1930. — **(292)** Klopstock: DM 1926, p. 1460. — **(293)** Landsteiner et al: JEM *45*, 465, 1927. — **(294)** Blumenthal et al: CB *121*, 85, 1931 (B). — **(295)** Donath et al: MM 1904, p. 1590; EH *7*, 184, 1925. — **(296)** Oe. Fischer: K 1928, p. 2061. — **(297)** Nanba: DM 1925, p. 594. — **(298)** Sunami: Tohoku JEM *16*, 277, 1930. — **(299)** Weil: BaR *5*, 293, 1941. — **(300)** Lewis: JI *41*, 397, 1941 (B). — **(301)** Schwentker et al: JEM *60*, 559, 1934. — **(302)** Parks et al: Pr *35*, 418, 1936. — **(303)** Schwentker et al: JEM *70*, 223, 1939. — **(304)** Halpern: ZI *11*, 609, 1911. — **(305)** Krusius: AA *67*, 6, 1910 (B). — **(306)** Hektoen et al: JID *34*, 433, 1924. — **(307)** Metalnikoff: AP *14*, 577, 1900. — **(308)** Weil: ZI *47*, 316, 1926. — **(309)** Irwin et al: JID *70*, 119, 1942. — **(310)** Benjamin et al: Z. Kinderheilk. *3*, 73, 1912. — **(311)** Heimann: ZI *50*, 525, 1927. — **(312)** Klopstock: ZI *55*, 304, 1928. — **(313)** Sachs: CB *104*, Beih., 128, 1927. — **(314)** Il'ina et al; Vashkov: CA *36*, 3840, 1942. — **(315)** Plaut et al: ZI *74*, 333, 1932 (B). — **(316)** Rudy: BZ *245*, 431, 1932. — **(317)** Sachs et al: BZ *159*, 491, 1925. — **(318)** Dienes: JI *17*, 137, 1929. — **(319)** Prüsse: ZI *78*, 437, 1933. — **(320)** Hazato: ZI *89*, 1, 1936. — **(321)** Weil: ZI *80*, 75, 1933. — **(322)** Ramon: SMW 1941, p. 1366. — **(323)**

Schäfer: AIF H.37, p. 16, 1939.—(324) Stitz: CB *143*, 437, 1939.—(325) Doerr et al: BZ *131*, 13, 1922.—(326) Gibson et al: JEM *12*, 411, 1910.—(327) Witebsky: ZI *59*, 139, 1928.—(328) Wheeler et al: JI *36*, 349, 1939.—(329) Andersen: APS Suppl. 37, 82, 1938.—(330) Morgenroth et al: BZ *131*, 525, 1922.—(331) Stuart et al: AJ *30*, 775, 1940 (B); JI *37*, 159, 1939.—(331a) Culbertson: AJH *22*, 190, 1935.—(332) Witebsky et al: ZI *54*, 181, 1927.—(333) Tsuneoka: ZI *22*, 567, 1914.—(334) Doerr et al: ZI *47*, 291, 1926.—(335) Sachs: Acta Soc. med. Fenn. *A*, 15, 1932 (B).—(336) Cesari: AP *44*, 534, 1930.—(337) Eagle: JEM *55*, 667, 1932.—(338) Doerr et al: ZI *81*, 132, 1933.—(339) Sachs et al: DM 1925, p. 589; HP *13*, 433, 1929.—(340) Gonzalez et al: SB *106*, 1006, 1931; AJH *17*, 277, 1933 (B), *19*, 184, 1933.—(341) Zozaya et al: JEM *55*, 325, 1932; Pr. *30*, 47, 1932.—(342) Freund: S *75*, 418, 1932.—(343) Landsteiner et al: Pr *30*, 1055, 1933; JEM *59*, 479, 1934 (B).—(344) Plaut et al: ZI *81*, 87, 1933.—(345) Zozaya et al: JEM *57*, 21, 1933.—(345a) Misawa: ZI *79*, 80, 1933.—(346) Mutsaars: SB *120*, 263, 1935.—(347) Meyer: ZI *57*, 42, 1928.—(348) Klopstock et al: K 1927, p. 119.—(349) van der Scheer: ZI *71*, 190, 1931.—(350) Fränkel et al: K 1927, p. 1148, 2473.—(351) Guggenheim: ZI *61*, 361, 1929.—(352) Hedén: ADV *21*, 181, 1940.—(353) Ramon et al: SB *103*, 1202, 1930; AP *40*, 1, 1926.—(354) Glenny et al: JP *34*, 267, 1931.—(355) Prigge: CB *145*, 241, 1940.—(356) Freund: JI *40*, 437, 1941.—(357) Seibert: JI *28*, 425, 1935.—(358) Caulfeild: JA *7*, 451, 1936.—(359) Holford et al: JI *46*, 47, 1943 (B).—(360) Swift et al: JEM *63*, 703, 1936.—(361) Sabin et al: JEM *68*, 659, 1938.—(362) Rudy: K 1934, p. 4.—(363) Meyer: ZI *57*, 42, 1928 (B).—(364) Pedersen-Bjergaard: ZI *82*, 258, 1934.—(365) Chargaff et al: AP *54*, 708, 1935, *64*, 301, 1940.—(366) Dienes et al: JI *12*, 137, 1926.—(366a) Ninni: AP *52*, 502, 1934.—(367) Avery et al: JEM *38*, 73, 81, 1923, *42*, 347, 355, 367, 1925; AM *6*, 1, 1932.—(368) Tillett: JEM *45*, 713, 1927.—(369) Avery et al: JEM *42*, 367, 1925.—(370) Goebel et al: JEM *54*, 431, 437, 1931.—(371) Spassky et al: Br *17*, 38, 1936.—(371a) Partridge et al: Br *21*, 180, 1940.—(372) Francis et al: JEM *52*, 573, 1930; Pr *31*, 493, 1934.—(373) Zozaya et al: Pr *30*, 44, 1932.—(374) Finland et al: JI *29*, 285, 1935 (B).—(375) Avery et al: JEM *58*, 731, 1933 (B).—(376) White: The Biology of Pneumococcus, Commonwealth Fund, New York, 1938.—(376a) Pochon: RI *3*, 136, 1937.—(377) Felton et al: BJH *62*, 430, 1938 (B); Pub. Health Rep. *53*, 1855, 1938; IC, p. 784.—(378) Chow: CJP *11*, 223, 1937; JEM *64*, 843, 1936.—(379) Pappenheimer et al: Pr *31*, 37, 1933.—(380) Enders et al: JEM *60*, 127, 1934.—(381) Felton et al: JBa *38*, 579, 1939 (B).—(382) Wong et al: Pr *40*, 357, 1939.—(383) Landsteiner et al: JEM *46*, 204, 1927.—(384) Horsfall et al: JI *31*, 135, 1936.—(385) Dubos: EE *8*, 135, 1939; JEM *66*, 113, 1937 (B).—(386) Boivin et al: SB 1934–36; CR 1934; Arch. Roumain. Path. *8*, 45, 1935 (B).—(387) Raistrick et al: Br *15*, 113, 1934 (B).—(388) Morgan et al: BJ *34*, 169, 1940 (B).—(389) Goebel et al: JBC *148*, 1, 1943.—(390) Chargaff et al: JBC *109*, XIX, 1935.—(391) Nishimura: JEM *50*, 419, 1929.—(392) Nozu: JJ VII, *2*, 55, 1934.—(393) Ikeda: JJ VII, *1*, 221, 1932; v. Giovanardi: CB Ref. *123*, 51, 1936.—(394) Fujimura: JB *21*, 371, 1935.—(395) Uhlenhuth et al: ZI *92*, 171, 1938 (B).—(396) Uhlenhuth et al: ZI *82*, 229, 1934.—(397) Campbell: Pr *36*, 511, 1937.—(398) Seideman: JI *38*, 237, 1940 (B).—(398a) Gelfand: JA *14*, 203, 1943.—(399) Rothschild: Enzymologia *5*, 329, 1939.—(400) Wedum: JID *52*, 203, 1933.—(401) Bashford: Arch. Int. Pharmacodyn. *9*, 451, 1901.—(402) Ford et al: J. Pharmacol. *4*, 235, 1913 (B).—(403) Adelung: AIM *11*, 148, 1913.—(404) Sachs et al: BZ *159*, 491, 1925; DM 1925, p. 1017.—(404a) Klopstock: ZI *48*, 97, 1926.—(405) Ornstein:

WK 1926, p. 785. — **(405a)** Folch: JBC *146*, 35, 1942. — **(406)** Uhlenhuth et al: ZI *4*, 780, 1910. — **(407)** Tokunoyama: Tohoku JEM *22*, 252, 1933. — **(407a)** Fierz-David et al: H *22*, 3, 1939; v. *24*, 5E, 1941. — **(408)** Levene et al: JEM *46*, 197, 1927. — **(409)** Plaut et al: ZI *73*, 385, 1932. — **(410)** Belfanti: ZI *56*, 449, 1928. — **(411)** Dessy: Boll. Ist. sieroter milan. 7, 599, 1928. — **(412)** Fujimura: JB *25*, 595, 1937. — **(413)** Weichsel et al: JI *32*, 171, 1937. — **(414)** Weil et al: K 1931, p. 1941. — **(415)** H. Maier: ZI *78*, 1, 1933. — **(416)** Klopstock: CB *104*, 435, 1927. — **(417)** Grün et al: BC *59*, 1350, 1926. — **(418)** Wadsworth et al: JI *26*, 25, 1934, *28*, 183, 1935. — **(419)** Kimizuka: JB *21*, 141, 1935. — **(420)** Tropp et al: ZI *83*, 234, 1934 (B); v. JBC *116*, 527, 1936. — **(421)** Plaut et al: ZI *66*, 152, 1930. — **(422)** Bisceglie: Biochimica e Ter. sper. *15*, 299, 1928. — **(423)** Aoki: C *110*, 2219, 1939. — **(424)** Brandt et al: K 1936, p. 1875. — **(425)** Thompson: Ph *21*, 595, 1941. — **(426)** Weil et al: ZI *76*, 69, 1932. — **(427)** Weil et al: ZI *76*, 76, 1932; Pr *36*, 238, 1937. — **(428)** Berger et al: K 1932, p. 158; ZI *76*, 16, 1932, v. *80*, 75, 1933. — **(429)** Caltabiano: ZI *95*, 260, 1939. — **(430)** Selter: ZI *68*, 409, 1930. — **(430a)** Wadsworth et al: JI *29*, 135, 151, 1935. — **(431)** Hahn et al: ZI *88*, 16, 1936. — **(432)** Durham: JEM *5*, 353, 1901. — **(433)** White: Great Britain Med. Res. Council Rep. Ser. #103, 127, 1926. — **(433a)** Tulloch: J. Roy. Army Med. Corps, 1927. — **(434)** Landsteiner et al: JEM *63*, 325, 1936. — **(435)** Landsteiner et al: JEM *71*, 445, 1940. — **(436)** Landsteiner et al: JI *17*, 1, 1929, *20*, 179, 1931. — **(437)** Akune: ZI *73*, 75, 1931. — **(438)** Andersen: ZR *4*, 49, 1931. — **(439)** Krumwiede et al: JI *10*, 79, 1925. — **(440)** Friedenreich et al: ZI *78*, 152, 1933; APS. Suppl. *37*, 163, 1938. — **(441)** Brockmann: ZI *9*, 87, 1911. — **(442)** Thomsen: ZR *7*, 1, 1935. — **(443)** Landsteiner et al: JI *33*, 19, 1937. — **(444)** Lattes et al: JI *9*, 407, 1924. — **(445)** Landsteiner et al: JI *11*, 221, 1926. — **(446)** Friedenreich: ZI *71*, 291, 1931. — **(447)** Landsteiner et al: JI *9*, 221, 1924. — **(448)** Landsteiner: Pr *28*, 981, 1931. — **(449)** Holzer: ZI *84*, 170, 1935. — **(450)** Landsteiner et al: JI *9*, 213, 1924. — **(451)** Irwin: JG *35*, 351, 1938; G *24*, 709, 1939; AN *74*, 222, 1940. — **(452)** Morgan: The Theory of the Gene, New Haven, Conn., Yale University Press, 1926. — **(453)** Landsteiner et al: Pr *30*, 209, 1932. — **(454)** Birkofer et al: ZPC *265*, 94, 1940 (B). — **(455)** Küster: ZPC *172*, 138, 1927. — **(456)** Schenck et al: ZPC *215*, 87, 1933. — **(457)** Trendtel: BZ *180*, 371, 1927. — **(458)** Aszodi: BZ *212*, 158, 1929. — **(459)** Lang: AEP *148*, 222, 1930. — **(460)** Schütze: DM 1902, p. 804. — **(461)** Cattabeni: BPh *123*, 376, 1941. — **(462)** György et al: MM 1929, p. 195. — **(463)** Friedberger et al: ZI *71*, 453, 1931. — **(464)** Gräfenberg et al: ZI *9*, 749, 1911. — **(465)** Picado: AP *44*, 584, 1930. — **(466)** Nattan-Larrier et al: SB *98*, 926, 1928, *110*, 510, 1932. — **(466a)** Cumley et al: JI *46*, 63, 1943. — **(467)** Castle et al: PNA *19*, 92, 1933. — **(468)** Glock: Biol. Zbl. *34*, 385, 1914. — **(469)** Sasaki: Nihon-Chikusan-Jakkwai-Zashi *3*, 88, 1928. — **(470)** Lühning: Inaug.-Diss., Bern 1914. — **(471)** Bruck: BK 1907, p. 793. — **(472)** Marshall et al: Philippine J. Sci. *3*, 357, 1908. — **(473)** FitzGerald: JM *21*, 41, 1909. — **(474)** Fischer et al: ZI *94*, 104, 1938. — **(475)** Sasaki: ZZ *38*, 361, 1937. — **(476)** Moritz: "Der Züchter" *6*, 217, 1934; BDB *51*, 52, 1933. — **(477)** Loeb: S *45*, 191, 1917. — **(478)** Cumley et al: PNA *27*, 565, 1941; G *27*, 228, 1942 (B). — **(479)** Guyer: S *71*, 175, 1930. — **(480)** Hertwig: Nw 1934, p. 425. — **(481)** Felton: IC, p. 784. — **(482)** Miller et al: IC, p. 800.

IV

THE NATURE AND SPECIFICITY OF ANTIBODIES

NATURAL ANTIBODIES. — The knowledge that antibodies are present in the serum of normal, non-immunized animals dates from the early work of Landois (1). Exploring for the cause of shock following transfusions of animal blood into human beings, he found an explanation in the clumping or lysis of the red blood corpuscles which frequently ensues when the serum of an animal is mixed with the blood of another species. Subsequent studies [1] confirmed and elaborated these observations. Similarly, normal sera contain agglutinins and lysins, and complement fixing antibodies for bacteria (7–11). The investigation of these antibacterial agents was initiated at a time when studies on immunity to infectious diseases had awakened interest in the properties of the blood serum, and the question of the relative importance of serum and cellular elements for protection against infectious agents was in the foreground of discussion.[2] In following this trend there were detected in the serum of apparently normal human beings and animals other antibodies: antitoxins, antiviral antibodies, opsonins, antilysins.

Some flocculation and complement fixation reactions, of indistinct specificity, with lipids and other substances (p. 130) [3] may be due to unstable globulins which are apt to form precipitable complexes. Whether these are to be considered as antibody reactions is indeed a question of definition.[4]

One could have thought that agglutination and lysis of bacteria and blood corpuscles by serum of normal animals is attributable to one substance, or a few substances, capable of acting on many sorts of cells. An experiment by Bordet (15) contradicted this simple assumption. When cholera vibrios were agglutinated by normal horse serum, and the bacteria after combination with agglutinins removed by centrifugation, the serum no longer acted on cholera vibrios yet still clumped typhoid bacilli as intensely as before. If the two sorts

[1] v. (2–6). On antibodies for spermatozoa v. (6a).
[2] Nuttall; Buchner (12).
[3] Cf. (13), (13a), (13b).
[4] v. Boyd (14).

of bacteria were added in reverse order, the serum separated from the typhoid bacilli agglutinated the vibrios and not the typhoid bacilli. Analogous experiments with agglutinins and lysins of normal serum and with various kinds of blood [5] and bacteria [6] in most cases yielded results in conformity with Bordet's observation, as illustrated in the following table. However, a distinct decrease in agglutinin titre for cells other than the one used for exhaustion of the serum was sometimes recognizable.[7] Special well established cases are the absorption of human isoagglutinins by the blood corpuscles of animals,[8] and the absorption of large fractions of the agglutinins acting on one sort of blood by the corpuscles of closely related species (man-chimpanzee, mouse-rat, Cavia porcellus-Cavia rufescens).

TABLE 17

	Unabsorbed Serum	Goat Serum absorbed with				
		Pigeon Blood	Rabbit Blood	Human Blood	Pigeon and Rabbit Blood	Pigeon and Human Blood
Pigeon Blood	+	o	+	+	o	o
Rabbit Blood	+	+	o	+	o	+
Human Blood	+	+	+	o	+	o

Malkoff (16), whose experiment is reproduced in Table 17, reached the conclusion, widely accepted (Ehrlich and others), that a normal serum contains as many specific agglutinins as there are sorts of cells that are agglutinated by the serum. Since normal sera, e.g. ox serum, may agglutinate numerous bacteria and blood corpuscles, and contain hemagglutinins that differentiate individuals of the same species,[9] this would imply the presence of an exceedingly large number of different active substances in the serum. Such a conclusion, unlikely at first sight, is reduced almost to absurdity when one considers that accord-

[5] (16–18).
[6] (19), (20). In experiments of Bürgi (8) on agglutination of bacteria normal sera of different animals could be arranged in a series according to their agglutinative strength (e.g. ox or horse sera were more active than rabbit or guinea pig sera and gave stronger flocculation also with gum mastic suspensions) and various bacteria showed the same gradation in agglutinability, regardless of the normal sera with which the tests were made.
[7] (21–25), (4), (6).
[8] von Dungern and Hirschfeld (26).
[9] (4); v. (27).

ing to Malkoff's hypothesis each of these antibodies should be specifically related to substances occurring in a certain species of animals or in bacteria.[10] Moreover, if the serum contains so many agglutinins the absolute amount of agglutinin acting on one kind of blood ought to be well-nigh infinitesimal; but this is not the case, as shown by experiments in which the agglutinins were set free from their union with erythrocytes by heating the agglutinated cells in saline solution, and the protein in the agglutinin solutions estimated by means of precipitins.[11] In addition the purified agglutinins when tested against the blood corpuscles of various animals acted most strongly on the red cells used for the absorption, but also agglutinated other sorts of blood, a result which would seem to prove that the agglutinins absorbed by a certain blood are not highly specific.

Although Thomsen (32) found that agglutinated cells may absorb other agglutinins than those reacting specifically, non-specific absorption is presumably not the sole cause of the phenomenon described. At any rate, there is an apparent contradiction between the results of absorption and splitting off of agglutinins, probably to some extent because the usual method of titrating agglutinins by determining the highest active dilution of serum is not very accurate; hence only gross differences are demonstrable.

Further investigation will be required to reconcile the inconsistencies. On the basis of the available results the most reasonable assumption would appear to be that natural antibodies are specific in so far as they react to a different degree on various cells.[12] Consequently, if one assumes that normal serum contains a sufficient number of agglutinins, each reacting distinctly only with a certain proportion of all bloods, a given sort of blood will absorb from a serum all those agglutinins for which it has affinity, and there will remain after absorption some that react with freshly added blood of other species. In a general way our view is supported by the existence of several kinds of substances which in their visible effects simulate typical antibodies but whose specificity is on a lower level; hemolysins of plant and animal derivation, plant hemagglutinins, and viruses, have previously been mentioned in this respect.

Normal sera contain, in addition to "cold hemagglutinins" and autoantibodies against spermatozoa (6a), other antibodies acting on materials

[10] v. (28–30).
[11] Landsteiner and Prašek (31).
[12] v. Browning (33).

derived from the same individual. Kidd and Friedewald (34) observed complement fixation — without demonstrable species specificity — with rabbit sera and extracts of the tissues of rabbits and other species, attributable to particles which can be separated by centrifugation at 20,000 r.p.m. (p. 81). Flocculation by chicken sera at low temperature of aqueous and alcoholic tissue extracts of all species tested was described by Duran-Reynals; these reactions become stronger after immunization with various antigens and — like certain complement fixation reactions of normal sera — they resemble the Wassermann reaction [Duran-Reynals (35)].

Concerning the origin of the antibodies in apparently normal sera, a subject of medical and epidemiological consequence — discussed clearly and in detail by Topley (36) — there are several possibilities: a purely physiological mechanism, past unapparent or frank infections, previous absorption of antigenic substances (microbes, food) from the intestinal tract [13] or otherwise; in the last two cases the resulting antibodies may be directed against serologically related bacteria as well as against the incitants themselves. To give an example of the conflicting opinions,[14] the occurrence in normal human serum of diphtheria antitoxin or antibodies which neutralize the virus of poliomyelitis has been attributed by most authors to contact with the infectious agents, by some to spontaneous formation. Significant evidence was provided by animal experiments [15] and tests with human sera. Hughes and Sawyer (44) showed that antibodies protecting mice against yellow fever virus are frequently found in serum of persons living in regions where the disease is prevalent, but are absent in individuals who had never been exposed to the virus. Similar in outcome are statistical studies on the distribution of diphtheria and scarlatinal antitoxins in populations and groups of individuals, exposed to infection in various degrees.

On the other hand, there can be no doubt about the physiological genetically determined [16] formation of antibodies. A direct proof is the regular presence of isoagglutinins in human serum, in strict correlation to inherited isoagglutinogens, and this argument is not weakened if one accepts the hypothesis that the isoagglutinins are found in consequence of "autoimmunization." [17] A similar origin may be pre-

[13] v. (37) (describes also experiments on oral immunization).
[14] Hirszfeld (38), Neufeld (39), Friedberger (40), Jungeblut et al. (41).
[15] Bailey et al. (42), Ramon (43).
[16] (45–47).
[17] A general discussion is given by Friedenreich (48). The suggestion (49) that isoagglutinins originate from oral immunization is entirely unsupported.

sumed for most normal hemagglutinins and hemolysins acting on blood of foreign species. The possibility that antibodies against blood corpuscles may be formed as a result of bacterial infection has been established by interesting observations of Bailey (42). This author found that sheep hemolysins appear in the serum of rabbits when the animals are infected with a strain of B. lepisepticus (and probably N. catarrhalis) containing Forssman's antigen, or when they harbour the bacteria in the nasal cavity. Yet it is not likely that such "heterogenetic" immunization, probably of importance in the stimulation of bacterial antibodies, plays more than a secondary role for the formation of natural antibodies to erythrocytes.

At birth the serum is low in natural antibodies [18] (and in globulins) which become established gradually. For instance, the development of hemagglutinins in chickens is completed after about a month, sometimes later; [19] the serum of adult cattle has much greater agglutinating activity than that of calves (11); and human isoagglutinins reach a maximum titer at the age-level of about 5 to 10 years (51). The parallelism in the development of natural hemagglutinins and antitoxins or bacterial antibodies has been used as evidence for the spontaneous origin of all normal antibodies,[20] but the argument works both ways since contacts with external stimuli will naturally accumulate with age.

Evidence for the spontaneous origin of normal hemagglutinins and hemolysins is afforded by certain regularities in distribution (and perhaps by their permanence). This question has not been investigated extensively and, on account of individual variations, it is necessary to examine a number of sera from one species. Some pertinent facts, however, have been established. There is, as Gürber [21] first noticed, a correlation between antibodies and cell antigens, somewhat similar to that obtaining with isoantibodies, namely an approximately reciprocal relationship between the range of activity of the serum and the sensitivity of the blood cells. Other observations concerning the agglutinins of certain species [22] are the following: Sera of Macacus rhesus and Cynomolgus philippensis agglutinate human A-blood more intensely than blood of groups O and B whereas sera of Cercopithecus pygerthyrus act mainly on human blood B; in a species of baboons

---

[18] v. (40), (19), (10). A certain amount of antibodies (e.g. human isoagglutinins) is transmitted through the placenta or in the colostrum.

[19] Bailey (50).

[20] Hirszfeld (52), Friedberger et al. (40).

[21] (53–55).   [22] (26), (56–58).

there were differences in this respect among individuals.[23] With the blood of a Cebus species (Ceb. hypoleucus?) and a lemur, the sera of several Cercopithecidae (Macacus rhesus, Cerc. pygerthyrus, Papio) regularly showed distinct agglutination, and conversely the Cebus serum agglutinated the bloods of the three above-named species of Cercopithecidae. According to these examples, which could be added to, the occurrence of natural antibodies appears commonly to be a species characteristic, with the qualification that both classes of substances may vary from individual to individual, owing to constitutional differences.

From the foregoing one may distinguish two kinds of antibodies in apparently normal individuals, acquired and physiological; and the conclusion may be drawn that normal serum contains substances which are formed independently of external antigenic stimuli and, though not adjusted to any one antigen, resemble the antibodies produced by immunization. One cannot yet estimate their number nor venture an opinion concerning their physiological function. In this connection it is perhaps of significance that antibodies which agglutinate the individual's own cells (spermatozoa,[24] blood corpuscles) seem to be in part identical with agglutinins acting on cells of foreign species. Information may possibly be derived from a study of plant agglutinins whose presence in seeds would suggest a function in the development of the plant.[25]

Finally, it should be noted that there are dissimilarities between true normal and immune antibodies, apart from the higher specificity of the latter and the usual disparity in titre, the immune antibodies having in general greater avidity and thermostability (p. 141).

In experiments on normal bacterial agglutinins conducted by Jordan (11) these were found to be not less resistant than immune agglutinins. The observations on the lability of normal hemagglutinins, however, are well

[23] Recent observations by Wiener, Candela and Goss (58a) indicate that these regularities are connected with the presence or absence of group-specific substances in tissues.

[24] (59).

[25] Dr. P. R. White, who obligingly agreed to carry out some experiments, found that the growth of tomato roots cultivated in vitro was inhibited by a concentration of 2 to 10 mg. phasin from Phaseolus vulgaris per liter of solution, and there was an indication of a stimulating effect of concentrations below 1 mg. per liter (personal communication). Since the agglutinins found in a number of Papilionaceae (beans, peas, etc.) were not toxic at least for animals [Landsteiner & Raubitschek (59a)], these preliminary results, if further substantiated, would be significant.

established, and it may be considered that the discrepancy is due to the dual origin of the antibodies in normal sera.

Only few data [26] are at hand on the chemical nature of natural antibodies but enough to indicate their globulin nature. One may conjecture that there exists a much greater variety of globulin molecules in a serum than would appear from physico-chemical examination, some of which by virtue of accidental affinity to certain substrates are picked out as antibodies.

IMMUNE ANTIBODIES

PHYSICO-CHEMICAL PROPERTIES.—The association of antibody activity with serum globulins has long been recognized. Not only could antibodies be concentrated in certain globulin fractions but in physical and chemical properties they were seen to resemble common proteins closely, particularly in deteriorating under the influence of denaturing agents and in being attacked by proteolytic enzymes. More precise evidence regarding their nature has accrued in the last decade from application of the newer physico-chemical tools — electrophoresis, ultracentrifugation — and through the use of quantitative analytical methods. The substance of this work is the demonstration that antibodies are not indiscriminately scattered over or adherent to the various serum protein fractions [27] but can be separated as preparations that are homogeneous in the ultracentrifuge and electrophoretically — irrespective of subtle serological diversity. However, not only may the response to an antigen be species characteristic, but different antigens may elicit antibodies of unlike physico-chemical properties in the same animal, and the antibodies in a serum may exist in more than one globulin fraction.[27a]

Examination of immune sera and purified antibodies (pp. 136, 137) in the electrophoresis apparatus of Tiselius [28] revealed that antibodies from rabbits and (with some exceptions, which possibly depend on the length of immunization) the antibodies in antibacterial horse immune sera were contained in the most slowly moving (at alkaline pH) globulin fraction, i.e. the γ-globulin. Conclusive proof of this has been afforded by the marked reduction in the γ-globulin component

[26] Landsteiner and Calvo (60), Gibson (19), Bleyer (60a).
[27] v. (61).   [27a] Cf. (227).
[28] Tiselius and Kabat; van der Scheer et al. (62), Pappenheimer et al. (63).
Fell et al. (64). On antibodies in the serum of allergic patients v. Newell et al. (65).

of anti-pneumococcus horse immune sera upon absorption with the homologous polysaccharide, as seen from the electrophoretic Longsworth diagram given here. In most antitoxic horse immune sera, however, a new globulin component (T), not present in normal serum, was detected and found to carry the antibody activity, alone or jointly

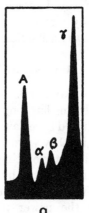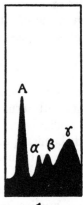

Serum components of a horse antipneumococcal serum (a) before absorption, and (b) after absorption of antibodies by means of pneumococcal polysaccharide. Serum albumin is shown by A, the globulins by α, β, and γ.

a   b

with the γ-globulin; [29] indeed, it will be seen that horse antitoxin is to be precipitated in a different serum fraction from that carrying horse antibacterial (antipneumococcal) immune bodies. According to Kekwick and Record (66) diphtheria horse antitoxin contains two antibodies moving with the β- (β_2, possibly identical with component T) and γ-globulins respectively; these antibodies moreover exhibit other differences. Data on sera of rabbits immunized to tuberculin protein are given by Seibert (67).

With regard to molecular weight, so far as the present evidence from the ultracentrifugal method goes, antibodies, in unaltered state, fall for the most part into one of two categories: those having approximately the same molecular weight (160,000–195,000) as the normal serum globulins, e.g. antibacterial antibodies produced by rabbit, man, monkey, and antibodies with a much larger molecular weight, more than 900,000,[30] typified by the antibacterial antibodies of the horse, cow and pig. This grouping of species, thought to indicate a general rule, does not hold invariably;[31] several non-conforming in-

[29] van der Scheer et al. (65a).
[30] Biscoe et al.; Wyckoff (69), Heidelberger and Pederson; Kabat et al. (70), Heidelberger et al. (71), van der Scheer et al. (72).
[31] In some anti-pneumococcus horse sera the antibody property was detected in

stances were encountered, such as diphtheria horse antitoxin, which has a molecular weight around 180,000 according to Petermann and Pappenheimer, and Rothen,[32] anti-sheep rabbit hemolysin with a sedimentation constant close to that of horse antibodies — and seemingly exceptional as to electrophoretic mobility (77) — and the syphilis reagin in human serum, apparently of great molecular size.[33]

The figures given for isoelectric points of antibodies [34] vary somewhat with the preparations and the experimental conditions. For horse pneumococcus antibodies [35] pH values of about 4.8, for pepsin treated horse diphtheria antitoxin [36] a pH slightly below 7, are recorded. Isoelectric points of rabbit pneumococcus antibodies [37] are given as 5.8 to 6.6, and those of various other rabbit sera [38] as about 6.2.

The axis ratios of antibody molecules (pneumococcus antibodies) have been computed by Neurath (86a) from the Svedberg dissymmetry constants and Perrin's equation, with the following results: horse 20.1, rabbit 7.5, man 9.2. The differences concern mainly the length of the molecules, these being respectively 950, 274, 338 Å. Since with some proteins x-ray analysis led to values conflicting with those obtained by the above calculation, the figures quoted may have to be revised [Crowfoot (86b)].

Ultrafiltration experiments on antibodies performed by Elford and coworkers, and Goodner, Horsfall and Bauer,[39] demonstrated not only the great disparity in molecular size between horse and rabbit pneumococcus antibodies but the presence of aggregates which could be dispersed by changes in the medium.

PURIFICATION METHODS. — For therapeutic ends, the concentration of antibodies and discarding of inert proteins has long been practised on a commercial scale, fractionation of serum proteins by salting-out methods, as with ammonium sulphate or sodium chloride, and isoelectric precipitation at low electrolyte concentration (including electrodialysis) having mainly been employed.[40] Since in general the

---

components of different sedimentation constants [Petermann and Pappenheimer (73)].
[32] Petermann and Pappenheimer (74), Rothen (75); cf. Fell (64), Païc (76), Kekwick et al. (66).
[33] Païc (78), Deutsch (79).
[34] v. Michaelis et al.; Szent-Györgyi et al.; Felton; (80–83); on the isoelectric point of sensitized antigens v. (80), (83a); (83b); (83c).
[35] (71), (84).
[36] (74), (75).
[37] (80–83).
[38] (85), (86).
[39] (87); v. (88).   [40] For literature v. (89–91), (80).

fractionation procedures, however, do not secure a separation of uniform, electrophoretically homogeneous proteins and the results are influenced by factors such as changes in technique, the derivation of the sera [41] and initial antibody concentration, it is easy to understand that antibodies of the same sort have been recovered, variously, in the pseudoglobulin or euglobulin fraction, or in both;[42] however, despite these difficulties, highly purified antibody preparations, largely free also from substances causing untoward reactions, have been obtained.[43] Certain regularities have been found: antitoxins in horse serum are almost consistently recovered in the water-soluble pseudoglobulin precipitate, horse pneumococcus antibodies are carried down with the water-insoluble euglobulin, pneumococcus antibodies produced by the rabbit remain in the water-soluble pseudoglobulin.[44]

For the concentration of antitoxic horse sera, the procedure of Gibson and Banzhaf (105) consists of adding ammonium sulfate and heating at about 60°, whereby inactive pseudoglobulin becomes insoluble and can be removed together with euglobulin; the active water-soluble globulin fraction is then precipitated by raising the ammonium sulfate concentration to nearly 50%, and further purification can be attained through removal of remaining water-insoluble globulins by isoelectric precipitation.[45] The antitoxin may also be recovered in the pseudoglobulin precipitate produced by sodium chloride (109).

In the comprehensive studies of Felton on pneumococcal antibodies, the water-insoluble euglobulin which contains these antibodies in the case of horse sera was brought down upon dilution with water (usually acidified slightly); preparations approximately 80% pure were secured by precipitation with low alcohol concentrations and removing inactive proteins by washing with water, and similar or better results were obtained with the use of aluminum or zinc chloride for precipitating inactive proteins. With horse pneumococcus anti-

[41] (92–93). (94) (reports also the separation of functionally different antibodies in the same serum).
[42] v. Ledingham (95); Gibson et al. (93).
[43] Prescriptions are given by Wadsworth (96) and Zinsser et al. (211). Processing of anti-pneumococcus rabbit sera to decrease the toxicity is described by Goodner, Horsfall and Dubos (97) (using heat and adsorption with kaolin). For fractionation methods v. (98) (antitoxins); (99), (100), (101), (antibacterial sera).
[44] Felton (102), Horsfall and Goodner (103); v. Chow (104).
[45] cf. (106). The "protéine visqueuse" of Doladilhe may well be related to γ-globulin (107); v. (108).

bodies an efficient method of purifying the antibody globulin was elaborated by Chow and Goebel (110) who found phthalate buffer particularly suitable to remove inert protein.

Attempts to purify antibodies by adsorption [46] after the pattern of the methods employed by Willstätter for enzymes have not led to practical results although some success was reported. Claims of Frankel and Olitzki of having separated protein-free antibodies by adsorption to kaolin followed by elution could not be verified by Rosenheim (118) and several other workers.

A high degree of purification is possible, owing to the reversibility of serum reactions, by splitting off antibodies from their combination with antigens.[47] Thus antibodies have been liberated from agglutinated blood stromata, bacteria or specific precipitates by various procedures — warming, extraction with acid, alkali, glycine buffer, salt or sugar solutions, enzymatic digestion [48] — the yield depending on the amount of bound antibody and the firmness of combination, the purity upon the more or less complete removal of antigenic material. Hemolysin preparations so obtained [49] were several hundred times as active as the original serum, and highly purified precipitins for polysaccharides were secured [50] containing up to 95–98 per cent specifically precipitable nitrogen; one preparation was even completely precipitable. The recovery upon dissociating certain immune precipitates or agglutinates in 15 per cent salt solution was 20 to 40 per cent of the total antibody content (Heidelberger).

The fact that by dissociating antigen-antibody complexes antibody solutions have been obtained in which no protein could be detected is obviously ascribable to weak concentration of antibody, and is no proof for the existence of protein-free antibodies, as has been claimed (112). Some examples of the higher sensitivity of serological reactions as compared with chemical protein tests may be cited. Of the hemolysin separated by Euler and Bru-

---

[46] (111–113), (106); v. (113a), (114–117) (experiments on adsorbents coated with soluble specific substances, and elution).

[47] v. Hahn et al.; Landsteiner (119), Kosakai (120), Huntoon (112), Munter (120a), Ramon (61), Velluz (121) (diphtheria antitoxin); Locke et al. (122), Felton (123); cf. Goldie (124), Saeki (125) (precipitins); Kirk and Sumner (126) (antiurease); Bier et al. (127) (syphilis reagins). Precipitins for azoproteins can be separated by absorption with "azostromata" and dissociation with dilute acetic acid (128).

[48] Pope and Healey (129), Petermann and Pappenheimer (74).

[49] Locke et al. (122), Euler and Brunius (130), (131), v. (131a).

[50] Felton (123), Heidelberger et al. (132); v. (133), (134), (135).

nius about 4×10^{-3} mg. were sufficient to dissolve, upon addition of complement, 1 cc of sheep blood, and 4×10^{-4} mg. of Felton's purified pneumococcus antibodies protected mice against 10^6 lethal doses of pneumococci. Plant agglutinins exhibit a degree of activity similar to the purified antibodies. A water-soluble protein fraction isolated by the author and van der Scheer from beans, precipitable at 0.6 to 0.8 saturation with ammonium sulfate,[51] agglutinated distinctly 1 cc of 0.5% washed horse erythrocytes in quantities of less than 10^{-5} mg.

Likewise using the antigen-antibody complex as a source, Northrop [52] was able to achieve an impressive result, the crystallization of an antibody (diphtheria antitoxin). The antitoxin preparation was made by digestion of the complex with trypsin at acid reaction and subsequent fractionation with ammonium sulfate; and on examination by ultracentrifugation, electrophoresis and solubility tests it appeared to be a homogeneous protein containing some carbohydrate [v. (74)].

By weighing particulate antigens before and after absorption of antibodies or by nitrogen determination on specific precipitates the quantity of antibody in an immune serum can be determined. In very potent precipitating antisera this was estimated to be of the order of 10 mg. per cc., i.e., a considerable proportion of the total globulin;[53] commonly the concentration is lower. The values found for hemolytic antibodies were about 1 to 2 mg. per cc. (140), and less in experiments of Euler and Brunius.[54]

ANTIBODIES AND NORMAL GLOBULINS.—Chemical analysis of antibodies has been unsuccessful in so far as no indication was found for the existence of specific prosthetic groups, and no sure difference could be detected with regard to amino acid composition [55] between normal agglutinins and purified antibodies or those contained in specific precipitates. This result is paralleled by serological evidence, and although negative in a sense, has definite significance when one contemplates the problems of antibody formation and specificity.

Apart from direct analytical methods, other attempts have been made [56] to establish differences between normal and antibody globu-

[51] v. (136).
[52] (137). Crystallization of an hemagglutinating substance (concanavalin) from jack beans was reported by Sumner and Howell (138). This substance, soluble only in strong salt solutions, differs also in other respects from typical phytagglutinins.
[53] Heidelberger and Kendall (139); (241), (139a).
[54] (130), v. (141).
[55] Hewitt (142), Chow and Goebel (110), Velluz (143), Breinl and Haurowitz
[56] v. (143), (147–148a). (144), Calvery (145), Banzhaf et al. (146).

lins and some apparently positive findings were reported concerning isoelectric points, titration curves, solubility, viscosity, adsorption, and precipitation by various colloidal solutions; in part these observations were not made with sufficiently pure antibody preparations and with exactly corresponding normal globulins. The results as a whole scarcely constitute positive proof of differences in chemical constitution. It should further be mentioned that in the case of diphtheria antitoxic sera no physico-chemical distinction could be made between active and inert pseudoglobulins.[57] On the other hand, antibodies were found (in horse sera) in protein fractions differing physico-chemically from those in normal serum.[58]

Antibodies, upon being employed as antigens, were found to be related to the normal serum proteins. Thus agglutinins and antitoxins could be precipitated by immune sera prepared against normal serum of the species,[59] and antibodies may engender immune sera that react with normal serum proteins. The latter demonstration was made by immunizing with washed specific precipitates, especially polysaccharide precipitates, whereby any antigenic material other than antibodies is all but eliminated.[60] It could then be anticipated that antibodies functionally different but derived from the same species would be antigenically alike, and this was actually found to be the case, with the reservation that in experiments of Ando, and of Treffers and Heidelberger, antibodies from the horse fell chiefly into two antigenic groups, associated with different globulins — those against bacteria being water insoluble, others, as diphtheria antitoxin and anti-egg albumin, soluble in water (p. 136). The members of each group were practically identical in antigenic properties, and, as in previous experiments, antibodies derived from different species although prepared with the same antigen showed no relationship in tests with precipitin sera. In other words, antibodies employed as antigens exhibit what

[57] (63); v. Reiner and Reiner (149); (75).

[58] (150), (62).

[59] Bordet; Dehne et al.; Kraus and Pribram; Eisler (151); Smith and Marrack (152); Eagle (153). Since such antisera contain precipitins for the various serum proteins, in different proportions, the effect on antibodies is not constant. This might explain negative results such as those with Northrop's highly purified diphtheria antitoxin which failed to precipitate with an immune serum against normal horse serum [v. (141), (154–157)]. Experiments of Wright (158) with anti-horse immune sera indicate that the results are connected with the presence of antibodies for γ-globulin in the precipitating antisera.

[60] Ando et al. (90), Marrack and Duff (159), Treffers and Heidelberger (157). In some of the experiments of these workers differences were noted between normal and immune globulins of the same electrophoretic mobility.

has been called "specificity of origin," but there is no evidence for the existence of true anti-antibodies which would bear a relation to the specific antibody function.

BINDING SITES OF ANTIBODIES.—While from the analysis of precipitates it is definitely known that antigens are multivalent,[61] the number of combining sites depending on their molecular size, the maximum valence of antibodies is still conjectural. Reflections on the process of specific precipitation and agglutination have led to the opinion that in order to flocculate antibodies must at least be bivalent, that is, possess two binding areas. For diphtheria antitoxin and ovalbumin combination of antibody with antigen in the ratio of one to two, deduced from ultracentrifugal analysis, has geen reported (63). Higher ratios seem not to have been directly established experimentally. On a priori reasoning multivalence would seem probable in the case of horse antibodies of large molecular size (159a).

Experiments on dysentery immune sera which react with sheep cells as well as with Shiga bacilli, and on antisera combining with lipids and polysaccharides of tubercle bacilli, were brought forward by K. Meyer [62] as evidence for the presence of specifically different groups in a single antibody (v. p. 269). On the other hand, on immunizing at the same time with a few or numerous antigens, antibodies reacting with more than one of them could not be detected showing, as Hektoen believes, "that the same globulin molecule may not have more than one precipitin" [63] (p. 58).

The assumption of several combining sites in antibodies adjusted to different determinants in the antigen has been made with the view of explaining cross reactions; and higher valence is held by some authors to be the reason for greater reactivity and wider range of cross reactions of antibodies formed upon prolonged immunization.[64]

It could have been supposed that combination of antibody with a precipitin would abolish the antibody function. That this is not necessarily true follows from the fact that horse antitoxin precipitated by an "antihorse" serum is still capable of neutralizing toxin.[65] Con-

---

[61] v. (p. 58); (159a). This serological "valence" is of course not the same as the valence of atoms. In still a different sense the expressions "polyvalent" or "multivalent" are currently applied to immune sera produced against several antigens.

[62] (160); v. Morgan (161).

[63] cf. (69), (159a), (161a), (227).

[64] v. (pp. 19, 144); (162-164); v. (165).

[65] Smith and Marrack (152), Eagle (166); v. Kraus et al. (166a); (151).

versely, antitoxin saturated with toxin is brought down by the corresponding precipitin. Hence, Marrack concludes that "the adsorbing sites of a globulin acting as an antibody appear different from those by which it is bound when acting as an antigen."

Upon adsorption to charcoal, antitoxins lose their capacity for neutralizing toxin (167), in distinction to the reactivity of adsorbed precipitable substances (pp. 144, 247) and of some other antisera, and in experiments of Freund (168) toxin adsorbed to collodion particles could be neutralized by antitoxin but collodion particles coated with antitoxin became toxic on treatment with the homologous toxin.

ALTERATIONS OF ANTIBODIES.—When subjected to various chemical or physical agents antibodies follow, generally speaking, as remarked before, the model of ordinary proteins. Inactivation of antibodies [66] by heat has, like denaturation, a high temperature coefficient and proceeds rapidly around the coagulation temperature of serum proteins. Different antibodies were found to vary in resistance to heat [67] (or acid and alkali),[68] for instance, flagellar are more resistant than somatic O-agglutinins. Kleczkowski (179) ascribes such differences in heat resistance to the formation of complexes between antibodies and unspecific serum proteins (v. p. 56) interfering with the flocculating effect, but to a different degree when various (e.g. O and H) antigens are tested. The conclusion that loss of the flocculating property of antibodies on mild heating can result from the formation of aggregates with other proteins was confirmed by Jennings et al. (184a) in experiments in which antibodies, i.e. the water insoluble globulin from horse anti-pneumococcus serum, were heated alone or in presence of other proteins (water-soluble serum globulin, casein). Kleczkowski's explanation, however, does not account for all cases, for example differences between sheep blood hemolysins (181) or the definitely greater heat resistance of immune in comparison to normal hemagglutinins, seen when the same antigen was employed [69] (p. 132). The rate of destruction by heat is influenced by changes in pH and can be reduced by sufficient concentrations of salt and non-electrolytes

[66] Fuller information on denaturation of antibodies is provided in the reviews by Hartley (89) and Marrack (80). On changes by irradiation v. (80), (169); on alcohol denaturation v. (170), (171). Immune sera can be precipitated by alcohol, acetone or alcohol-ether, at low temperature (172), or with high concentrations of the precipitants (173), without impairment of activity. Regeneration of antibodies after inactivation by acid has been claimed by Chow et al. (174).

[67] (92), (175–182).

[68] v. (89), (183), (184).   [69] Prašek (185).

(sugar, glycerol, urea, etc.).[70] The order of reaction was found in several cases, but not regularly, to correspond to a monomolecular reaction (175).

Of chemical reactions known to affect proteins several have been tried on antibodies for theoretical inquiry or in order to modify (for therapeutic use) the undesirable antigenic properties of immune sera; also, substances to be employed as preservatives (such as phenol or cresol) were examined for their injurious effect on antibodies. Treatment with formaldehyde, reversible under certain conditions, has been investigated repeatedly [71] and discussed with regard to the significance of NH_2 groups for specific combination and the mechanism of flocculation. The latter consideration was suggested by the observation that small concentrations of formaldehyde abolish the flocculating activity of antitoxins and anti-carbohydrate precipitins without destruction of their specific affinity,[72] a situation likewise encountered in the acetylation of antibodies by ketene.[73] Coupling with diazonium compounds [74] in not too large quantities, though altering the antigenic properties, diminishes the activity of antibodies but slightly; there was no indication that varying the azo components would influence the rate of degradation, or change the antibody specificity, which indeed has not been accomplished by any other means. Common to all these alterations is the difference in susceptibility of various antibody functions [75] and the progressive deterioration with increasing intensity of chemical treatment.

Regarding the action of enzymes [76] on immune sera, most reports agree in that antibodies are digested readily by pepsin and, less easily, by trypsin which in some experiments was found to act very slowly, a fact at times erroneously taken as argument against the protein nature of antibodies. Working with agglutinins for B. typhosus, Rosenheim (200) observed that after repeated injections of antigen

---

[70] v. (80), (185a), (185b).

[71] (80), (89), (110), (132), (186), (187); on the effect of various aldehydes v. (186). Other reactions studied are treatment with iodine (188); nitrous acid, phenyl isocyanate (130), (131); ninhydrin (189). Fluorescent conjugates were made by coupling with isocyanates of polynuclear aromatic hydrocarbons (190). Such an antibody conjugate made with fluorescin isocyanate was used by Coons, Creech et al. for specifically staining antigen in tissues.

[72] (186), (187).

[73] (191), (192); Tamura and Boyd (193), Chow and Goebel (110).

[74] Reiner; Bronfenbrenner et al.; Breinl and Haurowitz; Marrack (194), Eagle et al. (195); cf. Heidelberger and Kabat (196).

[75] (186), (188), (189), (195), (197–199).

[76] v. Marrack (80).

the flagellar agglutinins became more resistant to pepsin and trypsin — not to papain — while no such change took place with somatic agglutinins. As an explanation Rosenheim suggests that these phenomena may be connected with the number and distribution of the binding sites in the molecule and the different mode of cleavage by pepsin or trypsin and, on the other hand, by papain.

Of considerable interest in their bearing on antibody structure are recent studies utilizing enzymatic digestion for the purification of antitoxins. This subject was brought forward by Parventjev's patent for a process of antibody refinement consisting of partial peptic digestion between pH 4 and 4.5 and subsequent purification, which is aided by differential heat coagulation of the more readily denatured inactive protein, and results in a product with practically unimpaired antitoxic power but reduced antigenicity and altered reactivity as antigen.[77] Flocculation of this product with toxin, splitting of the complex, and fractionation with ammonium sulphate led to an antitoxin having 77% of the nitrogen precipitable by toxin, a molecular weight of 113,000 against 184,000 for the unchanged antitoxin, and an axis ratio of 5.3 instead of 7.[78] Thus "the increase in immunological potency of the molecule is directly proportional to the decrease in size,"[79] and the authors conclude that "the molecule has been split in a plane normal to the major axis." Furthermore, since an inactive portion of the molecule is split off by enzymatic action and the remaining part has full activity, it is suggested that the antitoxic sites are unsymmetrically distributed on the molecules.[80]

PLURALITY OF ANTIBODIES IN SERA. — The argued question[81] whether the various reactions shown by an antiserum (agglutination, lysis, precipitation, complement fixation, opsonic action, passive protection, sensitization) are brought about by a single antibody

---

[77] Weil et al. (201); Pope (202) (digestion by pepsin or trypsin); Hansen (106); Modern et al. (203); Coghill et al. (204) (Takadiastase); Schultze (205) (peptic digestion of antibodies and normal proteins, at pH 4.5); Huddleson et al. (206) (Brucella antiserum); Grabar (207); Treffers and Heidelberger (208) (pneumococcus antibodies); Caltabiano (209); Sandor et al. (210); Kass et al. (210a).

[78] Petermann and Pappenheimer (74), Rothen (75). A decrease in the sedimentation constant after partial digestion was demonstrated by these authors also with pneumococcus antibodies (73). Electrophoretic analyses of partly digested antibodies were conducted by van der Scheer, Wyckoff et al. (72).

[79] v. (75).

[80] Pappenheimer et al. (63).

[81] (211), (80), (212–215).

("unitarian" hypothesis) or by different ones can now be answered in general, if not in every special instance. It is certain that one antibody frequently causes reactions different in appearance. An acknowledged example, quantitatively verified,[82] is the agglutination of bacteria by the same antibodies that give precipitation and complement fixation with bacterial polysaccharides. Actually agglutination may be regarded as equivalent to a precipitin reaction on the surface. Thus in experiments of Jones (218), collodion particles coated with proteins were agglutinated by high dilutions of the corresponding precipitins.[83] Similarly blood corpuscles coated with soluble specific substances from cholera vibrios were found to be agglutinated by anti-cholera immune sera [84] and red cells treated with a subliminal amount of agglutinin were intensely clumped upon addition of a precipitin which combined with the agglutinin (Moreschi).[85] As an analogy illustrating the unitarian principle the various effects of tannic acid — agglutination, hemolysis, hemotropic action — have been offered (Reiner) (pp. 6, 255).

These facts are clearly in agreement with the unitarian hypothesis but it is not less certain that complex materials (sera, cells) can give rise to as many antibodies as there are antigenic components.[86] A further complication arises for the reason that one antigenic molecular species or even one determinant group does not engender just a single, uniquely defined counterpart of the antigen, but antibodies differing in specificity (p. 268), in the degree of affinity, and therefore in the shape of the reaction curves, and in other immunological manifestations also.[87] Related to this issue are changes in the immune sera during the course of immunization. A matter of common experience is the increase in strength and extent of cross reactions [88] upon con-

---

[82] Heidelberger et al. (216), Gerlough et al. (216a). On the identity of anaphylactic antibodies and precipitins v. Doerr and Russ (217).

[83] v. Mudd et al. (219), Nicolle (220), Arkwright (221). That precipitins are not effective in as high dilutions as are agglutinins and lysins depends largely upon the different dispersion of the antigens, the amount of antibody necessary for agglomeration being, *ceteris paribus*, a function of the total surface of the particles. [Zinsser (222), v. p. 259].

[84] A model experiment is the agglutination by H_2S of bacteria previously treated with lead salts [Neisser and Friedemann (223)].

[85] (224); v. (225).

[86] (226), (139), (227–229). Hetero- and isoagglutinins in normal sera likewise contain fractions that vary in avidity, i.e. with regard to the highest temperature at which combination still takes place [Bialosuknia and Hirszfeld; Friedenreich (230)]; (231).

[87] Goodner and Horsfall (227); (135).   [88] v. (232).

tinued injection of antigen. In part this is referable to higher antibody content (p. 19) but there are observations indicating the development of antibodies with greater combining capacity,[89] and a qualitative change was strikingly demonstrated by experiments of Hooker and Boyd (237) in which the precipitation by sera taken at a later stage was inhibited by a substance that showed no inhibition with the samples from earlier bleedings. Other findings concern the increased resistance to enzymes, just mentioned, the gradual development of heat resistance,[90] and avidity,[91] and variations in the distribution among the globulin fractions.[92]

Certain species differences in antibodies have been noted in the preceding pages; [93] many others (p. 246) were disclosed in the thorough work of Goodner and Horsfall,[94] which included examination of anti-pneumococcus sera from over ten mammalian species, classifiable into two groups. An outcome of their work was the introduction of rabbit antisera into the therapy of pneumonia.

THE FORMATION OF ANTIBODIES. — Before broaching the problem of antibody formation, it will be convenient to take cognizance of some pertinent experimental facts.[95]

Antibodies may be detectable in a few days following the injection of antigen (according to some statements within 24 to 48 hours); [96] after one to two weeks the antibody titre in the serum reaches its peak whereupon it gradually falls off as the result of elimination and diminishing production.[97] The rate and intensity of antibody formation is generally enhanced when antigen is administered again to a previously treated animal; after copious bleedings the antibody level is restituted shortly and the production may even be accelerated.[98] Upon long-continued immunization, dependent on

[89] Heidelberger et al. (233), (234), Malkiel and Boyd (235), Pappenheimer (236); v. (p. 140).

[90] Thiele and Embleton (238); (239), (185).

[91] Mueller (240); (240a), (239), (229).

[92] (229) (fractionation by dialysis); (95), (241).

[93] For individual differences v. pp. 104, 150, 64, 17 (Table 4); (234), (235).

[94] v. (242); Perlmann et al. (243).

[95] A comprehensive discussion of the subject, with bibliography, is presented in a monograph by Burnet (244); v. Browning (33).

[96] Oerskov and Andersen (245), Ramon (246); (263).

[97] The "lifetime" of antibodies has been estimated by means of introduction into antibodies of isotopic nitrogen (administration of isotopic glycine). A period of about two weeks "probably closely approximates the actual half lifetime of the antibody molecule." Passively injected antibodies did not take up the heavy nitrogen. [Schoenheimer et al.; Heidelberger et al. (247)].

[98] For comment on these matters, also for stimulation by non-specific substances, the "negative phase" in which the antibody level is temporarily depressed

the individual animal and the antigen, the response to antigenic stimuli finally declines or ceases.

Because the weight of antibodies is appreciable (p. 138), their titre can, of course, not be raised above a set limit but it is of interest that antibodies may be produced against many simultaneously injected antigens (p. 58); indeed, mixtures of several bacterial vaccines and toxoids have been recommended for practical use.[99] Concurrent with antibody formation there is an increase in the globulin content of the serum (at times along with diminution of the albumin fraction). While this increase was found by some authors to correspond to the amount of antibody globulin [100] it was, in the majority of investigations, considerably greater.[101] Thus, as shown by Pappenheimer et al. (63), immunization against diphtheria toxin calls forth the development of a protein that, but for its inactivity, is indistinguishable from the antibody globulin.

As to the site of antibody formation [102] it is currently accepted that the reticulo-endothelial system is chiefly involved, an opinion suggested by the phagocytic properties of these cells and strengthened by experimental and histological evidence.[103] The methods chiefly applied to this subject — extirpation of organs, impairment of the reticulo-endothelial cells by irradiation or blockade — gave in part fairly conclusive results.[104] Participation of the spleen in antibody production is indicated by experiments of Topley (261) in which spleen tissue, taken from rabbits treated with bacteria, gave rise to agglutinins upon injection into normal animals. MacMaster and his co-workers (262) obtained definite evidence pointing to lymph nodes as the source of antibodies in well-controlled experiments in which it was found that after injecting bacteria or vaccinia virus into the ears of mice, agglutinins were detectable in higher concentrations in the cervical lymph nodes than in the serum. Concordant results were secured by Burnet (244), and in experiments of Ehrich (263) lymph collected from an infected leg contained more antibody than the lymph of the opposite leg and was conspicuously rich in lymphocytes.

Sédallian (264) demonstrated the production of antibodies in surviving

---

following injection of much antigen, and the fate of injected antibody, v. (36). On the suppression of immunizing activity of antigens by combination with large quantities of antibodies v. Olitzki (247a); on immunization by toxin-antitoxin mixtures v. (211).

[99] Ramon et al. (248).

[100] Bjorneboe (249).

[101] Ionesco-Mihaiesti (250), Boyd et al. (241), van der Scheer (62); v. Marrack (80). Boyd and Bernard contemplate that the excess of globulin may include antibodies too weak to be demonstrable.

[102] Data on the distribution and the removal of injected antigen are given, for instance, in (251–253a). On the distribution of antibodies in tissues v. (254, 255).

[103] v. Sabin (256), Kyes (257), Doan (257a).

[104] (36), (258–260).

legs of immunized rabbits which were transfused by connecting the vessels with those of a normal animal; it remained undecided which tissues were responsible for the result.

Findings which suggest the possibility of local antibody formation in infected sites have been presented by Oerskov and Andersen (245), and others.[105] Roemer's report (267) on immunization of the conjunctiva against abrin, and experiments on induced resistance of red cells to the hemolysin of eel serum,[106] likewise dating far back, have not been confirmed. The latter observations could possibly depend on adsorption of antibodies from the serum or on non-specific resistance.

The resistance of lower animals against infections has been investigated by Metchnikoff (269) in connection with his theory of cellular immunity by phagocytosis, and acquired immunity and antibody formation have been observed in lower vertebrates, and in insects and other invertebrates.[107] If antibody formation is taken as a special case of adaptive biological processes, the induced fastness of bacteria and protozoa towards serological agents or chemicals would be in some degree related phenomena.

On natural and acquired immunity in plants and the alleged but wholly doubtful production of antibodies see (272–276).

THEORIES OF ANTIBODY FORMATION. — The antibody response following the introduction of immunizing substances, intimately connected with the physiological synthesis of proteins, is in various of its aspects still as puzzling today as at the time when this remarkable phenomenon was first discovered. Of the hypotheses that have been proposed the simplest, in a certain sense, is the assumption, first advanced by Buchner and still maintained at times, that the antigens enter into the composition of the antibodies. This offers an evident reason for the specificity of antibodies, although it would not by itself account for their affinity. To this conception are opposed strong arguments, chiefly the failure to demonstrate the presence of the antigen in immune sera, even when antigens detectable in minute amounts are used for immunization.

Doerr and Friedli, and Berger and Erlenmeyer [108] after immunization with a conjugated protein containing arsenic found no arsenic in the sera, or not more than in sera of untreated animals, and in the investigations of Heidelberger and co-workers [109] with a highly col-

---

[105] v. (36), (260), (244), (265), (266).
[106] Camus and Gley; Kossel (268).
[107] v. Bordet (270), Metalnikoff (271).
[108] (277), (278); v. (279).
[109] (280). Similarly, immune sera against hemoglobin contain no blood pigment (281).

oured antigen the antibody was practically free of dye. Likewise, experiments by Hooker and Boyd (282) with the arsanilic acid azo-protein, by Haurowitz et al. (283) with several "marked" antigens, and by Wollman and Bardach (284) who attempted to demonstrate antigen in immune sera by anaphylaxis, yielded results not in keeping with the hypothesis in question.

Another objection raised by Topley (261), Heidelberger et al. (285), and Pappenheimer (286) is based upon the quantitative relation between antigens and antibodies, i.e. the observation that antibodies combine with a much larger quantity of antigen than that necessary for their production.[110] According to a calculation made by Hooker and Boyd (290) the discrepancy may be so great that a single antigen molecule gives rise to a quantity of antibodies sufficient to agglutinate several hundred bacilli, and in determinations by Pappenheimer the weight of the resultant antibody was ten thousand times that of the antigen injected. Moreover, the occurrence of non-reciprocal reactions (p. 267) is difficult to understand if antigens incorporated in the antibodies are responsible for their specificity.

An indirect but significant argument is the consideration that the proposed hypothesis fails to explain the specificity of natural antibodies of the hemagglutinins and hemolysins in plants, and of enzymes.

Among the attempts to explain the phenomenon of immunization and the specificity of antibodies, an hypothesis propounded by Ehrlich had many adherents at one time. Ehrlich tried to account for the specificity of antibodies by the assumption that "antibodies are normal constituents of the body which in the cell protoplasm act as receptors and are responsible for the toxic action and the fixation of the antigen; as a result of this specific union, sometimes aided by a stimulating effect, the antibodies are regenerated in excess and enter the blood stream" (translated). Apart from the minor points that antibodies for proteins are usually not demonstrable in normal serum and that differences exist between normal and immune antibodies, this hypothesis is untenable on account of the unlimited number of physiological substances which it would presuppose.

In view of the arguments stated there remains hardly any other conclusion than to regard the production of immune antibodies as a synthetic function of the animal body and to assume that under the

[110] v. (287–289).

influence of antigens the formation of certain globulins (and perhaps of normal antibodies)[111] is modified in such a manner that the resulting globulins are closely adapted to the immunizing substance. Various ideas, necessarily conjectural, to explain the moulding effect of antigens, and the formation of antibodies in general have been advanced.[112] Considerable attention has been given to the hypotheses, concording in the main, of Breinl and Haurowitz, and Mudd, who assume that in contact with the antigen globulins are synthesized which spatially and in chemical affinity correspond to the antigenic pattern. According to this concept antibodies would be permitted to differ in their amino acid composition. A different theory submitted by Burnet (244) supposes that antigens do not influence the globulins directly during their formation, but modify the proteinases [113] operative in this synthesis, in analogy to the formation of new "adaptive" enzymes [114] for certain substances by bacteria upon being grown in media to which these substrates, e.g. carbohydrates, are added.

On the basis of a theory proposed by Mirsky and Pauling concerning the structure of the globular protein molecule the problem has been approached by Pauling (302) from a new point of view.

In considering the perplexing problem how a prodigious number of different antibodies, produced in the same species, can possibly exist which appear to be chemically identical or very similar and, except for some sub-divisions (p. 139), are serologically indistinguishable, it has been suggested that a great variety could result from folding of the same peptide chain in different ways.[115] This would be in harmony [116] with the fact that the activity of antibodies is completely destroyed by denaturation in which process, it is supposed, the bonds that stabilize the globular or elliptical shape of the protein molecule are broken up, at least in part.[117] Pauling's theory of antibody formation includes not only this general idea but several specific propo-

---

[111] (238), (240).

[112] Breinl and Haurowitz (144), Mudd (291), Alexander (292), Manwaring (293), Eastwood (294), Zinsser (295), Stearn (296); (297), (21). Concerning the question of the formation and physiological function of serum proteins v. Whipple (298), Mann (299), Landsteiner and Parker (299a).

[113] v. Bergmann and Niemann (300).

[114] v. Dubos (301).

[115] Rothen and Landsteiner (303).

[116] v. (304).

[117] Films of pneumococcus antibodies of 8–12 Å thick were found still to react specifically (303). It is possible, however, that in the extended molecule configurations due to folding of the peptide chain still persist; various antibodies might also behave differently.

sitions. It may best be rendered in the author's own words: "It is assumed that antibodies differ from normal serum globulins only in the way in which the two end parts of the globulin polypeptide chain are coiled, these parts, as a result of their amino acid composition and order, having accessible a very great many configurations with nearly the same stability; under the influence of an antigen molecule they assume configurations complementary to surface regions of the antigen, thus forming two active ends.[118] After the freeing of one end and the liberation of the central part of the chain this part of the chain folds up to form the central part of the antibody molecule, with two oppositely directed ends able to attach themselves to two antigen molecules." The stimulating arguments of the author deal with the antigen-antibody ratio, the postulated maximum bivalence of antibodies, antigenic activity, protein structure, and other items.

For the decision of the fundamental point whether all antibodies (from the same animal species) are strictly identical chemically, i.e. in the content of amino acids and the arrangement in the chain, the available analytical and serological data are not quite sufficient should the reactive groupings constitute only small portions of the molecule.[119] From a teleological point of view one could surmise that changes of the amino acid make-up, in conjunction with different folding, would afford the widest scope for variation, if the unfounded assumption of prosthetic, non-peptide groups is left out of account.

An obstacle to all theories based upon relatively simple premises is the complexity of the biological phenomena of immunization, which is already seen from the association of antibodies with particular protein fractions or the differences between individuals as regards antibody production,[120] the latter, moreover, being influenced by the nature of the antigens. Auxiliary hypotheses are thus required to explain the increase in serum globulins following immunization, and the enhanced and often modified response to repeated administration of antigen, indicating that an impression is left upon the antibody-producing cells which lasts after the disappearance of the antigen.[121]

[118] Cf. with this statement p. 143. In later papers of Pauling the contingency of a valence greater than two is accepted.

[119] v. (304), (208).

[120] The differences, probably constitutional, which cause these variations have been traced only in special cases to the presence or absence of substances related to the antigen, in the experimental animals (p. 104).

[121] Stimulation of antibody production by a second, unrelated antigen or other substance ("anamnestic reaction") is not unanimously accepted (36), (62).

This inference seems supported by experiments with tissue cultures that showed that cells can maintain their sensitivity to tuberculin for a few transplantations.[122]

It is not surprising that efforts should have been made to produce antibodies outside the animal body, in cell cultures or in cell-free solutions.

Antibody formation in cell cultures has been repeatedly reported but in careful work of Parker (310) the results were positive only when the tissue (spleen) was taken from an animal which had been injected with the antigen 2 or 3 days earlier.

As to the experiments made in the absence of active cells, the specificity, alleged to be equal to that of immune antibodies, of hemagglutinins which were recovered from normal serum by splitting the combination of natural agglutinins and erythrocytes can be readily explained by selective absorption of the agglutinins originally present in the serum (311). The claim made long ago by Osstromuislenskii (312) of having produced diphtheria antitoxin by keeping a mixture, containing 6% salt, of serum and toxin for 1–2 days at 37° is unsupported;[123] also Loiseleur's experiments (314) (electrodialysis of antigens) have apparently not been repeated by others, and the assertions of Mez on the production of artificial sera to plant proteins have been refuted by Eisler (p. 23). Quite recently provocative results were communicated by Pauling (315). Following up a plan suggested by his theory of antibody formation, Pauling mixed γ-globulin with dyes (aniline blue or an arsenic acid azodye) or the polysaccharide of Type III pneumococci and kept the mixtures under conditions in which partial denaturation and refolding of the proteins into a configuration complementary to the added substance might occur, and with the protein solutions so obtained he produced precipitation of the dyes or the polysaccharide which appeared to simulate antibody reactions in their relation to the substances employed. Before more extensive results on cross reactions and on the production of precipitins for proteins and cells are available, it is too early to judge to what extent the treated protein solutions are comparable in specificity to antibodies generated in the animal, but it goes without saying that,

---

[122] Moen and Swift (305). The fact that immunity can last for many years would be a decisive proof, but in the most striking case of virus infections (smallpox, measles, yellow-fever), the permanence of active virus cannot be excluded [v. Rivers (306)].

Pearce (307) found no correlation between immunity and the persistence of vaccinia virus up to 17 weeks, but Olitsky et al. (308) came to the opposite conclusion. On persistence of bacterial antibodies v. Kolmer et al. (309).

Certain allergies in which long persistence of the inciting agent can scarcely be assumed are possibly a relevant instance.

[123] Several similar allegations are quoted in (313).

if it were fully corroborated that proteins through artificial means can acquire, even to some degree, a predetermined configuration, this would be of fundamental importance for the understanding of antibody specificity, as well as for the knowledge of protein structure.

BIBLIOGRAPHY

(1) Landois: Die Tranfusion des Blutes, Vogel, Leipzig, 1875. — (2) Lüdke: CB *42*, 69, 1906 (B). — (3) Rissling: CB *44*, 541, 1907. — (4) Brockmann: ZI *9*, 87, 1911. — (5) Krainskaja: DZG *19*, 446, 1932; ZI *75*, 489, 1932. — (6) Shimidzu: Tohoku JEM *18*, 526, 1932. — (6a) Walsh: JI *10*, 803, 1925. — (7) Kolle et al: ZH *44*, 1, 1903. — (8) Bürgi: AH *62*, 239, 1907, v. *68*, 95, 1909. — (9) Hetsch et al: Festschrift f. Robert Koch, Fischer, Jena, 1903. — (10) Lovell: J. Comp. Path. & Therap. *45*, 27, 1932 (B). — (11) Jordan: JID *61*, 79, 1937. — (12) Buchner: CB *5*, 817, 1889. — (13) Sachs: HPM *2*, 884, 1929. — (13a) Wassermann et al: Ztschr. exp. Path. *4*, 273, 1907. — (13b) Kidd et al: JEM *76*, 557, 1942. — (14) Boyd et al: JI *33*, 111, 1937. — (15) Bordet: AP *13*, 225, 1899. — (16) Malkoff: DM 1900, p. 229. — (17) Ehrlich et al: Berl. Kl. W. 1900, p. 681; Croonian Lecture, RS *66*, 424, 1900. — (18) Neisser: DM 1900, p. 790. — (19) Gibson: JH *30*, 337, 1930 (B), JI *22*, 211, 1932. — (20) Finkelstein: JP *37*, 359, 1933 (B). — (21) Landsteiner et al: ZH *58*, 213, 1908. — (22) Gordon et al: JP *35*, 549, 1932, *37*, 367, 1933. — (23) Boissevain: SB *87*, 1255, 1922. — (24) Eisler et al: ZI *79*, 293, 1933. — (25) Streng: APS, Suppl. *38*, 152, 1938. — (26) von Dungern et al: ZI *8*, 526, 1911. — (27) Landsteiner et al: Pr *30*, 209, 1932. — (28) Pfeiffer et al: DM 1901, p. 834. — (29) Bordet: AP *15*, 303, 1901. — (30) Gruber: WK 1903, p. 1097. — (31) Landsteiner et al: ZI *10*, 68, 1911. — (32) Thomsen: ZI 70, 140, 1931. — (33) Browning: Antigens and Antibodies, SystB *6*, 202, 219, 1931. — (34) Kidd et al: JEM *76*, 543, 557, 1942. — (35) Duran-Reynals: YJB *12*, 361, 1940. — (36) Topley et al: PB. — (37) Ingalls: JI *33*, 123, 1937. — (38) Hirszfeld: Konstitutionsserologie etc., Springer, Berlin, 1928. — (39) Neufeld: K 1929, p. 49. — (40) Friedberger et al: ZI *64*, 294, 1929, *67*, 67, 1930; DM 1929, p. 132. — (41) Jungeblut et al: Pr *29*, 879, 1932; JI *23*, 35, 1932. — (42) Bailey et al: AJH *7*, 370, 1927, *8*, 398, 477, 485, 723, 1928. — (43) Ramon: RI *2*, 305, 1936. — (44) Hughes et al: JAM *99*, 978, 1932. — (45) Hirszfeld: K 1932, p. 950. — (46) Landsteiner et al: JI *20*, 179, 1931. — (47) Schermer et al: K 1932, p. 335; ZZ *24*, 103, 1932. — (48) Friedenreich: ZI *71*, 314, 1931. — (49) Dupont: AI *9*, 133, 1934. — (50) Bailey: AJH *3*, 370, 1923. — (51) Thomsen et al: ZI *63*, 67, 1929. — (52) Hirszfeld: EH *8*, 367, 1926. — (53) Gürber: Beitr. z. Physiol., Vieweg, Braunschweig, 1899, p. 121. — (54) Landsteiner : Oppenheimer's Handb. d. Biochemie *2*, 395, 1909. — (55) von Toth: ZI *75*, 277, 1932. — (56) Landsteiner: JI *15*, 589, 1928. — (57) Hirano: Philippine J. Sci. *47*, 449, 1932. — (58) Buchbinder: JI *25*, 33, 1933 (B). — (58a) Wiener et al: JI *45*, 229, 1942. — (59) London: Arch. des Sci. Biol., St. Petersburg *9*, 84, 1903. — (59a) Landsteiner et al: CB *45*, 660, 1908. — (60) Landsteiner et al: CB *31*, 781, 1902. — (60a) Bleyer: ZI *53*, 386, 1927. — (61) Ramon: SB *116*, 917, 1934. — (62) van der Scheer et al: JI *44*, 165, 1942 (B). — (63) Pappenheimer et al: JEM *71*, 247, 1940. — (64) Fell et al: JI *39*, 223, 1940. — (65) Newell et al: JA *10*, 513, 1939. — (65a) van der Scheer et al: JI *39*, 65, 1940 (B). — (66) Kekwick et al: Br *22*, 29, 1941; JC *60*, 486, 1941. — (67) Seibert et al: Pr *49*, 77, 1942; JAC *65*, 272,

1943. — **(68)** Moore et al: JI *38*, 221, 1940. — **(69)** Wyckoff: S *84*, 291, 1936 (B). — **(70)** Kabat et al: S *87*, 372, 1938; JEM *69*, 103, 1939 (B). — **(71)** Heidelberger et al: N *138*, 165, 1936. — **(72)** van der Scheer et al: JI *41*, 209, 1941. — **(73)** Petermann et al: S *93*, 458, 1941. — **(74)** Petermann et al: JPC *45*, 1, 1941. — **(75)** Rothen: JGP *25*, 487, 1942. — **(76)** Paić: CR *208*, 1605, 1939. — **(77)** Paić: CR *207*, 1074, 1938. — **(78)** Paić: BCB *21*, 412, 1939. — **(79)** Deutsch: CR *208*, 603, 1939. — **(80)** Marrack: CAA. — **(81)** Boyd: FI. — **(82)** Joffe et al: JGP *18*, 599, 1935. — **(83)** Liu et al: CJP *11*, 211, 1937. — **(83a)** Abramson: JGP *14*, 163, 1930. — **(83b)** Smith et al: Br *11*, 494, 1930. — **(83c)** McCutcheon et al: JGP *13*, 669, 1930. — **(84)** Tiselius et al: JEM *69*, 119, 1939 (B). — **(85)** Girard et al: SB *116*, 1010, 1934. — **(86)** Lourau-Dessus: J. Chim. Phys. *34*, 149, 1937. — **(86a)** Neurath: JAC *61*, 1841, 1939. — **(86b)** Crowfoot: ChR *28*, 215, 1941. — **(87)** Goodner et al: Pr *34*, 617, 1936 (B). — **(88)** Went et al: ZI *91*, 157, 1937. — **(89)** Hartley: SystB *6*, 249, 1931. — **(90)** Ando et al: JI *34*, 295, 303, 1938 (B). — **(91)** Baecher: HPM *2*, 203, 1929. — **(92)** Pick: BCP *1*, 351, 393, 445, 1902. — **(93)** Gibson et al: JBC *3*, 233, 253, 1907 (B). — **(94)** Lewin: ZI *87*, 289, 1936. — **(95)** Ledingham: JH *7*, 65, 1907. — **(96)** Wadsworth: SM. — **(97)** Goodner et al: JI *33*, 279, 1937. — **(98)** Gerlough et al: JI *22*, 331, 1932. — **(99)** Green et al: JI *36*, 245, 1939. — **(100)** Murdick et al: JI *28*, 205, 1935. — **(101)** Scherp et al: JEM *63*, 547, 1936. — **(102)** Felton: JID *37*, 199, 1925; BJH *38*, 33, 1926. — **(103)** Horsfall et al: JEM *62*, 485, 1935. — **(104)** Chow: Pr *34*, 651, 1936. — **(105)** Gibson et al: BJH *22*, 106, 1911. — **(106)** Hansen: BZ *299*, 363, 1938. — **(107)** Doladilhe et al: CR *208*, 1439, 1939; SB *126*, 557, 1937 (B). — **(108)** Bierry et al: CR *206*, 785, 1938. — **(109)** Banzhaf et al: Studies Lab. City of New York *8*, 208, 1914/15. — **(110)** Chow et al: JEM *62*, 179, 1935. — **(111)** Klobusitzky: JI *35*, 329, 1938 (B). — **(112)** Huntoon et al: JI *6*, 117, 123, 185, 1921 (B). — **(113)** Helferich et al: ZPC *267*, 23, 1940. — **(113a)** Bleyer: ZI *33*, 477, 1922. — **(114)** Yung Tsü: ZI *70*, 223, 289, 1931. — **(115)** d'Alessandro et al: ZI *84*, 237, 1935, *85*, 410, 1935 (B). — **(116)** K. Meyer et al: AP *56*, 401, 1936. — **(117)** Schmidt: APS, Suppl. *16*, 400, 1933. — **(118)** Rosenheim: JP *40*, 75, 1935. — **(119)** Landsteiner: MM 1902, p. 1905. — **(120)** Kosakai: JI *3*, 109, 1918 (B); v. Japan Med. World *1*, Oct. 1925. — **(120a)** Munter: ZH *94*, 152, 1921. — **(121)** Velluz: SB *118*, 745, 1935. — **(122)** Locke et al: JID *39*, 126, 1926 (B); NK p. 1049. — **(123)** Felton: JI *22*, 453, 1932. — **(124)** Goldie: SB *121*, 761, 1936. — **(125)** Saeki: BPh *68*, 388, 1932. — **(126)** Kirk et al: JI *26*, 495, 1934. — **(127)** Bier et al: JI *40*, 465, 1941. — **(128)** Landsteiner et al: JEM *63*, 325, 1936. — **(129)** Pope et al: Br *20*, 213, 1939. — **(130)** Euler et al: ZI *68*, 124, 1930. — **(131)** Brunius: Chemical Studies, etc., Fahlcrantz, Stockholm, 1936. — **(131a)** Heidelberger et al: JGP *25*, 523, 1942. — **(132)** Heidelberger et al: JEM *68*, 913, 1938 (B). — **(133)** Lee et al: Pr *43*, 65, 1940. — **(134)** Chow et al: CJP *11*, 139, 1937. — **(135)** Lee et al: Pr *37*, 462, 1937. — **(136)** Schneider: JBC *11*, 47, 1912. — **(137)** Northrop: JGP *25*, 465, 1942; S *93*, 92, 1941. — **(138)** Sumner et al: JBa *32*, 227, 1936; S *87*, 395, 1938. — **(139)** Heidelberger et al: JEM *62*, 697, 1935. — **(139a)** Schmidt: APS, Suppl. *16*, 400, 1933. — **(140)** Heidelberger et al: JGP *25*, 523, 1942. — **(141)** Landsteiner et al: ZI *10*, 68, 1911. — **(142)** Hewitt: BJ *28*, 2080, 1934. — **(143)** Velluz: SB *113*, 684, 1933, *116*, 981, 1934; VI Congr. Chim. biol. 1937, p. 173. — **(144)** Breinl et al: ZPC *192*, 45, 1930. — **(145)** Calvery: JBC *112*, 167, 1935. — **(146)** Banzhaf et al: JI *2*, 125, 1916. — **(147)** Mutzenbecher: BZ *243*, 100, 1931 (B). — **(148)** Felton: JI *30*, 381, 1936. — **(148a)** Marrack et al: RS B, *106*, 1, 1930. — **(149)** Reiner et al: JBC *95*, 345, 1932. — **(150)** Green et al: JI *36*, 245, 1939. — **(151)** Eisler: CB *84*, 46, 1920. — **(152)** Smith et al: Br *11*, 494, 1930. — **(153)** Eagle: JI *29*, 41, 1935. — **(154)**

Maloney et al: Proc. Roy. Soc. Canada 3rd ser., Sect. V, *19*, 1925. — **(155)** Doladilhe: CR *206*, 1150, 1938. — **(156)** Salfeld et al: JID *61*, 37, 1937. — **(157)** Treffers et al: JEM *73*, 125, 293, 1941 (B), *75*, 135, 1942. — **(158)** Wright: JID *70*, 103, 1942. — **(159)** Marrack et al: Br *19*, 171, 1938. — **(159a)** Hooker et al: JI *45*, 127, 1942. — **(160)** Meyer et al: AP *59*, 477, 594, 1937. — **(161)** Morgan: BJ *31*, 2003, 1937. — **(161a)** Haurowitz et al: Br *23*, 146, 1942. — **(162)** Burnet: Br *15*, 354, 1934. — **(163)** Heidelberger et al: JEM *59*, 519, 1934, *62*, 697, 1935, *71*, 271, 1940; BaR *3*, 49, 1939. — **(164)** Morgan: JH *37*, 372, 1937. — **(165)** Kendall: ANY *43*, 85, 1942. — **(166)** Eagle: JI *30*, 339, 1936. — **(166a)** Kraus et al: CB *39*, 72, 1905. — **(167)** Eisler: BZ *150*, 350, 1924 (B). — **(168)** Freund: JEM *55*, 181, 1932. — **(169)** Wells: CAI, p. 33. — **(170)** Mellanby: RS B *80*, 399, 1908. — **(171)** Merrill et al: JGP *16*, 243, 1932. — **(172)** Hartley: Br *6*, 180, 1925 (B). — **(173)** Merrill et al: Pr *29*, 799, 1932. — **(174)** Chow et al: CJP *11*, 175, 1937. — **(175)** Madsen et al: ZP *70*, 263, 1910. — **(176)** Jones: JEM *46*, 291, 1927. — **(177)** Streng: ZH *62*, 281, 1909. — **(178)** Meyer et al: AP *59*, 282, 1937; SB *123*, 935, 1936. — **(179)** Kleczkowski et al: Br *22*, 192, 1941, *23*, 178, 1942. — **(180)** Felton et al: JI *11*, 197, 1926. — **(181)** Klingenstein: ZI *66*, 99, 1930. — **(182)** Olitzki: ZI *72*, 498, 1931. — **(183)** Weil et al: JI *37*, 413, 1939. — **(184)** Selter: ZI *54*, 113, 1927. — **(184a)** Jennings et al: JI *45*, 105, 111, 1942. — **(185)** Prašek: ZI *20*, 146, 1913. — **(185a)** Goldie: SB *125*, 861, 1937. — **(185b)** Silber et al: ZI *63*, 506, 1929. — **(186)** Eagle: JEM *67*, 495, 1938. — **(187)** Mudd et al: JGP *16*, 947, 1933. — **(188)** Breinl et al: ZI *77*, 176, 1932. — **(189)** Eggerth: JI *45*, 303, 1942. — **(190)** Coons et al: Pr *47*, 200, 1941; JI *45*, 159, 1942. — **(191)** Goldie et al: SB *129*, 391, 1938 (B). — **(192)** Sandor et al: BCB *20*, 1130, 1938; IC, p. 811. — **(193)** Tamura et al: Pr *38*, 909, 1938. — **(194)** Marrack: N *133*, 292, 1934. — **(195)** Eagle et al: JEM *63*, 617, 1936 (B). — **(196)** Heidelberger et al: JEM *65*, 885, 1937. — **(197)** Braun ZI *78*, 46, 1933. — **(198)** Iwanoff: ZH *118*, 197, 1936. — **(199)** Olitzki: CB *106*, 267, 1928. — **(200)** Rosenheim: BJ *31*, 54, 1937. — **(201)** Weil et al: JI *35*, 399, 1938. — **(202)** Pope: Br *19*, 245, 1938, 20, 201, 1939. — **(203)** Modern et al: SB *133*, 158, 1940, *134*, 290, 1940; BZ *305*, 405, 1940. — **(204)** Coghill et al: JI *39*, 207, 1940. — **(205)** Schultze: BZ *308*, 266, 1941. — **(206)** Huddleson et al: S *90*, 571, 1939. — **(207)** Grabar: CR *207*, 807, 1938. — **(208)** Treffers et al: JEM *73*, 125, 1941. — **(209)** Caltabiano: C *112*, 903, 1941. — **(210)** Sandor et al: SB *131*, 461, 1939. — **(210a)** Kass et al: JI *45*, 87, 1942. — **(211)** Zinsser et al: JI *6*, 289, 1921; Immunity Principles, etc. Macmillan Company, New York, 1941. — **(212)** Doerr: HPM *1*, 838, 1929. — **(213)** Neufeld: HPM *2*, 964, 1929. — **(214)** Heidelberger et al: Pr *31*, 595, 1934. — **(215)** Teale: JI *28*, 241, 1935. — **(216)** Heidelberger et al: JEM *67*, 181, 1938 (B). — **(216a)** Gerlough et al: JI *40*, 53, 1941. — **(217)** Doerr et al: ZI *3*, 181, 1909. — **(218)** Jones: JEM *48*, 183, 1928. — **(219)** Mudd et al: JEM *52*, 313, 1930. — **(220)** Nicolle: AP *12*, 161, 1898. — **(221)** Arkwright: JH *14*, 261, 1914. — **(222)** Zinsser: JI *18*, 483, 1930. — **(223)** Neisser et al: MM 1904, p. 827. — **(224)** Moreschi: CB *46*, 49, 1908. — **(225)** Bordet et al: CB *58*, 330, 1911. — **(226)** Heidelberger et al: JEM *61*, 563, 1935, *64*, 161, 1936. — **(227)** Goodner et al: JEM *66*, 425, 437, 1937. — **(228)** Lee et al: Pr *38*, 101, 1938. — **(229)** Raffel et al: JI *39*, 317, 337, 349, 1940. — **(230)** Friedenreich: ZI *71*, 283, 1931. — **(231)** Landsteiner et al: JI *12*, 441, 1926. — **(232)** Hooker et al: JI *26*, 469, 1934 (B), *33*, 57, 1937. — **(233)** Heidelberger et al: JEM *62*, 697, 1935. — **(234)** Heidelberger et al: JEM *71*, 271, 1940. — **(235)** Malkiel et al: JEM *66*, 383, 1937. — **(236)** Pappenheimer: JEM *71*, 263, 1940. — **(237)** Hooker et al: Pr *47*, 187, 1941. — **(238)** Thiele et al: ZI *20*, 1, 1913. — **(239)** Makino: J. Chosen Med. Ass. *29*, 76, 1939. — **(240)** Mueller: AH *64*, 62, 1908. — **(240a)**

Landsteiner et al: CB *39*, 712, 1905. — **(241)** Boyd et al: JI *33*, 111, 1937. — **(242)** Horsfall: JBa *35*, 207, 1938. — **(243)** Perlmann et al: JI *43*, 99, 1942. — **(244)** Burnet: Monographs Hall Institute No. 1, MacMillan, Melbourne, 1941. — **(245)** Oerskov et al: ZI *92*, 487, 1938. — **(246)** Ramon: SB *99*, 1295, 1928; CR *179*, 514, 1924. — **(247)** Heidelberger et al: JBC *144*, 541, 555, 1942. — **(247a)** Olitski: JI *29*, 453, 1935. — **(248)** Ramon et al: RI *6*, 5, 1940. — **(249)** Bjorneboe: JI *37*, 201, 1939. — **(250)** Ionesco-Mihaiesti et al: Arch. Roum. Path. Microbiol. *8*, 269, 1935. — **(251)** Opie: JI *8*, 55, 1923. — **(252)** Haurowitz et al: ZPC *239*, 76, 1936 (B). — **(253)** Sullivan et al: JI *26*, 49, 1934 (B). **(253a)** Culbertson: JI *28*, 279, 1935. — **(254)** Kahn: S *79*, 172, 1934; Tissue Immunity, Thomas, Springfield, 1936. — **(255)** Wang et al: CJP *13*, 417, 1938. — **(256)** Sabin: JEM *70*, 67, 1939. — **(257)** Kyes: JID *18*, 277, 1916. — **(257a)** Doan: JLC *26*, 89, 1940. — **(258)** Cannon et al: JI *17*, 441, 1929. — **(259)** Sachs: HP *13*, 447, 1929. — **(260)** NK p. 881, 1035. — **(261)** Topley: JP *33*, 339, 1930. — **(262)** MacMaster et al: JEM *61*, 783, 1935, *66*, 73, 1937. — **(263)** Ehrich et al: JEM *76*, 335, 1942. — **(264)** Sédallian et al: RI *5*, 227, 1939. — **(265)** Seegal et al: JEM *54*, 249, 1931; IC, p. 761. — **(266)** Cannon et al: Pr *29*, 517, 1932. — **(267)** Roemer: Graefe's Arch. f. Opthal. *52*, 72, 1901. — **(268)** Kossel: BK 1898, p. 152. — **(269)** Metchnikoff: L'immunité dans les maladies infectieuses, 1901, Paris, Masson. — **(270)** Bordet: Traité de l'immunité, Masson, Paris, 1939. — **(271)** Metalnikoff et al: AP *44*, 273, 1930. — **(272)** Price: Am. Natural. *74*, 117, 1940. — **(273)** Chester: Q *8*, 129, 1933. — **(274)** Frémont: SB *113*, 775, 777, 1933. — **(275)** Wilhelm: CB 2. Abt., *89*, 107, 1933. — **(276)** East: HL 1930/31, p. 112. — **(277)** Doerr et al: 14. Kongr. Dtsch. Dermat. Ges. 1925. — **(278)** Berger et al: ZH *113*, 79, 1931; BZ *252*, 22, 1932. — **(279)** Haurowitz et al: ZPC *205*, 259, 1932. — **(280)** Heidelberger et al: S *72*, 252, 1930. — **(281)** Heidelberger et al: JEM *38*, 561, 1923. — **(282)** Hooker et al: JI *23*, 465, 1932. — **(283)** Haurowitz et al: JI *43*, 327, 1942. — **(284)** Wollman et al: SB *118*, 1425, 1935. — **(285)** Heidelberger et al: JEM *58*, 137, 1933. — **(286)** Pappenheimer: JEM *71*, 263, 1940. — **(287)** Knorr: MM 1898, p. 321, 362. — **(288)** Roux et al: AP 7, 65, 1893. — **(289)** Vincent: SB *113*, 340, 1933. — **(290)** Hooker et al: JI *21*, 113, 1931. — **(291)** Mudd: JI *23*, 423, 1932. — **(292)** Alexander: Protoplasma *14*, 302, 1932. — **(293)** Manwaring: The Newer Knowledge of Bacteriology and Immunology, University of Chicago Press, 1928, p. 1078. — **(294)** Eastwood: JH *33*, 259, 1933. — **(295)** Zinsser: Resistance to Inf. Dis., Macmillan, New York, 1931, p. 100. — **(296)** Stearn: JBa *29*, 52, 1935. — **(297)** Sahli: SMW 1920, p. 1129. — **(298)** Whipple: AJM *196*, 609, 1938. — **(299)** Mann et al: XVI Internat. Physiol. Kongr. 1938, p. 201. — **(299a)** Landsteiner et al: JEM *71*, 231, 1940. — **(300)** Bergmann et al: S *86*, 187, 1937. — **(301)** Dubos: BaR *4*, 1, 1940; HL 1939–40, p. 223. — **(302)** Pauling: JAC *62*, 2643, 1940. — **(303)** Rothen et al: S *90*, 65, 1939; JEM *76*, 437, 1942. — **(304)** Landsteiner et al: JEM *67*, 709, 1938. — **(305)** Moen et al: JEM *64*, 339, 943, 1936. — **(306)** Rivers: S *95*, 107, 1942. — **(307)** Pearce: JID *66*, 130, 1940. — **(308)** Olitsky et al: JEM *50*, 263, 1929. — **(309)** Kolmer et al: JID *70*, 54, 1942. **(310)** Parker: S *85*, 292, 1937 (B). — **(311)** Landsteiner: ZI *18*, 220, 1913. — **(312)** Osstromuislenskii: CA *10*, 214, 1916. — **(313)** S *96*, 181, 1942. — **(314)** Loiseleur et al: SB *134*, 212, 214, 1940 (B). — **(315)** Pauling: JEM *76*, 211, 1942.

V

ARTIFICIAL CONJUGATED ANTIGENS. SEROLOGICAL
REACTIONS WITH SIMPLE CHEMICAL COMPOUNDS

FOR a considerable period of time after discovery of serological phenomena and despite an abundance of observations, a method was yet wanting for the systematic investigation, along chemical lines, of specificity in serum reactions. It was indeed clear that serological reactions must somehow be dependent upon the chemical properties of the substances involved, though this was doubted,[1] but insufficient chemical knowledge concerning the available antigens (proteins) and still more so concerning the nature of antibodies made a closer analysis impossible. With this state of affairs it is comprehensible that even in 1917 Morgenroth (3), although a follower of Ehrlich's chemical theories, was led to comment on serology as a field to which there leads no bridge from chemistry.

A workable approach was found when it proved possible, by attaching simple chemical compounds to proteins, to prepare conjugated antigens containing specifically reacting components of known constitution, chosen at will. At the outset, the prospects of realizing this plan were slight. It is true that, as the investigations of Obermayer and Pick had shown (p. 48), the two defining properties of antigens — their capacities to immunize and to combine with antibodies — may persist when the structure of the protein molecule is altered by rather drastic chemical treatment, as iodination of the tyrosine groups, oxidation or nitration. Yet these properties were destroyed by other chemical changes (treatment with enzymes, alkali), and on the basis of all existing evidence there was an almost dogmatic belief that the two above named functions of antigens are inseparable, and that a special chemical constitution, peculiar to proteins and not even to all of them, is necessary for the production of antibodies, and accordingly for the reactions in vitro as well. It could therefore not be foreseen that antibodies would react with chemical compounds entirely unrelated to proteins, just as no other substances but proteins were known to be susceptible to peptic or tryptic digestion. Actually

[1] v. (1), (2).

early experiments in this direction proved unsuccessful and were not continued (4). Years later, however, an investigation conducted by the author and his colleagues (5) gave promising results in a first series of experiments. These consisted of the introduction of acyl groups into proteins by treatment with anhydrides or chlorides of acids (butyric, isobutyric, mono-, di- and trichloroacetic, anisic and cinnamic acids). As in the case of the previously investigated acetyl- and alkylproteins (p. 51) the original specificity of the protein was changed, but in addition the various products were serologically clearly differentiated and showed cross reactions in the case of chemically related acyl radicals.[2] The procedure was a step forward since it became possible to prepare by the same general chemical process a number of serologically different protein derivatives whose specificity was undoubtedly ascribable to the nature of the acyl groups that had been introduced. There still remained the question, however, whether these are capable of reacting by themselves or only in conjunction with neighbouring portions of the protein molecule. It was therefore of importance that in the coupling of proteins with diazonium compounds [3] an easily applicable method for the preparation of "synthetic antigens" was found (11–13) which removed this doubt and, as will be seen, proved to be of general application.

The reaction is that used for the preparation of ordinary azocompounds; e.g., phenol and benzenediazonium chloride give p-hydroxyazobenzene:

$$\langle\bigcirc\rangle OH + \langle\bigcirc\rangle N=NCl \rightarrow \langle\bigcirc\rangle -N=N-\langle\bigcirc\rangle OH + HCl$$

The coloured products formed by coupling with diazonium compounds (designated as azoproteins), when prepared in a suitable manner, give but weak reactions with immune sera for the unchanged protein and elicit the formation of antibodies in rabbits even when the serum of this species is used for the preparation. Nevertheless, the protein specificity is retained to a certain extent, depending on the azocomponent and the degree of coupling,[4] for immune sera

[2] The tests with the insoluble substances were made by complement fixation. Medveczky and Uhrovits (6) who repeated these experiments with some modifications (employing soluble acyl proteins and precipitin or anaphylaxis tests) found no marked specificity due to the acyl groups. This can be accounted for by their use of acyl residues which were similar chemically and, in some experiments, of too closely related proteins, possibly also by insufficient acylation [v. (7)].

[3] German patent quoted in (8); v. Heumann (9); Pauly (10).

[4] v. (14).

against azoproteins precipitate not only the homologous azoantigen but also mostly the original protein as well as various azoderivatives thereof; and there is evidence to show that these reactions are not to be explained solely by some uncoupled protein admixed with the immunizing antigen [Heidelberger and Kendall (15)], but to antibodies stimulated by the azoproteins. Further, on testing azoantigens prepared with the same azocomponent but with different proteins the reactions, as a rule, vary in intensity according to the relationship of the proteins to that contained in the immunizing antigen. The amount of precipitate diminishes, for example, in the order horse, ox, man, birds and rabbit when the antiserum employed is produced in the rabbit with an azoantigen from horse serum. Evidently in these reactions determinant structures comprising the azogroups and adjacent parts of the proteins are implicated. Therefore, to exclude overlapping reactions caused by the protein part of the antigen, it is essential to carry out the immunization and the precipitin tests with azoantigens made from widely different proteins, e.g., the immunization with azoproteins from horse serum and the reactions in vitro with azoproteins prepared from chicken serum (or serum of the immunized species), or using serum globulin and egg albumin in a similar way. Indeed, this artifice together with the choice of reactive azocomponents was decisive in definitely establishing the specific reactivity of haptenic groups in artificial compound antigens. Under proper conditions, then, the specificity is directed towards the azocomponents only and is independent of the protein portion of the antigen.[5] This was proved by the agreement of the reactions of azoantigens made with the same diazonium compounds but different animal or plant proteins (21) — casein, hemoglobin, zein, legumin, etc. — or even gelatin [6] or histone which were not known to be suitable for serum reactions. One notices also that blood stromata and red cells, after treatment with diazonium compounds, are agglutinated by the corresponding anti-azoprotein sera (21a). Hence, for most purposes it

---

[5] For this reason the objection which has been made occasionally against the method, that it changes the specificity of the protein regardless of the nature of the introduced group, is beside the point.

[6] v. Hooker and Boyd (16).

The production of immune sera with azogelatin has already been noted (p. 62); according to Adant (17) the sera would precipitate unchanged gelatin [v. Bruynoghe et al. (18)]. Reactions in vitro have been seen also with other gelatin derivatives [Clutton, Harington and Yuill (19)]. Protein coupled with diazonium compounds after it had been deprived of its antigenic activity by treatment with alkali was found by Doerr and Girard (20) to have no antigenic effect.

does not matter whether protein mixtures, such as blood serum, or purified proteins are chosen for the preparation of the azoantigens.[7] The reactions are performed most conveniently by precipitation but as Klopstock and Selter (25) showed, complement fixation tests are applicable as well.

It is not always easy and may require a number of animals to obtain immune sera specific for the azocomponents and often, contingent upon the response of the individual animal and the nature of the introduced groups, sera are produced that precipitate strongly the immunizing azoprotein and others containing kindred proteins,[8] but precipitate not at all or only slightly azoantigens made from widely distant proteins. Of course such sera are not suitable for the studies here under discussion on serological specificity in relation to chemical constitution, a point which has sometimes been overlooked.

Tests with two immune sera tabulated in a recent paper by Haurowitz (27) may be quoted as example: The author's statement that "the determinant group of the azoproteins includes a portion of the proteic molecule" applies to the tests he considers but not to reactions one may secure with azoantigens whose protein moiety is sufficiently foreign to that in the immunizing antigen.

Those immune sera which possess the desired hapten specificity contain commonly, as shown by adsorption or quantitative determination of the precipitates, three or more sorts of antibodies, directed towards the protein used for preparing the antigen, towards azoproteins made from this or somewhat related proteins containing the homologous or even other azocomponents, and towards the uncombined azocomponents.[9] The existence of antibodies of the second type adjusted more closely either to azotyrosine or to azohistidine groups has been demonstrated by Hooker and Boyd (31) in inhibition tests with azocompounds of phenol and imidazol. That the conjugates are presumably not entirely homogeneous is of little consequence;[10] special antibodies adjusted to antigens having many or few haptenic groups respectively do not seem to exist (29).

It appeared from experiments with amino acids (10) that the for-

---

[7] v. Erlenmeyer and Berger (22). Also bacterial proteins [Heidelberger (23)] and proteoses (24) have been used.
[8] (11), (26).
[9] Landsteiner and van der Scheer (28), Haurowitz (27), (29). Similar observations were made with various chemically altered proteins (29), (30).
[10] It is here worthy of notice that Reiner (32) was able to prepare crystallized azoinsulin.

mation of coloured azocompounds involves the tyrosine and histidine groups [11] in the proteins, these amino acids being capable of combining with one or two molecules of diazonium compounds to form azodyes, as for example tyrosine-diazo-benzoic acid,

$$\text{HOOC} - \text{C}_6\text{H}_4 - \text{N} = \text{N} - \underset{\underset{\text{CH}_2 - \text{CH}_2(\text{NH}_2) - \text{COOH}}{\big|}}{\overset{\overset{\text{OH}}{|}}{\bigcirc}} - \text{N} = \text{N} - \text{C}_6\text{H}_4 - \text{COOH}$$

From the analytical data one can calculate, on an estimate of a molecular weight of 70,000, that a molecule of serum albumin contains about 20 tyrosine and more than 10 histidine residues. Accordingly, if all these are occupied the protein is studded with many foreign groups, but attachment of about 10 groups (arsanilic acid) was found sufficient [12] to render conjugates, made with heterologous proteins, specifically precipitable, and in experiments of Haurowitz only a few groups were needed for the production of antibodies against the hapten. Overloading with foreign groups reduced the antigenic effect; antigens with one to two per cent As were optimal.[13]

In preparations made with a large amount of the diazonium compound Haurowitz [14] found figures for the arsenic content up to 10.6%, corresponding to more than 200 azogroups; in glucoside-azoproteins a sugar value of 10% was found by Goebel and Avery (40).

With azoproteins prepared by intense treatment the analytical values are too high to be accounted for by the coupling of each tyrosyl and histidyl group with two molecules of diazonium compounds [15] which must mean that other parts of the protein molecule participate in the combination.

The assertion of Klopstock and Selter (25) that mixtures of diazo solutions and proteins immunize just as well as azoproteins was corrected by Heidelberger and Kendall [16] who found that coupling takes

---

[11] Kapeller-Adler and Boxer (33) obtained azodyes also with phenylalanine, tryptophane, proline and oxyproline and conclude that these amino acids take part in the coupling of proteins. Aliphatic amino acids were found to form unstable colourless compounds with diazonium salts [Busch et al. (34); v. Boyd and Mover (35), Eagle and Vickers (36), Fierz-David et al. (37), Landsteiner (21)].
    On electrochemical properties of azoproteins cf. (37a).
[12] Hooker and Boyd (38), (35), Haurowitz (29).
[13] (12), (29).
[14] (29); v. (39).
[15] Boyd and Hooker (8), (35), Haurowitz (29).
[16] (41); (42).

place under the conditions employed by these authors. With simple compounds loose attachment to proteins by adsorption, serviceable in the case of certain lipoid substances (pp. 101, 111), apparently does not suffice for producing antigenic complexes; antibody production against dyes by injecting dyed proteins has so far not succeeded. [Experiments which perhaps suggest that the same bacteria treated with salts of various heavy metals can be differentiated by immune sera have been reported (43).]

Direct proof that the groups attached to protein are of themselves capable of combining with antibodies, and that the protein part which affords the antigenic stimulus in the immunization is unnecessary for the specific reactions, has been brought in the case of numerous azoproteins by inhibition reactions (p. 182). The heterologous protein contained in the test antigens serves, therefore, essentially as a carrier for the specifically reacting azocomponents, aiding precipitation or rendering it possible. The formation of precipitates is certainly furthered by the colloidal state of proteins.

On account of some inexact statements in the literature it will not be superfluous to point out the essential difference, in typical cases, between proteins whose original serological properties are changed by chemical treatment, and conjugated antigens in which a protein is compounded with substances capable of reacting — as haptens — independently of the protein molecule, with the antibodies incited by the conjugated antigen. Thus inhibition reactions showed that antibodies produced, for example, by a protein coupled with a certain α-glucoside, combined with this simple substance and differentiated it from the β-glucoside, whereas sera for nitroprotein are not specific for and do not react with the NO_2 group as seen from the negative tests with the three nitrophenols or nitrobenzoic acids [Mutsaars (44)] (p. 195). (In this respect the role of NO_2 groups — as, similarly, iodine in iodoproteins — is comparable to that of oxygen in the serological reactions of an oxidized protein.) Or, as Heidelberger (45) remarks, "The specificity conferred upon egg albumin by the introduction of phosphoryl groups is not entirely a hapten specificity . . . for the anti-PEa (anti-phosphorylated-egg-albumin) sera are not precipitated by phosphates nor is the specific precipitation . . . inhibited by as much as 0.15 M phosphate."

The principle of the method of preparing synthetic conjugated antigens having been established (12), it was rather obvious and indeed predicted that other chemical reactions besides coupling with diazonium compounds can be utilized for the preparation of conjugate antigens, for any method which results in a firm chemical attachment of the substance chosen as antigenic determinant, to protein, should be effective. Up to now, treatment with anhydrides, acylchlorides, azides, isocyanates, aldehydes, halogen compounds, oxazolones and a quinone have been subjected to examination. Whether with these methods antigens exhibiting hapten specificity are secured is largely a question of the attached substances, that is of their capacity to serve as determinants.[17] Coupling with diazonium compounds has been used preferentially because of the facility with which widely varied conjugated antigens can be prepared; for special purposes other methods of combination may be of advantage and with some compounds possibly may yield more active antigens. That a "more natural" linkage like the peptide bond is preferable to azo-linkages, as suggested by Clutton and Harington,[18] is not evident from the results so far obtained and, when the same independently reactive substance is coupled to protein in different ways, the corresponding immune sera, if containing antibodies for the attached group, will necessarily react with the uncombined hapten in similar fashion. Experiments of Mutsaars (49) with ureido- and azo-proteins, made to test this point, do not carry conviction since the residues (phenyl-, naphthyl-) employed are not effective determinant groups, but the postulate was fulfilled in the cross reactivity of glucosidophenyl azo-protein and glucosidocarbobenzyloxytyrosylprotein.[19]

SEROLOGICAL REACTIONS OF AROMATIC COMPOUNDS. — The experiments on conjugated antigens were at the start made with easily available aromatic amino compounds — generally with acids (carboxylic, sulfonic and arsonic acids) — on the assumption that salt forming groups would have greater reactivity. Afterwards other compounds were used, and it became evident that a great variety of chemical structures foreign to proteins are capable of giving serum reactions with antibodies.

The immune sera prepared to the various substances had their own specificities and reacted most intensely and in some cases only with the antigen bearing the homologous azocomponent; the cross reactions encountered showed definite regularities, in close accordance with chemical relationships.

---

[17] Positive results, apart from those discussed in the following, have been obtained with β-naphthoquinone-sulfuric acid [Obo (46)].

[18] (47); Harington (48) (Lecture on "Synthetic Immunochemistry").

[19] Clutton et al. (50); v. Creech and Franks (51).

The number of group reactions was increased, though without material alteration to the results, when the precipitation was intensified by taking a larger proportion of immune serum, or on longer standing. Similarly, as with natural antigens, there were differences in the cross reactions of sera procured from individual animals.[20] With regard to technique it should be mentioned that with too high antigen concentrations, particularly of those azoproteins which are not easily soluble, non-specific precipitation may occur.

The principal results of numerous precipitin tests with azoproteins were the following, as illustrated in Tables 18–20:[21]

1. First of all, the nature of the acid groups was of decisive influence. Sulfonic acid immune sera reacted markedly with several sulfonic acids, but little, if any, with carboxylic acid antigens, and carboxylic acid immune sera only exceptionally gave distinct reactions with azoproteins containing sulfonic acid groups. The determining influence of the arsenic acid radicals was still more pronounced, as is indicated by the fact that arsanilic acid serum precipitated all of the six substances tested containing the group AsO_3H_2 and none of the other antigens.

2. In contrast to acid groups, substitution of the aromatic nucleus by methyl, halogen, methoxyl and nitro groups was of less influence on the specificity.

Thus in the tests presented in Table 19 the immune sera acted with varying intensity on almost all of the antigens possessing mono-, or di-substituted benzene rings; the more strongly polar group NO_2 appeared to change the specificity to a somewhat greater extent than halogen and CH_3. The radicals containing a carbonyl or CONH group constituted an exception, for, with the same sera, antigens prepared from acetyl p-phenylenediamine and p-aminoacetophenone did not give any, or but slight precipitation. In experiments of Hopkins and Wormall on phenylureido proteins the introduction of bromine into the phenyl group did not alter the specificity considerably.

On testing antisera to azoproteins prepared from aniline, p-aminodiphenyl and β-aminoanthracene with the respective homologous and β-naphthylamine [22] antigens, distinct specificity was shown by "ani-

---

[20] (12) (Table II, lines 10, 11, 12, 13, 22, 23); (v. p. 17, Table 4).

[21] The antigens and immune sera are designated, for brevity, with the name of the simple compound used in the preparation of the antigen. The other abbreviations are self-explanatory.

[22] v. (52).

TABLE 18

[after Landsteiner et al. (12)]

Immune Sera:	Aniline	o-Aminobenzoic acid	Amino-m-toluic acid	m-Aminobenzoic acid	Amino-o-toluic acid	3-Amino-4-methyl benzoic acid	4-Chloro-3-aminobenzoic acid	4-Bromo-3-aminobenzoic acid	p-Aminobenzoic acid	o-Aminocinnamic acid	m-Aminocinnamic acid	p-Aminocinnamic acid	o-Aminobenzene sulfonic acid	4-Bromoaniline-2-sulfonic acid	m-Aminobenzene sulfonic acid	4-Aminotoluene-2-sulfonic acid
Aniline	+++	o	o	o	o	o	o	o	o	o	o	o	o	o	o	o
o-Aminobenzoic acid	o	+++	+++	o	o	o	o	o	o	o	o	o	+	o	o	o
m-Aminobenzoic acid	o	o	o	+++	o	o	o	o	o	o	o	o	o	o	o	o
4-Methyl-3-amino benzoic acid	o	o	o	o	o	+++	o	o	o	o	o	o	o	o	o	o
4-Chloro-3-aminobenzoic acid	o	o	o	o	o	o	+++	+	o	o	o	o	o	o	o	o
4-Bromo-3-aminobenzoic acid	o	o	o	o	o	o	+	+++	o	o	o	o	o	o	o	o
p-Aminobenzoic acid	o	o	o	o	o	o	o	o	+++	o	o	o	o	o	o	o
o-Aminocinnamic acid	o	o	o	o	o	o	o	o	o	+++	o	o	o	o	o	o
m-Aminocinnamic acid	o	o	o	o	o	o	o	o	o	o	+++	o	o	o	o	o
p-Aminocinnamic acid	o	o	o	o	o	o	o	o	o	o	o	+++	o	o	o	o
o-Aminobenzene sulfonic acid	o	o	o	o	o	o	o	o	o	o	o	o	+++	+	+	o
4-Bromoaniline-2-sulfonic acid	o	o	o	o	o	o	o	o	o	o	o	o	+++	+++	+++	+++
m-Aminobenzene sulfonic acid	o	o	o	o	o	o	o	o	o	o	o	o	+++	+++	+++	+++
4-Aminotoluene-2-sulfonic acid	o	o	o	o	o	o	o	o	o	o	o	o	+++	+++	+++	+++
4-Chloroaniline-3-sulfonic acid	o	o	o	o	o	o	o	o	o	o	o	o	o	o	o	o
1,3-Dimethyl-6-aminobenzene-4-sulfonic acid	o	o	o	o	o	o	o	o	o	o	o	o	o	o	o	o
p-Aminobenzene-sulfonic acid	o	o	o	o	o	o	o	o	o	o	o	o	o	o	+	o
6-Aminotoluene-3-sulfonic acid	o	o	o	o	o	o	o	o	o	o	o	o	o	o	+	+
5-Bromo-6-aminotoluene-3-sulfonic acid	o	o	o	o	o	o	o	o	o	o	o	o	o	o	+	+
1-Naphthylamine-4-sulfonic acid	o	o	o	o	o	o	o	o	o	o	o	o	o	o	o	o
Aminoazobenzene disulfonic acid	o	o	o	o	o	o	o	o	o	o	o	o	+	o	o	o
p-Aminophenylarsenic acid	o	o	o	o	o	o	o	o	o	o	o	o	o	o	o	o

TABLE 18 (continued)

Immune Sera: / Antigens from:	4-Chloroaniline-3-sulfonic acid	1,3-Dimethyl-6-aminobenzene-4-sulfonic acid	2,4,6-Tribromoaniline-3-sulfonic acid	p-Aminobenzene sulfonic acid	6-Aminotoluene-3-sulfonic acid	5-Bromo-6-aminotoluene-3-sulfonic acid	2,6-Dibromoaniline-4-sulfonic acid	3-Nitroaniline-4-sulfonic acid	1-Naphthylamine-4-sulfonic acid	Aminoazobenzene disulfonic acid	o-Aminophenylarsenic acid	4-Nitroaniline-2-arsenic acid	4-Chloro-3-aminophenyl-arsenic acid	p-Aminophenylarsenic acid	6-Aminotoluene-3-arsenic acid	3-Chloro-4-aminophenyl-arsenic acid
Aniline	o	o	o	o	o	o	o	o	o	o	o	o	o	o	o	o
o-Aminobenzoic acid	o	o	o	o	o	o	o	o	o	o	o	o	o	o	o	o
m-Aminobenzoic acid	o	o	o	o	o	o	o	o	o	o	o	o	o	o	o	o
4-Methyl-3-aminobenzoic acid	o	o	o	o	o	o	o	o	o	o	o	o	o	o	o	o
4-Chloro-3-aminobenzoic acid	o	o	o	o	o	o	o	o	o	o	o	o	o	o	o	o
4-Bromo-3-aminobenzoic acid	o	o	o	o	o	o	o	o	o	o	o	±	o	o	o	o
p-Aminobenzoic acid	o	o	o	o	o	o	o	o	o	o	o	o	o	o	o	o
o-Aminocinnamic acid	o	o	o	o	o	o	o	o	o	o	o	o	o	o	o	o
m-Aminocinnamic acid	o	o	o	o	o	o	o	o	o	o	o	o	o	o	o	o
p-Aminocinnamic acid	o	o	o	o	o	o	o	o	o	o	o	o	±	o	o	o
o-Aminobenzene sulfonic acid	o	o	o	o	o	o	o	o	o	o	o	o	o	o	o	o
4-Bromoaniline-3-sulfonic acid	+	o	o	o	o	o	o	o	o	o	o	o	o	o	o	o
m-Aminobenzene sulfonic acid	+	o	o	o	o	o	o	o	o	o	o	o	o	o	o	o
4-Aminotoluene-2-sulfonic acid	±	o	o	o	o	o	o	o	o	o	o	o	o	o	o	o
4-Chloroaniline-3-sulfonic acid	++++	+	o	o	o	o	o	o	o	o	o	o	o	o	o	o
1,3-Dimethyl-6-aminobenzene-4-sulfonic acid	+	++	o	o	o	o	o	o	o	o	o	o	o	o	o	o
p-Aminobenzene sulfonic acid	o	+	o	+++	±	o	o	o	o	o	o	o	o	o	o	o
6-Aminotoluene-3-sulfonic acid	o	o	o	++	+++	±	o	o	o	o	o	o	o	o	o	o
5-Bromo-6-aminotoluene-3-sulfonic acid	o	o	o	o	±	+++	+	o	+	o	o	o	o	o	o	o
1-Naphthylamine-4-sulfonic acid	o	o	o	o	o	o	o	o	++	+	o	o	o	o	o	o
Aminoazobenzene disulfonic acid	o	o	o	o	o	o	o	o	o	++	o	o	o	o	o	o
p-Aminophenylarsenic acid	o	o	o	o	o	o	o	o	o	o	+	±	+	+++	++++	++

The compounds are arranged according to chemical constitution, particularly as to the nature of the acid groups and their position relative to the amino group. Substances alike in these respects are grouped together, as indicated by braces. (The present table corrects some errors contained in the original paper, caused by the mislabeling of commercial preparations.) Concentration of antigens 0.01%. The degree of precipitation is indicated as in Table 6.

TABLE 19

[after Landsteiner et al. (57), with some correction]

Immune Sera: Antigens from:	Aniline	o-Toluidine	o-Anisidine	o-Nitroaniline	o-Chloroaniline	m-Toluidine	m-Nitroaniline	m-Chloroaniline	m-Bromoaniline	p-Toluidine	p-Anisidine	p-Nitroaniline	p-Chloroaniline	p-Bromoaniline	p-Iodoaniline
Aniline	++±	++	++	++	+±	++	+++	++	++	++±	++	+	++±	++	+±
o-Chloroaniline	++	+++±	++	++±	+++	+++	+++	+++	+++	+++	+++	+≡	+++	++±	++
p-Toluidine	++±	+±	±	±≡	+±	+++	+++	+++	++±	+++	+++	+++	+++	+++	++
p-Nitroaniline	++	±	0	0	++	++	++	+++	++	+++	+++	+++	+++	+++	+++
p-Chloroaniline	++	±	0		+±≡	++	+±	++	++	+±	++±	+++	++±	+++	+++

Immune Sera: Antigens from:	3-Nitro-4-methyl-aniline	4-Nitro-2-methyl-aniline	asymm. m-Xylidine	p-Xylidine	Acetyl-p-phenylene-diamine	p-Amino-aceto-phenone	Monomethyl-p-phenylene-diamine
Aniline	++	++±	+±	+±	0	±≡	+±
o-Chloroaniline	++±	+++	+++	++	0	0	
p-Toluidine	+++±	++±	+±	+++	0	+	
p-Nitroaniline	++±	++	+±	++≡	0	+	++±
p-Chloroaniline	+	±			0	0	

Concentration of antigens 0.01%.

line" immune sera while cross reactions were observed between aminodiphenyl or aminoanthracene sera and β-naphthylamine antigen [Jacobs (53)]. Sera for p-aminoethyl- and p-aminobutylbenzene showed some crossing and reacted faintly with aminohexylbenzene, not with aminooctylbenzene antigens.[23] Quite possibly the hydrocarbon residues are responsible for the specificity of the reactions only in conjunction with adjacent parts of the protein.

The prominent influence of acid groups, in contrast to that of other substituents, is also apparent from the fact that aniline sera do not precipitate antigens with acid groups and that the immune sera for azoproteins containing acid groups give negative or quite weak reactions with "neutral" antigens [24] (Table 20).

In this regard an azoprotein prepared with the methyl ester of p-aminobenzoic acid displayed a characteristic behaviour (57). Like other "neutral para-antigens" it gave definite reactions with aniline and p-toluidine immune sera, and very faint ones with a serum for p-aminobenzoic acid. If, however, the ester was hydrolyzed by gentle treatment of the azoprotein with NaOH, the precipitability by "neutral" immune sera gradually disappeared almost entirely, with the concomitant appearance of strong reactions with p-aminobenzoic acid serum. Immune sera prepared with the ester antigen precipitated the homologous azoprotein but not that made from p-aminobenzoic acid.

Furthermore, specificity is more sharply defined in antigens with acid groups in that their reactions are, for the most part, more strongly influenced by substituents than are those of the neutral antigens (Tables 18, 19).

3. Another rule, seen from the very distinctive reactions of the three isomeric aminobenzoic acids and aminocinnamic acids, is that the relative position of the acid radical to the azogroup has a pronounced effect on specificity and the occurrence of cross reactions. Thus, Table 18 shows clusters of positive reactions, namely with antigens prepared from m-aminobenzoic acid and its derivatives, a group of meta- and one of para-aminobenzene sulfonic acid antigens

[23] (54). Azoconjugates with higher hydrocarbons — anthracene, butylbenzene — appeared to be poor antigens and those containing hexyl- or octylbenzene failed to elicit antibodies. Antisera for anthranyl-carbamido protein, apparently with hapten specificity, were obtained by Creech and Franks (51, 55), and an analogous conjugate was made with the carcinogenic hydrocarbon 1,2,5,6 — dibenzanthracene.

[24] The discordant results of Adant (56) must be ascribed to unsuitable technique.

TABLE 20

[after Landsteiner et al. (57)]

Immune Sera:	p-Aminobenzoic acid	m-Aminobenzoic acid	o-Aminobenzoic acid	p-Aminophenylarsenic acid	Sulfanilic acid	o-Aminocinnamic acid	Aniline	p-Nitroaniline	p-Toluidine	m-Toluidine
p-Aminobenzoic acid	+++±	±	o	o	o	o	o	o	o	o
o-Aminobenzoic acid	o	o	++±	o	o	o	±	±	o	±±
p-Aminophenylarsenic acid	o	o	o	+++++	o	o	o	o	o	±±+++
Aniline	o	o	o	o	o	o	+++±±	++	++±	++
p-Nitroaniline	o	o	o	o	o	o	++	++±	+++±	++
p-Toluidine	o	o	o	o	o	o	++±	++	+++±	±±

Concentration of antigens 0.01%.

and derivatives of these acids. The p-aminobenzene sulfonic acid sera gave slight reactions with meta- and at a later reading also with o-sulfonic acid antigen (12), and the reactions of the p-aminophenyl-arsenic acid serum showed similar gradations. Worthy of note is the action of o-aminobenzoic acid sera on o-aminobenzene sulfonic acid antigen, which may be comparable in strength to the homologous preparation in spite of the difference in the acid groups, and evidently depends on the identical position.[25]

A striking illustration of the role played both by the nature of the acid groups and their position in the benzene ring is given in Table 21 from an experiment with an m-aminobenzene sulfonic acid serum (28).

TABLE 21. — ANTIGENS TESTED WITH IMMUNE SERUM FOR META-AMINOBENZENE SULFONIC ACID (METANILIC ACID)

Antigens	NH_2 — R ortho-	NH_2 — R meta-	NH_2 — R para-
Aminobenzene sulfonic acid	$+\pm$	$++\pm$	$\pm$
Aminobenzene arsenic acid	o	$+$	o
Aminobenzoic acid	o	$\pm$	o

R designates the acid groups ($COOH$ or SO_3H or AsO_3H_2).

In the case of less effective neutral substituents the position was of greater influence than their nature. Thus immune sera for para compounds acted with almost equal strength on most "para"-antigens (which in part could not be differentiated), not so intensely on meta-, and still less on ortho-antigens.

Passing on to subsequent work by several investigators, notice may first be taken of experiments by Haurowitz (27) on conjugates made from a diazotized ammonium base, m-amino-phenyltrimethyl-ammonium chloride. The results indicated that strongly basic groups are probably as effective as acid groups in directing specificity. Antigens containing a pyridine base were first prepared by Berger and Erlenmeyer; [26] in later studies pyridine and carboxy-pyridine conjugates were found to compare satisfactorily in specificity with benzene de-

[25] Other examples can be found in (12).
[26] (58); v. Hooker and Boyd (59).

rivatives.[27] In the examination of other heterocyclic substances —
pyrazolone compounds [28] — sharp distinction was seen between two
isomeric amino antipyrines,[29] a case of interest with regard to the
question whether the antibodies are adjusted to the whole ring system
or to circumscribed reactive groups.

A number of azoantigens were prepared by Berger and Erlenmeyer
in order to demonstrate the serological similarity of compounds which
are closely related in chemical properties and physical constants, and
which often are able to replace each other in crystals. Of the groups
of substances (p. 265) tested by these authors, benzene and thiophene,[30] and pyridine and thiazole,[31] may here be cited as pairs which
proved to be serologically nearly equivalent.

CONJUGATED ANTIGENS WITH ALIPHATIC SIDE CHAINS.[32] — By
combination with diazotizable aromatic compounds the azoprotein
method can be applied over a wide range of aliphatic substances.
Thus, to prepare antigens from aliphatic acids, dibasic acids were
fused with p-nitroaniline, and the nitroanilic acids formed in this way
were coupled to protein after reduction of the nitro groups and diazotization. For comparison, antigens were made from aminophenylacetic acid and aminoacetanilide and their homologues. (In other
cases aminobenzoyl derivatives and the like were used to obtain diazotizable compounds.)

As shown in Table 22 the immune sera against the lower anilic
acids (oxanilic and succinanilic acids) were distinctly specific, so that
lengthening or shortening of the chain by only one carbon atom produced a marked difference, whereas the other two immune sera (to
adipanilic and suberanilic acids) show much stronger overlapping
reactions with the neighboring members of the series and with the
"aminophenyl" preparations.[33] Similar but less selective were the
reactions of the antigens made from aminophenylacetic acid and its
homologues, and of the anilides. The more pronounced specificity of
the anilic acid antisera would seem to indicate that, like acid groups,
the polar CONH group is of significance; according to preliminary
experiments it seems to enhance the immunizing activity.

[27] (60); v. (p. 185).
[28] Berger and Erlenmeyer (61); v. (62).
[29] Harte (63).
[30] (52).
[31] (64); v. p. 187.
[32] Landsteiner and van der Scheer (65).
[33] cf. the comment of Haurowitz (14).

TABLE 22
[after Landsteiner et al. (54)]

Immune Sera:	p-Aminooxanilic acid	p-Aminomalonanilic acid n=1	p-Aminosuccinanilic acid n=2	p-Aminoglutaranilic acid n=3	p-Aminoadipanilic acid n=4	p-Aminopimelanilic acid n=5	p-Aminosuberanilic acid n=6	p-Aminobenzoic acid	p-Aminophenylacetic acid	p-Aminophenylbutyric acid	p-Aminophenylcaproic acid
p-Aminooxanilic acid	+±	o	o	o	o	o	o	o	o	o	o
p-Aminosuccinanilic acid ...	o	o	+++±	+	++	++±	++±	o	o±	o	o
p-Aminoadipanilic acid	o	o	±	++±	++±	+++±	+++±	o	+±≡	±	±
p-Aminosuberanilic acid	o	o	±	+±	++	+±	+++	±	+	±±	±±
p-Aminophenylacetic acid ...	o	o	o	o±	o±	±±≡	±±	o	±±≡	+±	o
p-Aminophenylbutyric acid .	o	o	o±	±	±	±	++	o	±	±+	+±
p-Aminophenylcaproic acid .	o	o	±≡	±	±	+	++	o	±	+	±±

Concentration of antigens 0.01%.
General formula of amino anilic acids: $NH_2 \cdot C_6H_4 \cdot NH \cdot CO \cdot (CH_2)_n \cdot COOH$.

Concerning the cross reactions it may be argued (also in other instances) that the compounds attached to protein are disintegrated in the animal body with subsequent production of antibodies to haptenic groups. Such an assumption could apply at any rate only to the cross reactions of compounds with shorter side chains and, as the substances differing from the homologous one by an even number of carbon atoms are not singled out, is inconsistent with the experimental evidence that in vivo oxidation of aliphatic chains proceeds by the loss of two carbon atoms at a time.

A new way of attaching aliphatic compounds was devised by Pillemer and Ecker (66), namely the interaction of organic halogen compounds with SH groups. The combination was effected with reduced keratin, rich in SH groups, and a series of halogenated fatty acids (or halides of alkyl-benzenes), feather keratin being used for the immunizing antigens and wool keratin antigens for the tests. The reactions paralleled those shown by azoantigens, especially in the greater specificity of the shorter chains.

Further observations concerning the specificity of aliphatic compounds will be mentioned in the following sections.

SPECIFICITY OF STEREOISOMERIC COMPOUNDS. — The marked differences among aromatic compounds, isomeric with respect to the position of substituents in the benzene ring, and the gradation in their cross reactions described above have already given a definite indication that spatial structure, as well as chemical constitution in the ordinary sense, plays an important role in serum reactions, as has been shown for enzymatic processes by E. Fischer (67), and repeatedly confirmed. Fischer's conception had, in a hypothetical way, been applied by Ehrlich to the specificity of serum reactions. To prove the matter, the investigation of stereoisomeric substances was undertaken by the author. Suggested by the work of Ingersoll and Adams (68) on staining with optically isomeric dyes, d- and l-para-aminobenzoyl phenylaminoacetic acids were first used (69):

On immunizing with the corresponding azoproteins, two sorts of sera were obtained which distinguished the two isomeric antigens. This demonstrated that a change in the spatial arrangement of atoms or radicals linked to one asymmetric carbon atom suffices to alter serological specificity.

The work was continued with the stereoisomeric tartaric acids (dextro, laevo, meso) which, after being converted into amino-tartranilic acids and diazotized, were coupled to protein.[34] Again the serum reactions showed a distinct difference between the stereo-isomers (Table 23). The stronger cross reactions of the meso-antigen, in comparison to those between the d- and l-antigens, are plausibly attributable to the circumstance that in the first case the difference in configuration involves one, in the other two asymmetric carbon atoms.

TABLE 23
[after Landsteiner et al. (70)]

Immune Sera:	Antigens from:					
	l-Tartaric acid		d-Tartaric acid		m-Tartaric acid	
	COOH HOCH HCOH COOH		COOH HCOH HOCH COOH		COOH HCOH HCOH COOH	
l-Tartaric acid	+++	++±	±	o	+	±
d-Tartaric acid	o	o	+++	++±	+	±
m-Tartaric acid	±	±	o	o	+++	+++

Concentration of antigens 0.05% (first column), 0.01% (second column).

Tartaric acid sera also react with malic acid antigens,[35] the d-serum chiefly with the d-, the l-serum with the l-compound, and vice versa, in agreement with the configurational correspondence demonstrated by Freudenberg and Brauns between those optically active malic and tartaric acids which rotate polarized light in the same direction. This showed that, in principle, serum tests, like enzyme reactions (E. Fischer), may be used for the determination of spatial configuration.

That steric configuration must be of importance also for the specificity of bacterial polysaccharides could hardly be doubted, especially on account of the relationship of tartaric acid to sugar acids. This presumption was substantiated by Avery and Goebel, who applied very successfully, as discussed below, the azoprotein method to the serological examination of carbohydrates.

[34] Landsteiner and van der Scheer (70).   [35] (13), (71).

The demonstration of immunological specificity in the typical case of cis-trans isomerism — maleic and fumaric acids — was attained indirectly (p. 192).

CARBOHYDRATE-AZOPROTEINS. — With the purpose of elucidating the serological reactions of bacterial polysaccharides, an extensive study on artificial conjugated carbohydrate-proteins was carried out by Avery and Goebel. First, the simple sugars glucose and galactose were converted into p-amino-phenol-β-glycosides which, after diazotization, were coupled to protein. The antisera produced with the resulting antigens differentiated the two substances sharply, showing that in sugars, too, interchange of H and OH on one carbon atom alters the immunological properties (72). Acetylation changed the specificity of the β-glucoside; the stereoisomeric α- and β-glucosides were distinguishable, but exhibited strong cross precipitation (73, 74).

p-Aminophenol α-glucoside   p-Aminophenol β-glucoside   p-Aminophenol β-galactoside

Antibodies for O-β-glucosido-N-carbobenzyloxytyrosyl protein were found by inhibition tests to be directed towards the β-phenolic glucoside grouping and the carbobenzyloxy group as well.[36]

Rather complicated relationships — represented in Table 24 — were encountered between the four disaccharides, lactose, gentiobiose, cellobiose, maltose, and the monosaccharides glucose and galactose. The occurrence and intensity of cross reactions depending on similarities in structure could not be definitely explained in all cases; of primary significance was the configuration of the terminal hexose.[37]

Extension of the work to uronic acids (76, 77) was indicated because of the prominence of these sugar acids in the serologically active polysaccharides of pneumococci, Friedländer bacilli and plant gums. While it could be expected that the azoproteins made by

[36] (19); v. (74a), (75).
[37] (75a); (p. 268, cf. p. 178); cf. the discussion by Marrack (75b).

coupling the diazotized p-aminobenzyl glycosides of glucuronic and galacturonic acids, and of glucose and galactose, would be quite different serologically, a novel fact emerged, namely immunological relationships of natural and artificial antigens; such reactions were shown by Goebel et al., who demonstrated the precipitation, in high dilutions, of glucuronic acid azoprotein by sera for pneumococci of types II, III, and VIII, and by Woolf, Marrack and Downie (78) with an azoprotein made from a natural glucuronide, euxanthic acid, and Pneumococcus type II immune serum.

The method of Goebel of synthesizing aminobenzyl glucosides is generally applicable to uronic acids. Woolf's method consists of introducing an aromatic amino group by coupling with a diazonium compound, followed by reduction cleavage of the $N=N$ linkage.

Cross reactions, thus far unique, of compounds unrelated except for the presence of acid groups were observed by Goebel and Hotchkiss (77) when they found that p-aminobenzene sulfonic and carboxylic acid antigens are precipitated by various pneumococcus horse immune sera. The precipitation of galacturonic acid antigen by antisera for pneumococci which do not contain this substance may be a somewhat similar case.

A still closer approach to the serology of bacterial polysaccharides was to be anticipated through the use of antigens prepared from aldobionic acids, of more complex structure than simple uronic acids.[38]

The experiments were begun with cellobiuronic acid which is a disaccharide consisting of one molecule of glucuronic acid and one molecule of glucose bound in glucosidic linkage, and is a chief constituent of the polysaccharide of Pneumococci III and VIII. Immune sera against cellobiuronic acid reacted with the disaccharide itself and with both portions of the molecule, glucose and glucuronic acid. More remarkably, not only did cellobiuronic acid antigen precipitate with Pneumococcus III, VIII (and II) sera, but cellobiuronic im-

[38] Goebel (79).

TABLE 24. — REACTIONS OF ANTISERA TO
WITH HOMOLOGOUS AND HETEROLOGOUS
1 : 50000 [GOEBEL, AVERY AND BABERS

Test antigens. ↓	α-glucoside antigen.	β-glucoside antigen.	β galactoside antigen.	β-cellobioside antigen.
α-glucoside	+++	+	O	±
β-glucoside	++	++++	O	+++
β-galactoside	O	O	+++	O
β-cellobioside	±	++±	O	++++
β-maltoside	+++	+++	O	+++
β-gentiobioside	±	++±	O	+++
β-lactoside	O	O	++	++±

mune serum agglutinated Pneumococci III, precipitated the capsular polysaccharide, and immunized mice against infection with the three pneumococcus types — the first time an immune serum produced by means of a synthetic substance acted upon a natural antigen and protected against an infectious disease. Another aldobionic acid used for the preparation of conjugated antigens was gentiobiuronic acid (80), likewise a compound of glucose and glucuronic acid, but differing from cellobiuronic acid in the position of the glucuronosidic linkage. The two acids cross-react, evidently owing to the glucuronic acid part which is common to both, and gentiobiuronic acid cross-reacts with gentiobiose. Gentiobiuronic acid immune sera protected mice against pneumococci of type II (not against type III or VIII), and since protection was then found to be afforded by sera for glucuronic acid, one may suppose (that is, on immunological grounds) that the uronic acid, not yet chemically identified, in the capsular polysaccharide of the microbes is glucuronic acid.

Applying the azoprotein method to a substance of large molecular size, Avery and Goebel (81) converted the type specific polysaccharide of type III pneumococci into a complete antigen by way of diazotizing an aminobenzyl ether of the carbohydrate and coupling, after diazotization, with serum globulin (p. 108).

PEPTIDE-AZOPROTEINS. — The serological examination of pep-

Mono- and Di-Saccharide Antigens
Antigens; Antigen Concentration
(75a)]; [after Marrack (119)]

β-maltoside antigen.	β-gentiobioside antigen.	β-lactoside antigen.
+ + ±	O	O
±	+ + +	O
O	O	+ ±
±	+ + ±	±
+ + + +	+ + ±	±
±	+ + + +	±
±	±	+ + + ±

tides [39] was undertaken with a desire to gain information on the specificity of proteins. The peptides were nitrobenzoylated and reduced to amino compounds, and these were converted into antigens by coupling to protein. Azoproteins were prepared in the same manner from the amino acids glycine, leucine, glutamic acid, and tryptophane.

The azoproteins from amino acids proved to be distinctly specific in their reactions with the corresponding antisera, strong cross reactions occurring only in cases of the related compounds glycine and alanine, valine and leucine, and asparaginic and glutamic acid, which however could be differentiated without difficulty (83).

Conjugated antigens containing tyrosine (p. 162) in peptide linkage were synthesized by Clutton, Harington and Yuill.[40]

Lettré and Haas (84) prepared an antigen containing benzoylalanin residues by treating protein with the corresponding oxazolone. Their tests, made only with products from somewhat related proteins (cattle and horse sera), are insufficient to form a final opinion about the hapten specificity of the antigens.

To explain the results of Dujarric de la Rivière and Kossovitch (85) on the distinction of the optical isomers of leucine and histidine by antisera,

[39] Landsteiner and van der Scheer (26), (82).
[40] (19), (75).

one would have to assume that azoproteins were formed by the interchange of azogroups between the protein and the azodyes used (cf. 180a).

Antisera against the (optically inactive) dipeptides glycyl-glycine, glycyl-leucine, leucyl-glycine and leucyl-leucine A gave the strongest precipitation with the homologous antigens (Table 25); they also gave overlapping reactions, chiefly when the terminal amino acid of the peptides was the same as in the immunizing antigen. Hence, in conformity with other instances where acid groups have predominant influence, the specificity was determined by the amino acid carrying the free carboxyl group, to a lesser degree by the second amino acid.[41]

TABLE 25
[after Landsteiner et al. (26)]

Immune Sera:	Antigens from:			
	Glycyl-glycine	Glycyl-leucine	Leucyl-glycine	Leucyl-leucine
Glycyl-glycine	++±	o	o	o
Glycyl-leucine I	o	++±	o	±
Glycyl-leucine II	+	+++	o	+
Leucyl-glycine I	+	o	+++	o
Leucyl-glycine II	++	o	+++	±
Leucyl-leucine	o	+	o	++

Concentration of antigens 0.01%.

In later experiments antisera for tri- and pentapeptides (made from glycine and leucine) and for two pentapeptide amides (also for glutathione) [42] were included and tested against various peptides built up from the two amino acids. Here again cross reactions occurred, but not invariably, with peptides having the same amino acids at the free end of the peptide chain; besides, overlapping reactions, particularly with an amide serum, were evoked by reason of other portions of the molecule, and the cross reactions were definitely correlated to similarities in constitution. For instance, a G_5 serum [43] precipitated G_2, not LG antigen, and precipitated LG_2 much less than G_3, and the precipitates produced by G_4L serum increased in the sequence L, GL, G_2L, G_3L, G_4L. G_2LG_2Am serum reacted with G_2L-

[41] Accordingly the antigenic imprint must be accomplished before any cleavage of the peptides takes place in the body; the possibility of preparing antibodies to a readily saponifiable ester is corresponding evidence (p. 168).

[42] Glutathione antiserum gave only weak reactions with the other peptides.

[43] G stands for glycine, L for leucine, Am for amide (e.g., LG_2=leucylglycylglycine).

Am and G_4LAm, not with LG_2Am despite the identity of the terminal part. It may further be noted that mere shifting of the position of one amino acid induced a marked change in the specificity of tripeptides (G_2L, GLG, LG_2), pentapeptides (G_4L, G_3LG_2, LG_4) and pentapeptide amides.

Sharper distinctions than those in direct precipitin tests were brought out through tests with partially absorbed sera and by inhibition reactions (p. 192). Thus upon absorption of G_2LG_2 serum with G_5 antigen an antibody fraction with selective affinity for the homologous peptide was separated which reacted only slightly with LG_2 and with none of eighteen other antigens, all containing glycine or leucine, or both (Table 26).

From the experiments described one can predict that the amino acids present in proteins permit the synthesis of a very large number of serologically different peptides; their pronounced specificity may be connected with the polarity of the CH_2CONH groups.[44] As yet only relatively simple compounds have been examined, and a closer approach to the conditions obtaining in proteins could well be reached by examining the behavior of peptides built up from a greater variety of amino acids, and having larger molecular size and more complex structure. The investigation of such substances may enable one to determine how complicated a chemical pattern can be fitted by antibodies, and to draw from these models inferences upon the size and nature of the determinant groups in proteins. To be sure, the work involved in the synthesis of more than a few higher polypeptides is cumbersome to the point of being almost prohibitive, but there is hardly any other class of synthetic chemical compounds that could furnish such complex and diversified substances possessing serological activity.

A connection between proteins and synthetic compound antigens by cross reactivity in analogy to the findings with carbohydrates has only been obtained in the very special case of thyroglobulin (p. 195). A result touching this subject remotely is the reaction of anthrax immune sera with glutamic acid azoproteins [45] which is due to the presence in the bacilli of a colloidal substance composed of glutamic acid (p. 226).

PHARMACOLOGICALLY ACTIVE SUBSTANCES, HORMONES. — On the

[44] Cohn (86). Comparison may be made with the indifferent specificity of long fatty acid chains.
[45] Ivanovics and Bruckner (87).

TABLE 26
[after Landsteiner et al. (82)]

Immune Serum	G	G₅	G₅	G'₂	G₅	LG₅	G₂LG₅	L₅G₅	LG'	Glut. G₅	Tyrosyl G₅	Glutathione
G₂LG₂ absorbed with G₆ (azostromata)	0	0	0	0	0	±	++	0	0	0	0	0
	0	0	0	0	0	+	+++±	+	0	0	0	0
G₂LG₂ unabsorbed	±	+	++	++	++	++	+++	+	++±	+	++	0

The readings were made after 1 hour (top line), and after being held in the icebox overnight (second line); the third line shows the relative intensities of the reactions of the unabsorbed serum after 1 hour. Eight other antigens gave negative results also with unabsorbed serum.

pattern of antitoxin production it seemed not out of the question that non-antigenic poisons after combination with protein would call forth antibodies capable of neutralizing their toxic effect. This idea, which would hold out hopes for practical application, was put to a test by Berger and Erlenmeyer (88) with an antiserum to a dimethyl-N-pyrrol-phenylarsenic acid (icterogen) and by Hooker and Boyd (59) with antibodies for strychnine,[46] but was unsuccessful in both trials. The authors considered the reasons for these failures to be dissociation of the antibody union in the animal, non-identity of the toxic and serologically binding portions of the drug molecule, and the large quantity of antibody which would be required for neutralization on account of the disproportions in molecular weight of antibody and drug.

On the other hand, Clutton, Harington and Yuill (121), working along similar lines with thyroxine, obtained positive results. Antigens were synthesized, by an ingenious method, in a series of reactions. Diiodothyronine methylester was converted into N-carbobenzyloxydiiodothyronyl azide; this was coupled to protein and the compound was iodinated. In the last reaction the introduced diiodothyronyl and the original tyrosyl residues were changed into thyroxyl and diiodotyrosyl residues respectively (v. p. 194). The antisera for the conjugated protein had positive haptenic specificity and, when injected into rats, suppressed the physiological action of administered thyroglobulin.[47] The neutralization extended also to thyroxine, but the authors think that this could be an indirect effect inasmuch as thyroxine may exert its activity after having entered into the composition of thyroglobulin. A similar exception cannot be taken against the neutralization of the pharmacological effect of aspirin by antisera against this compound, reported by Butler, Harington and Yuill (90). As Marrack (91) remarks, the result is surprising "considering the small amount of haptens bound by antibodies in vitro."

Tests in vitro may, in general, give a better chance for demonstrating neutralization of simple toxic substances, than in vivo experiments. For instance, attempts to obtain immune sera by means of conjugate antigens which would inhibit the powerful hemolytic activity of saponins may well be worth while.

---

[46] Antimorphine sera could not be obtained.
[47] Concordant results were obtained by active immunization against thyroglobulin and with anti-thyroglobulin immune sera [Kestner; Went et al. 89)].

Azoproteins containing sulfanilamide and related chemotherapeutic compounds have been made mainly in view of the allergic conditions which sometimes ensue after the administration of these drugs, and data on the serological distinction and cross reactions of these substances have been recorded.[48]

Histamine azoprotein has been used for the treatment of allergic patients (94), and Cohen et al. (94a) seem to have succeeded in neutralizing in vitro the physiological effect of small amounts of histamine with the serum of patients treated with histamine azoprotein. Antisera for conjugates made by coupling proteins with diazotized aminoadrenalin were examined by Went (95); the results presented do not quite suffice as proof that the sera were specific for adrenalin itself.

Mooser and Grilichess (96) state that they were unable to produce antibodies directed against androstendiol by immunizing with corresponding azoproteins; similar earlier experiments with an antigen supposed to contain cholesterol in azolinkage had a similar negative result (97).

SERUM REACTIONS WITH SIMPLE SUBSTANCES OF KNOWN CONSTITUTION. — Although the specificity of sera for azoproteins evidently is due to the reaction between antibodies and the substances linked to protein, precipitation was observed only with the protein compounds. When the tests were performed with the uncoupled azocomponents, or with dyes in which the diazotized substances were coupled to tyrosine or other phenols instead of to protein, there was no perceptible reaction.[49] This result did not appear extraordinary, since precipitin reactions were known to occur only with substances having high molecular weight and giving colloidal solutions. In order to render the supposed reactions visible, the author had recourse to the fact that precipitation by immune sera is diminished or prevented when the antigen is present in excess [v. Halban and Landsteiner (98)]. Accordingly it was possible that addition of the azocomponents (or simple azodyes) containing the specific groups would, by virtue of their combination with the antibodies, prevent precipitation (or complement fixation) of the homologous azoprotein. Experiments performed in this way indeed showed the effect sought for, namely specific inhibition [50] by the compounds corresponding to the immune serum, and related ones.[51] Thus it was demonstrated that simple

---

[48] Mingoia and Rocha e Silva (92), Wedum (93).

[49] See however p. 183.

[50] The reaction can also be carried out, less conveniently, by testing for a dissolving effect on antigen-antibody precipitates.

[51] Landsteiner et al. (99), (100), (69), (70), (26); Haurowitz and Breinl (101). Similar reactions have since been performed with various sorts of compound

substances which are lacking in antigenic power combine specifically with antibodies, and that serological reactivity in vitro is altogether independent of the power to immunize.

The phenomenon of specific inhibition in which a participation of proteins in the reaction with the antibodies is entirely excluded widened the field of serum reactions, limited originally to complex biological substances, for with the aid of the reaction a great variety of synthetic compounds of simple composition became directly accessible to serological examination. That the reaction results from union of the inhibiting substances with antibodies [52] seems the only plausible explanation and is supported by the analogy to the specific inhibition by an excess of antigen. Unquestionable proof was afforded by Marrack and Smith (103), and Haurowitz and Breinl (101) who found that the diffusion of an azodye made from p-arsanilic acid through collodion membranes was hindered by the corresponding antiserum.[53] Even more direct evidence was the finding that certain azodyes in high dilutions may actually give specific precipitin reactions with azoprotein immune sera, precisely like proteins or bacterial polysaccharides, and elicit anaphylactic shock in guinea pigs sensitized with the corresponding azoproteins.[54] The specifically precipitable dyes were derived from the aforementioned anilic acids by coupling with resorcinol or tyrosine; the strongest reactions were obtained with the anilic acid dyes having the longest aliphatic chains [resorcinoldiazo-p-suberanilic acid, $(OH)_2C_6H_2(N=N-C_6H_4-NH-CO-(CH_2)_6-COOH)_2$, and the analogous adipanilic and pimelanilic acid compound]. The results showed that (irrespective of the possible formation of aggregates in the solutions) it is not necessary to have a compound of high molecular weight for precipitation by immune sera or for the production of anaphylactic shock. In following up these

antigens, viz. conjugates made with isocyanates, azides, carbobenzoxy chloride, halogenated fatty acids (p. 172). Also, an analogous experimental procedure has been applied to enzymes, in particular for the investigation of the specificity of peptidases [v. von Euler and Josephson; Balls and Köhler (102)].

[52] On the supposition of bi- or multivalent antibodies the existence of compounds which contain all three constituents is also to be considered.

[53] v. (104).

Complement fixation tests, with the aid of lecithin, reported by Klopstock and Selter (105) are perhaps relevant. Mutsaars (44) did not succeed in separating hapten-antibody complexes in the ultra-centrifuge but considers his tentative experiment as indecisive. With univalent haptens, at any rate, the increase in molecular size over uncombined antibody would be negligible.

[54] Landsteiner and van der Scheer (106); (Table 27); see p. 198.

TABLE 27
[after Landsteiner et al. (106)]

Immune Sera	Azodyes prepared by coupling resorcinol with diazotized amino acids:					
	p-Amino-malonanilic acid	p-Amino-succinanilic acid	p-Amino-glutaranilic acid	p-Amino-adipanilic acid	p-Amino-pimelanilic acid	p-Amino-suberanilic acid
p-Aminosuccinanilic acid	0 / 0	+± / ++	0 / 0	0 / 0	0 / 0	0 / 0
p-Aminoadipanilic acid	0 / 0	0 / 0	0 / 0	+ / ++	±± / ±	±± / ±
p-Aminosuberanilic acid	0 / 0	0 / 0	0 / 0	± / ++	+± / +++±	++ / +++

Concentration of the dye 1/25000 millimol in 1 cc.
First line: reading after 2 hours; second line: reading at a later time.

experiments, weak precipitin reactions were seen with other dyes (106), and later many such observations [including complement fixation tests (106a)] were made in connection with the question of the insolubility of antigen-antibody complexes (pp. 249, 246, 243).

The specificity of inhibition reactions is essentially of the same order as that of precipitin reactions; accordingly, just as with conjugated antigens, simple, closely related compounds, for example the isomeric tartaric acids, can be differentiated serologically.

The inhibition reaction furnished new information not only on account of the ease with which it was possible to investigate numerous compounds but also because of the appearance of additional group reactions. This probably depends upon the simple composition of the substances used. Indeed, a wide range of cross reactions was observed with simply constituted substances (and the corresponding immune sera) which, so to speak, possess few distinctive characteristics. Drawing upon E. Fischer one may appropriately illustrate this state of affairs by the analogy of opening a variety of locks with a simply constructed key. With immune sera adjusted to more complex structures (aminobenzoyl phenylaminoacetic acid, tartranilic acid, polypeptides) the inhibitions were as specific as the corresponding precipitin tests, and in some instances apparently even more so (p. 192).[55]

The tests were made mostly with acids (neutral solutions of the sodium salts) but other substances such as glucosides, pyridine bases and pyrazolone derivatives have been found to be equally suitable. Because of considerable dissociation of the hapten-antibody compounds, high concentrations of the haptens are often required, yet there are differences depending on the nature of the substances. In inhibition tests with immune sera for aromatic amino acids their azocompounds with tyrosine or m-hydroxybenzoic acid were active in smaller concentrations than the amino acids themselves.[56] The reason lies most likely in the greater similarity of the dyes to azoproteins, on the assumption that cyclic amino acids, to which the haptens are linked in the proteins by the $N=N$ group, take part in the antibody reaction.[57] This explanation receives support from the experiments of Hooker and Boyd referred to above. That simple sugars are not

---

[55] (100); v. (99).
[56] (99); v. (107), (22). Similarly, sera for aspiryl protein were better inhibited by aspirylglycine than by aspirin (90).
[57] v. (107a).

inhibiting, while glucosides are, can be understood on analogous considerations, and there are other similar observations. As far as the studies on specificity are concerned, the question of combination with antibodies at other sites than the azocomponent is of secondary importance, because the inhibiting effect of the uncoupled haptens could be established with most azoprotein sera which have come under examination.[58]

Aromatic bases, such as aniline or toluidine, have not been shown to possess inhibitive capacity; that free acids impede the precipitation of acylated proteins would hardly be anticipated (6).

It is obvious that the degree of inhibition is dependent on the relative affinity of the antibody to the antigen and hapten. Consequently, as a rule, heterologous precipitation is more readily inhibited than the homologous reaction, and heterologous haptens are less effective than homologous ones. Quantitative studies of the reaction were made by several workers.[59] As shown by Woolf, while in the precipitin reaction the quantity of precipitate is little changed by dilution, inhibition depends on the concentration, not on the total amount of inhibitor; the effect was not much influenced by variation in the antigen-antibody ratio. The concentrations of hapten necessary for inhibition were greatly different in homologous and heterologous systems (v. p. 272, 250) and with sera of individual rabbits.

Selected results of inhibition reactions with azoprotein antisera are reproduced in Tables 28–31. The following may be noted:

1. The influence of position in the reactions of substituted benzene derivatives is so pronounced that independently of their nature the position of the groups (CH_3, Cl, Br, OH, NO_2) in mono-substituted benzoic acids can usually be determined with the help of the three aminobenzoic acid antisera, and regularly a difference is demonstrable between o- and D-compounds. A similar dependence on the position of aromatic substituents in the case of enzyme reactions is of interest, as in the oxidation by tyrosinase (110), the inhibition of this enzymic process by aromatic acids (111) and the action of carboxy-polypeptidases on the isomeric chloroacetyl aminobenzoic acids.[60]

2. The importance of the nature of acid groups (AsO_3H_2, SO_3H,

---

[58] Inhibition by the homologous azodyes of non-specific precipitation of sulfathiazol azoprotein was described by Bukantz and Abernethy (108).

[59] Marrack and Smith (103), Haurowitz and Breinl (101), (29), Erlenmeyer and Berger (104), Woolf (78), Pressman, Brown and Pauling (107a).

[60] Waldschmidt-Leitz and Balls (109); v. (102).

COOH) can be seen from Table 28. Again, as in precipitin tests, arsanilic acid sera reacted with all aromatic arsonic acids tested and, remarkably, even with an inorganic substance, namely arsenic acid. Arsenious acid, aliphatic [61] or secondary aromatic arsonic acids, acetaminophenylstibinic acid and arsenic oxides showed no inhibition.[62]

A study of the inhibiting activity of numerous substituted phenylarsonic acids on the precipitation of arsanilic acid azoprotein was conducted by Pressman, Brown and Pauling (107a). The effects depended, as in the above-mentioned tests, on the position of the substituents. The inhibiting power was improved over that of phenylarsonic acid by various substituents in the descending order: NO_2; CH_3CONH; $\langle\bigcirc\rangle$$-N=N-$; $\langle\bigcirc\rangle$$-CONH$; Cl, Br, I, CH_3; OH; NH_2; COOH. Most striking was the increase in the "bond strength constant" of the hapten reaction produced by the NO_2 group, which the authors relate to the significance of nitro groups in the formation of molecular compounds. This result deviates from those obtained in inhibition tests with sera for aminobenzoic acids.[63]

The effect of acids on the precipitation of kerateine conjugates, prepared by the substitution of SH groups (66), parallels the observations made with azoprotein antisera; the reason for the inhibition of benzylkerateine serum by benzoic acid is not quite evident.

Pyridine bases were found to inhibit the precipitation of aminopyridine azoprotein; piperidine was inactive (58), nicotine inhibited to some degree; with pyridine acids the relative position of carboxyl and the ring nitrogen was of significance (60). Precipitation of immune sera to 2-aminothiazole was inhibited almost equally well by aminothiazole and aminopyridine (64). Anti-strychnine sera gave a cross reaction with brucine, not with morphine or quinine.[64]

A number of pyrazalone derivatives were examined by Erlenmeyer and Berger for the purpose of finding a possible correlation between the portions of the molecule participating in the serological reactions and those which are responsible for the pharmacological effects.[65]

Several substances (veronal, salicylic acid, etc.) which form molecular compounds with antipyrine or pyramidon did not block their inhibitory

[61] Aliphatic arsonic acids are also ineffective as chemotherapeutic agents (112).
[62] (99), (101).
[63] (100); see also (107b).
[64] Hooker and Boyd (59).
[65] Erlenmeyer and Berger (61), Harte (63).

TABLE 28
[after Landsteiner

Substances used for

Immune Sera:	Control	Pro-pionic acid	Iso-valeric acid	Chloro-acetic acid
p-Aminophenylarsenic acid	++++	+++±	++++	++++
m-Aminobenzene sulfonic acid	++++	++++	++++	++++
p-Aminobenzoic acid	+++±	++++	+++	+++

Substances used for

Immune Sera:	Aminobenzene sulfonic acid			Benzoic acid
	o-	m-	p-	
p-Aminophenylarsenic acid	+++±	++++	+++±	++++
m-Aminobenzene sulfonic acid	o	o	+±	+++
p-Aminobenzoic acid	+++±	+++	+++±	o

0.2 cc. of 0.01% antigen solutions +0.05 cc. 1/10 molar solutions of the sodium salts of the

TABLE 29
[after Landsteiner

Immune Sera:	Control	Amino-benzoic acid			Methyl-benzoic acid		
		o-	m-	p-	o-	m-	p-
o-Aminobenzoic acid	+±	o	±	+	o	o	±
m-Aminobenzoic acid	++	±	±	+	+	o	+
p-Aminobenzoic acid I	+±	±	±	o	+	±	±
II	+++±				++±	+±	o

Tests as in Table 28; for p-aminobenzoic acid serum II 1/20 molar solutions were used.

et al. (99)]

inhibition tests:

Glycine	Tartaric acid	Fumaric acid	p-Amino-phenyl-arsenic acid	p-Hydroxy-phenyl-arsenic acid	Benzene-sulfonic acid
++++	++++	++++	o	o	++++
++++	++++	+++±	+++±	++±	±
+++±	+++±	+++±	+++±	+++±	+++±

inhibition tests:

Aminobenzoic acid			Hippuric acid	Pyro-mucic acid	Phenyl-acetic acid
o-	m-	p-			
+++±	+++±	++++	++++	++++	++++
+++±	+++	+++	+++	++++	+++±
+±	o	o	+++±	++	++±

substances used for the inhibition tests; addition of immune serum to the mixture.

et al. (99) (100)]

Substances used for inhibition tests:												Benzoic acid
Chloro-benzoic acid			Bromo-benzoic acid			Hydroxy-benzoic acid			Nitro-benzoic acid			
o-	m-	p-	o-	m-	p-	o-	m-	p-	o-	m-	p-	
o	o	±	o	o	o	o	±	+	o	o	±	o
+	o	±	+	o	±	±	±	+	+	o	+	±
+	±	o	+	±	o	±	±	o	+	±	o	±
			++±	+	o				+++	+±	o	

capacity; it would be unsafe, however, to draw conclusions as to the serologically reactive groups from this result without reliable knowledge about
the degree of dissociation of the molecular compounds in solution.

3. Inhibition of the precipitation by antisera to tartaric acid,
succinic acid or peptides is considerably stronger on use of nitro- or
amino-tartranilic (or succinanilic) acids and nitro- or aminobenzoyl
derivatives of the peptides than with the acids and peptides themselves. As in the case of azodyes this may be referred to a participation in the reaction of the aromatic groupings in the immunizing protein. But that choloracetylglycine and -leucine react more strongly
than do glycyl-glycine and glycyl-leucine with immune sera corresponding to these peptides, cannot well be explained in that manner.
This behavior is due perhaps to the higher acidity of the chloroacetyl
compounds, as is, in the opinion of Waldschmidt-Leitz,[66] the stronger
action of carboxypolypeptidases on chloroacetyl tyrosine than on
glycyltyrosine. Other cases, already mentioned, in which an increase
in inhibiting activity of haptens does not depend on similarity in constitution of hapten and the immunizing antigen are recorded by
Pressman, Brown and Pauling (107a).

4. Precipitation of the homologous antigen by succinanilic acid
immune serum (Table 30) (65) is specifically inhibited by succinic
acid — which in this way can be distinguished from malonic or glutaric acids — and also by higher dicarboxylic acids (pimelic, suberic,
sebacic acids) and, as far as the experiments go, the more intensely
the longer the chain. The phenomena must be attributed in part,
therefore, to a general property of salts of higher aliphatic acids.[67]
Nevertheless, the inhibitions are in part dependent upon the specificity of the antibodies, for although higher dicarboxylic acids inhibit
the precipitation of azoantigens having fatty acid chains, this is not
true of all antigen-antibody systems; these cases, therefore, form,
as it were, a transition between specific and non-specific union.

Similar phenomena were observed among the reactions of aminobenzoic acid sera (100). Such sera react with various aromatic and
cyclic acids, especially strongly with benzoic, the closely related
thiophene carboxylic, and naphthoic [68] acids, but not significantly

---

[66] (113); v. (102).

[67] v. Pillemer (66).

[68] See also Hooker and Boyd (220). Tests with antigens containing naphthyl
and phenyl residues, without a carboxyl group, showed only weak cross precipitation. (52), (53).

TABLE 30

[after Landsteiner et al. (54)]

Immune Sera:	Control	Substances used for inhibition tests:							
		Malonic acid	Succinic acid*	Glutaric acid	Adipic acid	Fumaric acid*	Maleic acid*	Mesaconic acid	Citraconic acid
Aminosuccinanilic acid	++±	+	0	±	+±	++	0	++	0
Aminoadipanilic acid	++±	+++	++	+±	±	++	+±	++	+

* Succinic acid COOH·CH₂·CH₂·COOH, Fumaric acid $\begin{matrix} \text{HOOC-CH} \\ \| \\ \text{HC-COOH} \end{matrix}$ Maleic acid $\begin{matrix} \text{HC-COOH} \\ \| \\ \text{HC-COOH} \end{matrix}$

o.1 cc. of o.4 molar solutions of the tested substances, otherwise as Table 28.

with lower fatty acids. From this, and because benzoic acid has very little effect on the precipitation by arsenic or sulfonic acid sera, one must consider the interaction of the aromatic acids as group reactions, connected with the presence of carboxyl attached to aromatic rings. Precipitation by amino benzoic acid sera was furthermore inhibited by higher fatty acids (caproic, heptylic and caprylic acids), and here the reactions are likewise to some extent specific, for such reactions do not occur with arsanilic acid sera or ordinary protein precipitins; other factors, probably the surface activity of soaps, appear to play a part (100). Salts of cholic and desoxycholic acids were found to impede precipitin reactions in general.

5. Upon testing the influence of substituents in aliphatic compounds (54) by means of antisera to succinanilic, phenylacetic and phenylbutyric acids, it was found that replacement of H by the polar groups OH or NH_2 caused marked changes in the immunological behavior, apparently more than halogen.[69] With succinanilic serum it was also seen that removal of both carboxyl groups of succinic acid, as in succinimide, abolishes completely the reactivity with the serum while the monoester and monoamide in which one carboxyl group is left intact still react distinctly.

6. The reactions of unsaturated stereoisomeric dicarboxylic acids with succinanilic acid serum showed a sharp difference between cis- and trans-forms (54). While fumaric acid is practically inactive, maleic acid inhibits the precipitation by this serum just as well as does succinic acid (Table 30). The same difference as that between the acids is evident in the monoesters and the methyl derivatives, citraconic and mesaconic acids. Accordingly, one could suppose that the succinic acid molecule can exist in a form corresponding to the cis configuration,[70] or that the antibodies adjust themselves to this.

7. With peptide azoproteins it was found that inhibition reactions may be a more effective means for detecting the serological diversity of chemically similar compounds than precipitin tests. In contrast to the latter, which often failed clearly to distinguish azoproteins made from structurally related peptides, practically each peptide could be singled out by inhibition tests with the acylated compounds [71] (Table

---

[69] v. Pillemer (66). Cf. the comment by Marrack on polar substituents [(119), pp. 119, 120].

[70] v. (114).

[71] Inhibition was also seen with high concentrations of a non-acylated pentapeptide.

31). Marked cross reactions occurred only between peptides of very similar structure. Observations of Goebel, Avery and Babers on glucosides of disaccharides present a parallel case. As explanation it may be considered that in inhibition tests with homologous antigens and heterologous haptens the latter are competing with the stronger

TABLE 31
[after Landsteiner et al. (82)]

Immune Sera for	Nitrobenzoyl peptides used for inhibition tests:							
	GG	GL	LG	LL	GGG	GGL	GLG	LGG
L	+	o	±	o	±	o	±	±
	++	+	+±	±	++	±	+±	+
GL	++	o	++	+	++	±	++	++
	++±	o	+++	++	+++	+	++±	++±
LG	+±	++	o	+±	+±	++	±	+±
	++±	+++	o	++±	++±	+++	+±	++±
GGG	+±	++	++	++	o	++	++	+±±
	++	+++	++±	+++	±	++±	++±	++
LGG	+±	++	+±	++	+	++	+±	o
	++±	+++	+++	+++	++	+++	+++	+

TABLE 31 (continued)
[after Landsteiner et al. (82)]

Immune Sera for	Nitrobenzoyl peptides used for inhibition tests:							
	GGGG	GGGL	GLGG	GGGGG	GGGGL	GGLGG	LGGGG	Control
L	+	o	±	+	o	±	±	+
	+±	+	+±	+±	±	+±	+±	++
GL	+±	±	+±	+±	±	+±	++	++
	++±	+	++±	++±	+	++±	+++	+++
LG	+±	++	+±	+±	++	++	+±	++
	++±	+++	++±	++±	++±	++±	++±	+++
GGG	±	+±	+±	±	++	+±	±	++±
	±	++	++	+	++±	++	+	+++
LGG	+±	++	±	+±	+±	±	+±	++
	++±	+++	+±	+++	+++	+±	++	+++

Tests as in Table 28; made with 0.05 cc. of neutralized 1/40 molar solutions of the nitrobenzoyl peptides.
First line, readings after 5 minutes; second line, after 1 hour.
GG = glycyl-glycine, GLG = glycyl-leucyl-glycine, etc.

affinity of the antigen while there is no such interference in the precipitin reaction. Besides, there may be present in the sera weak antibodies which do not precipitate by themselves but are carried down in precipitates formed by antibodies of higher activity (pp. 257, 256).

8. As the author has remarked,[72] the inhibition test provides a possibility of determining the specifically reacting groups in antigens of unknown composition. This suggestion was taken up by Wormall (115) in the investigation of iodinated proteins. His studies, confirmed by Jacobs (116), showed that the reactions of immune sera for iodoproteins were not inhibited by iodocompounds chosen at random such as potassium iodide or o-iodophenol but by 3.5 dihalogenated tyrosine, the effect diminishing in the order I, Br, Cl, whereby tyrosine disubstituted with halogen was proved to be the reacting group. Pursuing this inquiry Snapper [73] found that the precipitation of iodo-

proteins was prevented by diiodotyrosine $HO\langle\bigcirc\rangle C_3H_6O_2N$, with I substituents,

thyroxine $HO\langle\bigcirc\rangle-O-\langle\bigcirc\rangle C_3H_6O_2N$, with I substituents,

3,5 diiodo-4-hydroxybenzoic acid, or the corresponding sulfonic acid, by 3,5 diiodo-4-hydroxynitrobenzene and other compounds, all containing the group $I\langle\bigcirc\rangle I$ with OH. Diiodothyronine $HO\langle\bigcirc\rangle-O-\langle\bigcirc\rangle-$ with I substituents,

$C_3H_6O_2N$, 3,5 diiodo-2-hydroxybenzoic acid, or 3,5 diiodo-4-aminobenzoic acid, however, were inactive. These reactions show that, in addition to iodine or bromine, hydroxyl ortho to both halogens is essential for the serological specificity of iodo- or bromo-proteins; methylation of the hydroxyl group abolished the activity altogether.

Somewhat weaker inhibition reactions were also obtained with the acetic acid ester of 3,5-diiodo-4-hydroxybenzoic acid and 3,5-diiodopyridone; the latter reaction may be due to enolization (119).

Snapper and Grünbaum failed to find cross precipitation between iodoprotein and thyroglobulin (p. 23), and antithyroglobulin sera were not

[72] (99), (100).
[73] (117), (118).

interfered with by either thyroxine or diiodotyrosine.[74] However, the presence of thyroxine in the thyroglobulin molecule is indicated by the observation of Clutton, Harington and Yuill (121) that thyroglobulin is precipitated slightly — less than iodinated protein — but definitely by sera against artificial thyroxylglobulin. This precipitation was inhibited by diiodotyrosine or thyroxine, better by a mixture of both. The weak precipitation of diiodothyronylalbumin with thyroxylglobulin antisera was not affected by diiodotyrosine, an indication, as the authors conclude, that the reaction is in part determined by the diiodophenolic ether linkage of thyroxine.

The characteristic specificity of nitroproteins, discovered like that of iodoproteins by Obermayer and Pick, was carefully explored by Mutsaars (44) with the aid of inhibition tests on a number of nitrocompounds. Among three nitrated natural amino acids — arginine, phenylalanine, tyrosine — only the last substance prevented the precipitation by immune sera prepared against nitroproteins. The hydroxyl as present in tyrosine appeared to be indispensable, since neither nitrophenols nor nitrobenzoic acids were effective, in contrast to nitrosalicylic acids; and 3-nitro-4-hydroxybenzoic acid, itself strongly inhibitory, was rendered inactive by conversion into 3-nitro-4-methoxybenzoic acid. Esterification of the carboxyl in 3,5 dinitrosalicylic acid did not destroy the activity. From these and other results Mutsaars infers that the serological reaction of nitroproteins is attributable to nitrotyrosyl residues.[75] Some questions are still left for further study, especially why changes in the relative positions of OH, NH_2 and COOH seemed, from some tests, to be immaterial.

By inhibition tests with the phenylcarbamido acids from lysine and ε-amino-n-hexoic acids, Hopkins and Wormall (123) were able to demonstrate that in all probability phenylureido-lysine is the determinant group in the phenylcarbamido proteins, and that the free amino groups of the unaltered protein are the ε-amino groups of lysine. This is borne out by these authors' observation that zein, a protein lacking lysine, did not appear to combine with p-bromophenyl isocyanate (see, however, clupein, p. 52).

That the precipitation of formolized protein by the corresponding antiserum is inhibited by free formaldehyde is open to doubt. It is possible that

---

[74] This negative result is intelligible if groupings other than diiodotyrosine or thyroxine take part in the reactions, as may be deduced from the species specificity of thyroglobulin sera. A weak inhibitory effect of thyroxine has been reported by Adant and Spehl (120).

[75] Nitrogelatine inhibits the precipitation of diazoprotein, and diazogelatine that of nitroprotein (122).

the inhibition observed was brought about by an excess of formolized rabbit protein formed through the action of formaldehyde on the immune serum. Amino acids (lysine) which were combined with formaldehyde did not interfere with the reaction (124).

With regard to the tempting problem of tracing the determinant groups in unchanged proteins, inhibition effects of amino acids or simple synthetic peptides have thus far not been demonstrable as one could expect them to be, were the specific structures in natural proteins small and uniform as they are in iodoproteins and azoproteins. Yet, it was possible to obtain inhibition reactions with hydrolytic split products of proteins. The attempt was not successful in our hands when common antiprotein sera were used which, Michaelis showed (125), no longer give precipitation after brief peptic digestion (p. 45). Heteroproteose fractions, however, gave rise to antibodies that precipitated unaltered protein, peptic metaprotein and the proteose preparations used for the immunizing injections; and in the heteroproteose system definite inhibitions resulted from the addition of proteose fractions [76] which, themselves, gave no or very slight precipitation with the sera [v. Holiday (128)].

Better characterized inhibiting protein fragments were separated from silk.[77] Partial hydrolysis with hydrochloric acid, and fractionation with alcohol, yielded peptide preparations with inhibiting properties which appeared to have an average molecular weight of between 600 and 1000 and to contain at most 12 amino acids; the fractions were not homogenous, yet some had definitely crystalline appearance. One of these products was found to contain glycine, alanine, hydroxyamino acid and tyrosine in nearly the same proportions in which these amino acids are present in silk.

Inhibition of the precipitation of bacterial polysaccharides by hydrolytic products was observed by Heidelberger (130), Morgan (131), and Ivánovics, and such tests with substances of known composition have been used, in addition to direct precipitin reactions, to ascertain the determinant groups in polysaccharides (pp. 219–221).

For the purpose of differentiating antibodies in an immune serum, inhibition of complement fixation was applied by Schaefer (132), who tested antisera for tubercle bacilli with polysaccharides, lipids and proteins as inhibitors.

---

[76] Landsteiner and van der Scheer (126), Landsteiner and Chase (127).
[77] Landsteiner (129).

HYPERSENSITIVITY TO SUBSTANCES OF SIMPLE COMPOSITION. — The hypersensitivity, known as anaphylaxis, to a second injection of a protein can be transferred by injecting serum of hypersensitive animals into normal ones. Anaphylaxis then is certainly a consequence of the production of antibodies — assumed to be identical with precipitins — and for comprehension requires only the accepted tenets of serology and the assumption that the formation of antigen-antibody complexes in the animal gives rise to disturbances, such as injury to cell surfaces or liberation of toxic substances.[78] On the other hand, the explanation of certain other forms of (human and animal) hypersensitiveness [79] is not so obvious.

Serum sickness is mostly, and for plausible reasons,[80] believed to depend on antigen-antibody reactions (Pirquet and Schick) although the condition often occurs (after an incubation period) following a single injection of foreign serum. There is fair, though no regular, correspondence [81] between the appearance of symptoms and the presence of antibodies in the circulation and, as in ordinary immunization, the incubation period is shorter after repeated injections. Experiments suggesting passive transfer with serum of individuals convalescing from serum sickness seem to afford additional evidence.[82]

A group of allergic diseases, with hay fever and asthma as representatives, is characterized by special features which caused Coca to set them apart as "atopic" diseases from other forms of allergy. The main peculiarity is the pronounced influence of hereditary [83] disposition which manifests itself in that — given certain incitants and the same conditions of exposure — only a small minority of persons contracts the disease, and that to provoke hay fever experimentally in constitutionally not predisposed individuals is difficult, if not impossible (v. 147a). Antibodies, at first sought in vain, were detected in the serum of the patients by transfer to the skin of normal individuals,[84] and their titre may be increased by injection of the allergens.

[78] The pathogenesis of anaphylaxis, in particular the role of histamine, is reviewed by Dragstedt (133), Rocha e Silva (134).
[79] Hypersensitivity in man is often distinguished by the name of "allergy" but this term — originally meaning increased as well as diminished reactivity (Pirquet) — is also applied to hypersensitivity in general, usually with the exclusion of typical anaphylaxis. Some books and reviews dealing with the subject of human allergy are: Tuft (135), Coca, Walzer, Thommen (136), Rackemann (137); v. Topley and Wilson (138), Boyd (139), Sulzberger (140), Kallos et al. (141), Doerr (142), Landsteiner (143). The symptoms of anaphylactic shock in various animal species and in man are described in (139).
[80] v. Hooker (144).
[81] v. de Gara et al. (145).   [82] Voss; Karelitz et al. (146).
[83] Cooke et al. (147), Coca (136).
[84] Prausnitz and Küstner (148), De Besche (149), Coca and Grove (150).

These so-called reagins proved to be different in several respects from common anaphylactic antibodies, especially in being incapable of inducing passive anaphylaxis in animals and in failing to give in vitro precipitation. While there is indeed reason for speaking of a special type of allergy, a relation to ordinary immunization is indicated by the fact that also this form of specific hypersensitiveness comes into existence only after previous contact[85] and because there are incitants (such as p-phenylenediamine and parasitic worms) which elicit allergic asthma in many individuals and also evoke reagins (worms).[86]

Turning to allergy against simple chemical compounds[87] ("drug allergy") it will be seen that the mechanism of this condition presented an embarrassing problem. It is true, the old idea of spontaneous, innate idiosyncrasy had to be abandoned, and the necessity of previous contact with the incitants — not always demonstrable — had to be accepted.[88] And, naturally, after the impressive discovery of anaphylaxis one tried to bring all sorts of hypersensitiveness into line with this phenomenon. But it remained to be explained how simple chemical compounds could elicit an immunological response, in the face of common experience.

A clue was afforded[89] by the possibility of building up simple substances into antigens through attachment to proteins, as discussed in preceding pages. It could then be assumed that a non-antigenic substance can sensitize after combination with protein in the animal body.[90] In fact, animals injected with the said antigens become anaphylactic, are shocked by conjugate proteins which contain the corresponding azocomponent, and are desensitized by homologous azodyes.[91] Moreover, azodyes that are specifically precipitable were

---

Subsequently antibodies different in kind and of greater thermostability were detected in the patients' serum [Cooke et al; Loveless (151)].

[85] v. Grove; Coca (136).

[86] Rackemann; Brunner (151a), Fülleborn et al. (151b).

[87] The allergic symptoms, skin eruptions, asthma, fever, etc., although not always the same, are unrelated to the special pharmacological and toxic properties of the drugs; cases of increased susceptibility to the pharmacological effects belong to a different category.

[88] Persuasive evidence is the failure to find hypersensitivity to poison ivy among Eskimos (151c).

[89] v. Doerr (152).

[90] Such a mechanism had been early surmised [Wolff-Eisner (153)] on the ground that by chemical alteration proteins acquire special immunological properties.

[91] (154–157); v. (158) (absorption of antibodies in vivo); (159), (37).

Desensitization was not effected by uncoupled azocomponents like arsanilic acid.

shown to elicit anaphylactic shock in sensitized animals, and Schultz-Dale reactions [92] (p. 183).

The objection raised by Fierz and his collaborators against these experiments, implying that shock was actually produced by azoproteins that are formed in the body from the injected azodye has been disproved.[93]

The experiments cited showed indeed that simple substances can take part in anaphylactic manifestations; but inasmuch as sensitization was induced by conjugated proteins the results, although suggestive, still fell short of the goal. More direct information has accrued from experiments on sensitization with simple substances alone. Hypersensitiveness to such compounds was first experimentally produced in human beings with primrose extract [Nestler, Low (162)] and with an alkaloid from satinwood [Cash (163)], both materials being excitants of allergic skin affections, and later on with other compounds — arsphenamine, orthoform, etc. In animals definitely positive results were obtained by Bloch and Steiner (164), who were able to sensitize guinea pigs by treating the skin with ethereal extracts of Primula obconica, or with the crystallized active substance primulin ($C_{14}H_{18}O_3$ or $C_{14}H_{20}O_3$). Likewise chiefly in connection with a dermatological question, namely the etiology of eczema, several substances — p-phenylenediamine [94] and similar compounds, phenylhydrazine,[95] arsphenamin,[96] mesotan,[97] extracts of poison ivy,[98] nitrosodimethylaniline, chlorodinitrobenzene [99] — were subsequently tested in animals and proved to have sensitizing capacity.

A number of early and more recent papers on allergic effects of other substances are based on inconclusive evidence.[100] For inconsistencies encountered several factors seem to be responsible; differences in the susceptibility

---

[92] Landsteiner and van der Scheer (160).
[93] (161), (37).
[94] R. L. Mayer (165).
[95] Jadassohn (166).
[96] Frei, Mayer; Sulzberger and Simon (167).
[97] Silverberg (167a).
[98] Simon, Rackemann and Dienes (168). Poison ivy extracts and Japan lac, alike in allergic properties, contain several active components. A product — urushiol — separated by Majima, is a mixture of substituted catechols of the average formula $C_6H_3(OH)_2C_{15}H_{27}$, differing in the number and position of double bonds [v. (168a)].
[99] Landsteiner and Jacobs (169).
[100] Unconfirmed are reports concerning sensitization to pyramidon (170),

of individuals and breeds of animals are certainly of consequence. Experiments of Sulzberger and his colleagues may be quoted.[101] These workers succeeded in sensitizing guinea pigs in a German laboratory but were unable to reproduce these results in New York; in their opinion the diet was of influence.[102] In any event, it was shown later that the difficulties can be overcome, as with an improved technique guinea pigs could be sensitized anaphylactically with satisfactory regularity.[103]

For the sensitization with p-phenylenediamine and related bases [104] the "conjugation hypothesis" seems an adequate explanation. Phenylenediamine upon oxidation combines with proteins to form deeply colored products and because of this property is widely used as fur dye; it causes allergic conditions (asthma, skin eruptions) frequently in dyers, and in persons wearing furs. Here one may suppose that sensitization is brought about through the antigenic action of a protein conjugate formed in vivo, comparable to azoproteins. Strong evidence on the subject was provided by experiments [105] made with chloro- and nitro-substituted benzenes, chosen since it was known that 1,2,4 chlorodinitrobenzene is a frequent cause of allergy in factory workers handling this compound. The substance is excellently suitable for experimental work because on administration of solutions in organic solvents onto the intact skin, or intracutaneous injections of small fractions of a milligram, almost every guinea pig becomes hypersensitive and responds with an erythematous reaction when a drop of a solution is placed anywhere on the skin. Since there are theoretically more than 90 chloro- and nitro-substitution products of benzene, the following question presented itself. Are all the compounds sensitizers, and if not, is there a correlation between sensitizing capacity and any chemical characteristic? When the experiment was carried out there appeared to be sharp differences, sensitization being produced by only ten out of seventeen compounds tested. It turned out that the inactive ones were resistant to treatment with an organic base (aniline), and with one exception to treatment with

---

phenolphthalein (171), salicylic acid, aspirin, or the production of antibodies to thyroxine, adrenaline, quinine (172), (173), or melanine [cf. (174)]. On negative results with aspirin and barbital cf. (174a).

[101] (175), (167).

[102] v. (176).

[103] Landsteiner and Jacobs (177), Frei and Sulzberger (178).

[104] Nitti, Bovet and Depierre (179) investigated in guinea pigs the sensitizing capacity of various derivatives of p-phenylenediamine; o- and m-phenylenediamine were found to be inactive.

[105] Landsteiner and Jacobs (169).

sodium methylate and ethylate, while the ten sensitizing substances contained loosely bound Cl or NO_2 and formed substitution compounds with aniline by interacting with the amino group. In essence, these findings were corroborated in experiments on human beings.[106] Therefore, the conclusion is warranted that sensitization with 1,2,4 chlorodinitrobenzene, which is a typical incitant of contact dermatitis in man, and with related chemicals depends upon conjugation, probably with proteins, in the body. It could then be anticipated that compounds having similar chemical reactivity would likewise have sensitizing properties. In fact, following up this proposition, several classes of chemical compounds were found which render animals sensitive, such as acylchlorides, benzylchlorides (169) and acid anhydrides.[107] With all these substances formation of conjugates is easy to understand,[108] indeed beyond question in the case of rapidly reacting compounds (acylchlorides, acid anhydrides, diazomethane). In a number of other instances in which sensitization of animals was obtained — picric acid,[109] quinine [110] — a chemical interpretation is not immediately at hand.[111] It may be, however, that certain substances are converted in the animal into more reactive compounds; thus picric acid might combine after previous reduction of nitro to amino groups. Evidence on this point was contributed by Nitti and Bovet (183), who observed that treatment with p-sulfamido-chrysoidine sensitizes guinea pigs also to 1,2,4 triaminobenzene which is satisfactorily explained by formation of the amino compound in the animal through reductive cleavage.

The possibility that a substance may become effective through loose attachment to protein is perhaps indicated by Haxthausen's report on sensitization of human beings with mixtures of salts of heavy metals, and foreign serum (184).

From the evidence presented, the corollary would follow that the results are essentially the same whether the simple incitants or pro-

[106] Sulzberger et al. (179a).　　　　　　　　　　[107] Jacobs (180).
[108] By subcutaneous or intraperitoneal injection of aqueous solutions of diazonium salts Klopstock and Selter (157) [v. (156)] produced anaphylactic sensitization but not sensitivity to superficial skin application. These effects are apparently much the same as those resulting from immunization with azoproteins. Anaphylactic sensitization with a diazoamino compound would seem to be of similar significance [Fierz et al. (180a)].
[109] Landsteiner and DiSomma (181).
[110] Landsteiner and Chase (182).
[111] With these compounds a considerable degree of sensitivity was obtained by application to inflamed skin areas.

tein conjugates are employed for sensitization. Actually this expectation is not entirely fulfilled. Guinea pigs sensitized intracutaneously with an acyl chloride or picryl chloride reacted with skin inflammation to superficial application of the substances and with anaphylactic shock upon intravenous injection of acylated or picrylated protein;[112] intraperitoneal or cutaneous injection of picrylated stromata, on the other hand, produced anaphylaxis but at best very weak skin reactivity to contact (187). This shows that the two methods are not equivalent and withal that anaphylaxis and skin allergy to superficial application are distinct forms of hypersensitiveness, although the fact that both are produced by the same treatment strongly suggests related mechanisms.

The difference between the two types of sensitization mentioned appears also from other features. In contrast to anaphylaxis, contact dermatitis has not been transferred by serum; desensitization is not readily accomplished, and for producing sensitization treatment of the skin is far superior to other routes — which in general are of no avail — whereas in common immunization this is far less pronounced.[113]

As regards the discrepancies in the effects of simple chemicals and conjugates, and of sensitization by way of the skin or by other routes, the gap has been bridged in principle. Prompted by the experience that tubercle bacilli may increase or modify the response to antigenic stimuli (Lewis, Dienes, and others), it was found that picryl chloride or dinitrofluorobenzene, with killed tubercle bacilli as adjuvants, elicit skin sensitivity also when given intraperitoneally and much more effectively than the same substances injected alone by this route. Furthermore, the same technique led to a high degree of sensitivity when conjugates (picrylated stromata) were used instead of the uncombined chemical. In analogous manner, by extracutaneous injection, sensitivity could be induced against formaldehyde and iodine.[114] These experiments, in which skin hypersensitiveness of the contact dermatitis type was engendered by a full antigen, hardly per-

---

[112] (185), (186). The ease with which anaphylaxis is induced by a simple sensitizing compound depends upon its chemical properties; anaphylaxis was more regularly produced by acid chlorides or picryl chloride than by the less reactive chlorodinitrobenzene. With arsphenamine, sensitization as well as shock could be induced with the substance itself (177).

[113] Sulzberger; Landsteiner and Chase (188); v. (189).

[114] Landsteiner and Chase (190), (187) and unpublished preliminary experiments.

mit any conclusion other than that this form of allergy is intrinsically related to typical immunization processes.

Concerning the question of antibodies in drug allergy, substances belonging to this category — that is, specific substances formed in response to the incitants — must a priori be supposed to exist, if not in the circulation then in the sensitive tissues. Beyond that, the case is not without analogy. In the type of allergy to bacterial products,[115] of which the tuberculin reaction is the paradigm, passive transfer with serum could be achieved, if at all, only exceptionally,[116] and it is difficult to see what else than the action of antigens could be the cause of this allergic state;[117] Burnet inclines to the view that antibodies are formed but are at once taken up by tissues because of a special property of these antibodies (197). And even in common anaphylaxis, after a certain time, antibodies are not detectable in the serum although the animals and their isolated organs are still hypersensitive.

Hypotheses to explain the state of affairs in contact dermatitis were advanced to the effect that, without intervention of circulating antibodies, a specific change is set up in the skin tissues, brought about through spreading of the allergens along the skin.[118] Yet general sensitization could be attained by treatment of small skin islands that were completely isolated by a cut through the entire skin without severance of the lymph vessels on the surface of the skin muscle; [119] moreover, it seems impossible that sensitizing compounds which react so rapidly as diazomethane or certain acyl chlorides could travel very far without combining.

Transfer of sensitivity to simple compounds by intracutaneous injection of serum into the skin of human beings has been described,[120] but since some of the results are doubtful, and in general such attempts even with serum of highly sensitive individuals have been unsuccessful, the existence of antibodies in drug allergy is discounted in current texts. However, easily reproducible results have been obtained in animals. It having been established that cutaneous administration of acyl chlorides or picryl chloride produces at the same

[115] Its relation to other forms of hypersensitiveness is dealt with in Boyd (139); v. Bronfenbrenner (191).

[116] Zinsser and Müller (192); v. Seibert (193).

[117] Cf. the work on hypersensitiveness to streptococci and pneumococci by Derick and Swift (194), Julianelle (195), and the reactions of tuberculin type produced by Dienes with ovalbumin (196).

[118] (198), (199); v. (200).

[119] Landsteiner and Chase (188).   [120] (201–203), and others.

time cutaneous sensitivity and the anaphylactic state (p. 202), demonstrable by lethal shock following the injection of conjugates, or
by means of the Schultz-Dale method, it was shown that the serum of
the guinea pigs contains anaphylactic antibodies capable of passively
sensitizing normal animals [121] and, at times, precipitins. More than
this, it was found that sera from highly sensitive animals also sensitize the skin locally upon intracutaneous injection. Skin sites so
prepared react when the incitant chemicals (or conjugates) are injected into these areas or elsewhere, the reactions starting after a few
minutes in a manner altogether similar to the Prausnitz-Küstner
effect and fading within an hour or so.[122] Whether the antibodies in-
strumental in the skin reactions are identical with anaphylactic
antibodies is still undecided. Transfer sera, efficient in both respects,
were most readily secured with highly reactive compounds (acyl
chlorides and anhydrides) but effective sera were obtained also to
dinitrochlorobenzene and arsphenamine.[123]

The passively sensitized skin sites did not show the delayed erythematous reaction characteristic of contact dermatitis. On the other
hand, transfer of this form of skin sensitivity was achieved by injection of exudate cells [124] from animals sensitive to picryl chloride and
dinitrochlorobenzene, the exudates being produced by injection of
casein, or upon injection of killed tubercle bacilli (or tuberculin) into
guinea pigs made sensitive also to tuberculin. The mechanism of this
phenomenon, which possibly is the production of antibodies by cells
surviving in the recipient, has not yet been established; killed cells
appeared to be ineffective or very much less active.

It may be remarked in passing that since such simple substances as picryl
chloride or dinitrofluorobenzene incite anaphylactic antibodies one could
according to definition call them antigens. For the sake of a workable
terminology, even if a sharp definition can not be formulated, it would seem
preferable to reserve the term antigen, as customary, for substances of high
molecular weight, and to designate the simple compounds which are capable
of sensitizing — but probably only after combination with body constituents
— as allergens rather than antigens.

Several items concerning skin allergy require elucidation:[125] the
prominent role of the skin in sensitization and the fact that this tissue

---

[121] (186).   [122] (204), (204a).
[123] On similar reactions induced by common precipitin sera, see Chase (205).
[124] (204).
[125] Among the effects not yet reproduced in animals are sensitization by the

is often, as for instance in poison ivy allergy, the principal or only one that shows allergic effects; the chemical reactions by which antigenic complexes are formed in the body; the question whether tissues can become sensitive directly, without participation of free antibodies, for instance by distribution of allergens over the skin; the occurrence of localized sensitivity and the mechanism of "flare-up" reactions; [126] and yet other matters. Of particular interest is the experience that with certain substances only a few exposed individuals acquire hypersensitiveness. Evidence, however, is accumulating to indicate that no sharp line can be drawn on that score between drug allergy and other immunological phenomena.[127] Thus, while only a small proportion of persons who come in contact with primroses become allergic, treatment with the concentrated active principle, as shown by Bloch, eventually sensitizes almost every individual. From other sensitization experiments it appears that there exist all grades of sensitizing capacity. For instance, 1,2,4 chlorodinitrobenzene sensitized the great majority of persons when single drops of a strong solution were put once or twice on the skin,[128] while in our experiments with allylisothiocyanate (mustard oil) (209) only one out of seven volunteers became sensitive upon repeated application. Individual (heritable) differences, similar though less marked, in the response to antigens or allergens obtain also in animals, and Chase (210) succeeded in raising strains of guinea pigs of unequal susceptibility. In agreement with clinical knowledge, the differences in susceptibility did not run exactly parallel when two incitants were used (v. 147a).

The existence of species differences is evidenced by the experience that guinea pigs lend themselves well to the experiments, even though their sensitivity never reaches a degree comparable to that of highly allergic human beings, while with rabbits which otherwise are good antibody producers it was found possible but difficult to demonstrate sensitization to chemicals — perhaps owing to low skin reactivity.

When comparison is made with ordinary immunization it would seem that in the production of antibodies differences among indi-

---

oral route which certainly occurs in human beings, and sensitivity confined to limited areas of the skin, seen in the "fixed eruptions" of man [cf. Haxthausen (206)].

[126] A characteristic case is the reaction of old mustard gas burns upon exposure to slightly contaminated air.

[127] v. Doerr (207).

[128] Wedroff and Dolgoff; Landsteiner, Rostenberg and Sulzberger (208).

viduals are less pronounced than in drug hypersensitivity, particularly in certain forms of human allergy. Indeed, there are substances, so far ineffective in animal experiments (aspirin, phenolphthalein, pyramidon, etc.), to which relatively few persons become hypersensitive but some to a very high degree. One may hazard the guess that these are instances of a special sort, perhaps conditioned by unwonted modifications of the substances peculiar to certain individuals. On the other hand, considerable variation exists also in the formation of antibodies. Striking observations were made by Prigge [(211) (v. 211a)] who found that the quantity of antigen needed to immunize guinea pigs against diphtheria toxin showed enormous variations, and that there were differences between strains that had been inbred over a period of several years.[129]

Investigations on specificity [130] and cross reactions by testing the sensitivity of idiosyncratic human beings have not been carried out on an extensive scale, yet they serve to show that the specificity of drug allergy and of serum reactions with simple substances are of the same order. Marked specificity was seen in aspirin hypersensitiveness; the patients were unaffected by salicylic acid, methylsalicyate, or benzoic acid.[131] In experiments on hypersensitiveness to paraphenylenediamine and related substances, R. L. Mayer [132] found overlapping reactions with substances of quinoid structure. Dawson and Garbade (214) studied a case of hypersensitiveness to quinine and obtained skin reactions with a number of laevo-rotatory substances of the quinine group, but not with their dextro-rotatory isomers. Propyl-, isopropyl-, isobutyl- and isoamylhydrocupreine reacted positively, the higher alkyl derivatives negatively. The compound quitenine containing a carboxyl group was inactive, while a few esters of quitenine gave positive reactions. In several patients allergic to quinine considerable differences were found in the cross reactions with related substances.[133] The interesting question of specificity in quinine hypersensitiveness is open to study in animals. Guinea pigs sensitized to quinine gave strong skin reactions also to cinchonidine and optochin (ethylhydrocupreine), whereas cinchonine and quinidine were practically ineffective (unpublished experiments). The results of Nathan and Stern (216) in a case of idiosyncrasy to resorcinol are

---

[129] For inheritable differences in resistance to infection v. (138).
[130] Specificity tests with sensitized animals are described in (167), (169).
[131] Cooke (212).
[132] (165); v. (213).
[133] (215), (214).

similar to the observations on azoproteins containing substituted benzene rings. In their patient, in spite of strong hypersensitivity to resorcinol and resorcinol-monomethylether, the isomeric o- and p-compounds pyrocatechol and hydroquinone, and other phenols were without any effect. A patient of Urbach (217) although reacting most strongly with resorcinol responded weakly to both the other dihydroxybenzenes. Nickel and cobalt were found to be practically equivalent in skin tests [Haxthausen (184)].

The specificity of allergy to arsenicals is discussed by Frei (218), that to orthoform by Schwarzschild (219); specificity tests in individuals sensitive to 1,2,4 chlorodinitrobenzene were made by Haxthausen (206).

The observation by Bloch that individuals allergic to iodoform were sensitive to a number of compounds containing a methyl group has no analogue among reactions of conjugated proteins. Sensitization to compounds formed by decomposition of iodoform in the body might be the explanation. For clearing up this point a study on iodoform allergy in animals would be desirable.

BIBLIOGRAPHY

(1) Sleeswijk: Erg. Immtsf. *1*, 395, 1914. — (2) Traube: ZI *9*, 246, 1911. — (3) Morgenroth: BK 1917, p. 55; v. Festschr. P. Ehrlich, Fischer, Jena, 1914, p. 542. — (4) Obermayer et al: WK 1906, p. 327. — (5) Landsteiner et al: ZI *26*, 258, 1917. — (6) Medveczky et al: ZI *72*, 256, 1931. — (7) Kurtz et al: Pr *30*, 138, 1932, *31*, 265, 1933. — (8) Boyd et al: JBC *104*, 329, 1934 (B). — (9) Heumann: Die Anilinfarben, part 3, p. 1064, Vieweg, Braunschweig, 1900. — (10) Pauly: ZPC *42*, 508, 1904, *94*, 284, 1915. — (11) Landsteiner et al: ZI *26*, 293, 1917. — (12) Landsteiner et al: BZ *86*, 343, 1918. — (13) Landsteiner: K 1927, p. 103; Nw 1930, p. 653. — (14) Haurowitz: Fortschritte der Allergielehre, Karger, Basel, 1939, p. 19. — (15) Heidelberger et al: JEM *59*, 519, 1934, *58*, 137, 1933. — (16) Hooker et al: JI *24*, 141, 1933. — (17) Adant: AI *6*, 29, 1930; SB *103*, 541, 1930. — (18) Bruynoghe et al: SB *103*, 543, 1930. — (19) Clutton et al: BJ *32*, 1111, 1938. — (20) Doerr et al: ZI *81*, 132, 1933. — (21) Landsteiner: BZ *93*, 106, 1919. — (21a) Pressman et al: JI *44*, 101, 1942. — (22) Erlenmeyer et al: BZ *262*, 196, 1933. — (23) Heidelberger et al: JI *42*, 181, 1941. — (24) Landsteiner et al: ZH *113*, 1, 1931. — (25) Klopstock et al: ZI *55*, 118, 450, 1928. — (26) Landsteiner et al: JEM *55*, 781, 1932. — (27) Haurowitz: JI *43*, 331, 1942. — (28) Landsteiner et al: JEM *63*, 325, 1936. — (29) Haurowitz: ZPC *245*, 23, 1936. — (30) Kleczkowski: Br *21*, 98, 1940. — (31) Hooker et al: JI *25*, 61, 1933, *24*, 141, 1933. — (32) Reiner et al: JBC *139*, 641, 1941. — (33) Kapeller-Adler et al: BZ *285*, 55, 1936. — (34) Busch et al: J. prakt. Chem. *140*, 117, 1934. — (35) Boyd et al: JBC *110*, 457, 1935 (B). — (36) Eagle et al: JBC *114*, 193, 1936. — (37) Fierz-David et al: H *20*, 1059, 1937, *22*, 3, 1938. — (37a) Haurowitz: KZ *74*, 208, 1936. — (38) Hooker et al: JI *23*, 465, 1932. — (39) Marrack et al: Br *12*, 182, 1931. — (40) Goebel et al: JEM *50*, 521, 1929. — (41) Heidelberger et al: Pr *26*, 482, 1929. — (42) Landsteiner: ZI *62*, 178, 1929. — (43) Schwarz: ZI *1*, 77, 1908. — (44) Mut-

saars: AP *62*, 81, 197, 1939. — **(45)** Heidelberger et al: JAC *63*, 498, 1941. — **(46)** Obo: JB *33*, 241, 1941. — **(47)** Clutton et al: BJ *31*, 764, 1937. — **(48)** Harington: JCS 1940, p. 119. — **(49)** Mutsaars et al: SB *123*, 144, 1936. — **(50)** Clutton et al: BJ *32*, 1111, 1938; IC, p. 822. — **(51)** Creech et al: AJC *30*, 555, 1937. — **(52)** Erlenmeyer et al: H *16*, 733, 1933. — **(53)** Jacobs: JGP *20*, 353, 1937. — **(54)** Landsteiner et al: JEM *59*, 751, 1934. — **(55)** Creech et al: AJC *35*, 203, 1939; JAC *63*, 1670, 1941; CaR *3*, 133, 1943. — **(56)** Adant: AI *6*, 29, 1930. — **(57)** Landsteiner et al: JEM *45*, 1045, 1927. — **(58)** Berger et al: K 1935, p. 536. — **(59)** Hooker et al: JI *38*, 479, 1940. — **(60)** Landsteiner et al: JI *33*, 265, 1937. — **(61)** Berger et al: AEP *177*, 116, 1934; SMW 1936, p. 1309. — **(62)** Mulinos et al: Pr *35*, 305, 1936, *37*, 583, 1937. — **(63)** Harte: JI *34*, 433, 1938. — **(64)** Berger: SMW 1941, p. 1376. — **(65)** Landsteiner et al: JEM *59*, 751, 1934 (B). — **(66)** Pillemer et al: Pr *39*, 380, 1938; JEM *70*, 387, 1939. — **(67)** Fischer et al: ZPC *26*, 60, 1898; BC *27*, 2031, 2985, 1894. — **(68)** Ingersoll et al: JAC *44*, 2930, 1922, *47*, 1168, 1925. — **(69)** Landsteiner et al: JEM *48*, 315, 1928. — **(70)** Landsteiner et al: JEM *50*, 407, 1929. — **(71)** Landsteiner et al: Pr *29*, 1261, 1932. — **(72)** Avery et al: JEM *50*, 533, 551, 1929. — **(73)** Goebel et al: JEM *60*, 85, 1934. — **(74)** Avery et al: JEM *55*, 769, 1932. — **(74a)** Gaunt et al: BJ *33*, 908, 1939. — **(75)** Humphrey et al: BJ *33*, 1826, 1939. — **(75a)** Goebel et al: JEM *60*, 599, 1934. — **(75b)** Marrack: EE *7*, 281, 1938. — **(76)** Goebel et al: JEM *64*, 29, 1936; JBa *31*, 66, 1936; JBC *114*, XL, 1936. — **(77)** Goebel et al: JEM *66*, 191, 1937. — **(78)** Woolf et al: JC *55*, 156, 1936; RS *B*, *130*, 60, 1941. — **(79)** Goebel: JEM *68*, 469, 1938, *69*, 353, 1939. — **(80)** Goebel: JEM *72*, 33, 1940. — **(81)** Avery et al: JEM *54*, 431, 437, 1931. — **(82)** Landsteiner et al: JEM *59*, 769, 1934, *69*, 705, 1939. — **(83)** Van der Scheer et al: JI *29*, 371, 1935. — **(84)** Lettré et al: ZPC *266*, 31, 37, 1940, *267*, 108, 1940. — **(85)** Dujarric de la Rivière et al: RI *3*, 405, 1937. — **(86)** Cohn: HL 1938/39, p. 124. — **(87)** Ivanovics et al: ZI *93*, 119, 1938. — **(88)** Berger et al: Giorn. di Batter. e Immunol. *13*, 412, 1934. — **(89)** Went et al: AEP *193*, 312, 1939 (B). — **(90)** Butler et al: BJ *34*, 838, 1940. — **(91)** Marrack: ARB *11*, 629, 1942. — **(92)** Mingoia et al: Estr. Boll. Ist. sieroterap. Milanese *19*, 101, 1940. — **(93)** Wedum: JID *70*, 173, 1942 (B); Pr *45*, 218, 1940. — **(94)** Sheldon et al: JA *13*, 18, 1941. — **(94a)** Cohen et al: JA *14*, 195, 1943. — **(95)** Went et al: AEP *193*, 609, 1939. — **(96)** Mooser et al: Schweiz. Ztschr. *4*, 375, 1941. — **(97)** Berger: BZ *267*, 143, 1933. — **(98)** Halban et al: MM 1902, p. 473. — **(99)** Landsteiner: BZ *104*, 280, 1920. — **(100)** Landsteiner et al: JEM *54*, 295, 1931. — **(101)** Haurowitz et al: ZPC *214*, 111, 1933. — **(102)** Balls et al: BC *64*, 294, 1931 (B); Habilitationsschr., Prag, 1930. — **(103)** Marrack et al: N *128*, 1077, 1931; Br *13*, 394, 1932. — **(104)** Erlenmeyer et al: BZ *266*, 355, 1933; H *17*, 308, 1934. — **(105)** Klopstock et al: ZI *57*, 174, 1928. — **(106)** Landsteiner et al: Pr *29*, 747, 1932; JEM *56*, 399, 1932, *57*, 633, 1933. — **(106a)** Pressman et al: PNA *28*, 77, 1942. — **(107)** Berger et al: BZ *264*, 113, 1933. — **(107a)** Pressman et al: JAC *64*, 3015, 1942. — **(107b)** Pressman et al: JAC *65*, 728, 1943. — **(108)** Bukantz et al: Pr *47*, 94, 1941. — **(109)** Waldschmidt-Leitz et al: BC *64*, 45, 1931. — **(110)** Abderhalden et al: F *12*, 329, 1931. — **(111)** Landsteiner et al: Pr *24*, 692, 1927. — **(112)** Castelli: Arch. Schiffs. Tropenhyg. *16*, 605, 1912. — **(113)** Waldschmidt-Leitz et al: BC *61*, 299, 1928, *62*, 2217, 1929. — **(114)** Smyth et al: JAC *53*, 527, 4242, 1931. — **(115)** Wormall: JEM *51*, 295, 1930. — **(116)** Jacobs: JI *23*, 361, 375, 1932. — **(117)** Snapper: NT *79*, 2007, 1935; WK 1935, p. 1199. — **(118)** Snapper et al: Br *17*, 361, 1936. — **(119)** Marrack: CAA. — **(120)** Adant et al: SB *117*, 232, 1934. — **(121)** Clutton et al: BJ *32*, 1119, 1938. — **(122)** Mutsaars: SB *129*, 510, 1938. — **(123)** Hopkins et al: BJ *28*, 228, 1934. — **(124)** Horsfall: JI *27*,

553, 1934. — **(125)** Michaelis: DM 1904, p. 1240. — **(126)** Landsteiner et al: Pr *28*, 983, 1931. — **(127)** Landsteiner et al: Pr *30*, 1413, 1933. — **(128)** Holiday: RS *B*, *127*, 40, 1939. — **(129)** Landsteiner: JEM *75*, 269, 1942. — **(130)** Heidelberger et al: JEM *57*, 373, 1933. — **(131)** Morgan: BJ *30*, 909, 1936. — **(132)** Schaefer: AP *64*, 301, 1940. — **(133)** Dragstedt: PH *21*, 563, 1941. — **(134)** Rocha e Silva: ArP *33*, 387, 1942. — **(135)** Tuft: Clinical Allergy, Saunders, Philadelphia, 1937. — **(136)** Coca et al: Asthma and Hay Fever, Thomas, Springfield, 1931. — **(137)** Rackemann: Clinical Allergy, etc., MacMillan, New York, 1931. — **(138)** Topley et al: PB. — **(139)** Boyd: FI. — **(140)** Sulzberger: Dermatologic Allergy, Thomas, Springfield, 1940. — **(141)** Kallos et al: EH *19*, 178, 1937; Fortschritte der Allergielehre, Karger, New York, 1939, p. 5. — **(142)** Doerr: HP *13*, 650, 1929. — **(143)** Landsteiner: NEM *215*, 1199, 1936. — **(144)** Hooker: JI *9*, 7, 1924. — **(145)** de Gara et al: JI *44*, 259, 1942 (B). — **(146)** Karelitz et al: JI *44*, 271, 1942. — **(147)** Cooke et al: JI *1*, 201, 1916, *9*, 521, 1924. — **(147a)** Simon: AM *12*, 178, 1938 (B). — **(148)** Prausnitz et al: CB *86*, 160, 1921. — **(149)** De Besche: AJM *166*, 265, 1923. — **(150)** Coca et al: JI *10*, 445, 1925. — **(151)** Loveless: JI *38*, 25, 1940. — **(151a)** Brunner: JI *15*, 83, 1928. — **(151b)** Fülleborn et al: K 1929, p. 1988. — **(151c)** Heinbecker: JI *15*, 365, 1928. — **(152)** Doerr: HPM *1*, 808, 819, 1929. — **(153)** Wolff-Eisner: Dermat. Cbl. *10*, 164, 1907. — **(154)** Landsteiner: Kgl. Acad. Wet. Amsterdam *31*, 54, 1922, JEM *39*, 631, 1924. — **(155)** Meyer, K., et al: BZ *146*, 217, 1924. — **(156)** Landsteiner et al: JEM *52*, 347, 1930. — **(157)** Klopstock et al: ZI *63*, 463, 1929. — **(158)** Berger et al: BZ *255*, 434, 1932. — **(159)** Jadassohn et al: ADS *170*, 33, 1934. — **(160)** Landsteiner et al: JEM *57*, 633, 1933. — **(161)** Landsteiner et al: JEM *67*, 79, 1938. — **(162)** Low: Brit. J. Dermat. *36*, 292, 1924. — **(163)** Cash: Brit. Med. J. 1911, pt. 2, p. 784. — **(164)** Bloch et al: ADS *152*, 283, 1926, *162*, 349, 1930. — **(165)** Mayer, R. L.: ADS *163*, 223, 1931 (B). — **(166)** Jadassohn: K 1930, p. 551. — **(167)** Sulzberger et al: JA *6*, 39, 1934 (B). — **(167a)** Silverberg: AS *21*, 166, 1930. — **(168)** Simon et al: JI *27*, 113, 1934. — **(168a)** Mason et al: JAC *64*, 3058, 1942. — **(169)** Landsteiner et al: JEM *61*, 643, 1935. — **(170)** Golden et al: JI *36*, 277, 1939. — **(171)** Rosenthal: JI *34*, 251, 1938. — **(172)** Bauer et al: WK 1935, p. 1533; 1936, p. 1540. — **(173)** Hirose: Sei-I-Kwai Med. J. *53*, 31, 1934. — **(174)** Chorine: BP *35*, 491, 1937. — **(174a)** Frei: JD *4*, 111, 1941. — **(175)** Sulzberger et al: AS *24*, 537, 1931. — **(176)** Simon: JI *30*, 275, 1936. — **(177)** Landsteiner et al: JEM *64*, 717, 1936. — **(178)** Frei et al: JD *1*, 191, 1938. — **(179)** Nitti et al: RI *3*, 376, 1937. — **(179a)** Sulzberger et al: JD *1*, 45, 1938, *2*, 25, 1939. — **(180)** Jacobs et al: Pr. *43*, 641, 74, 1940. — **(180a)** Fierz et al: JEM *65*, 339, 1937. — **(181)** Landsteiner et al: JEM *72*, 361, 1940. — **(182)** Landsteiner et al: Pr *46*, 223, 1941. — **(183)** Nitti et al: BCB *19*, 837, 1937; RI *2*, 460, 1936. — **(184)** Haxthausen: ADS *170*, 378, 1934, *174*, 17, 1936; ADV *21*, 158, 1940; Congr. Dermat. Int. *9*, 1, 201, 1935. — **(185)** Landsteiner et al: JEM *64*, 625, 1936. — **(186)** Landsteiner et al: JEM *66*, 337, 1937. — **(187)** Landsteiner et al: JEM *73*, 431, 1941. — **(188)** Landsteiner et al: JEM *69*, 767, 1939. — **(189)** Haxthausen: ADV *20*, 396, 1939. — **(190)** Landsteiner et al: JEM *71*, 237, 1940. — **(191)** Bronfenbrenner: JLC *26*, 102, 1940. — **(192)** Zinsser et al: JEM *41*, 159, 1925. — **(193)** Seibert: JID *51*, 383, 1932. — **(194)** Derick et al: JEM *49*, 615, 883, 1929. — **(195)** Julianelle: JEM *51*, 643, 1930. — **(196)** Dienes: JI *17*, 531, 1929 (B). — **(197)** Burnet: Monographs Hall Inst. No. 1, MacMillan, Melbourne, 1941. — **(198)** Straus et al: JI *33*, 215, 1937. — **(199)** Schreus: K 1938, p. 1171. — **(200)** Anke: DW *109*, 1263, 1939. — **(201)** Kern: JA *10*, 164, 1939. — **(202)** Ensbruner: ADS *168*, 370, 1933. — **(203)** Biberstein: ZI *48*, 297, 1926. — **(204)** Landsteiner et al: Pr *49*, 688, 1942. — **(204a)** Chase et al: in preparation.— **(205)** Chase: Pr *52*, 238, 1943.— **(206)** Haxthausen: ADV *20*, 257, 1939.— **(207)** Doerr: SMW 1921, p. 937.— **(208)** Landsteiner et al: JD *2*, 25, 1939.— **(209)** Landsteiner et al: JEM *68*, 505, 1938.— **(210)** Chase: JEM *73*, 711, 1941.— **(211)** Prigge: ZH *119*, 186, 1937.— **(211a)** Stewart et al: Canad. Publ. Health J. *33*, 588, 1942.— **(212)** Cooke: JAM *73*, 759, 1919.— **(213)** Perutz: K 1932, p. 240.— **(214)** Dawson et al: JAM *94*, 704, 1930; v. *97*, 850, 930, 1931; J. Pharmacol. *39*, 417, 1930.— **(215)** Dawson et al: JI *24*, 173, 1933. **(216)** Nathan et al: DW *91*, 1471, 1930.— **(217)** Urbach: ADS *148*, 146, 1924. — **(218)** Frei: JD *5*, 29, 1942.— **(219)** Schwarzschild: ADS *156*, 432, 1928.— **(220)** Hooker et al: JI *45*, 127, 1942.

VI

CHEMICAL INVESTIGATIONS ON SPECIFIC NON-PROTEIN CELL SUBSTANCES

POLYSACCHARIDES.[1] — Even before accurate chemical data became available, adequate proof of the existence of specific non-protein substances in bacteria was supplied by Zinsser's residue antigens and some observations on acid-fast bacilli.[1a] As has been indicated already, the constituents of bacteria responsible for their specific properties are proteins, carbohydrates and complex substances containing polysaccharides and lipids. Polysaccharides, precipitable by immune sera, but not stimulating antibodies in rabbits, were discovered in the three first classified types of pneumococci by Heidelberger and Avery, and Goebel (6). These cocci bear capsules which in large part consist of colloidal carbohydrates. Convincing indeed was the body of evidence produced by the authors for the serological significance of the polysaccharides. Their activity increases with the degree of purity and was found to be the same with several methods of isolation; carefully purified preparations are free from protein, and the substances are resistant to pepsin or trypsin. The sensitivity of the reactions with antisera is of the same order as that of the specific precipitation of proteins, and the polysaccharides are recoverable from the specific precipitates. Finally, the carbohydrates turned out to be as sharply differentiated by their chemical composition and properties as by the serological reactions.

[1] Reviews are given by Heidelberger (1), Morgan (2), Mikulaszek (3), Marrack (4), Velluz (4a), Rudy (4b), Kimmig (5), Levinthal (5a).

[1a] See pp. 75, 76; 100, 101.

The polysaccharides precipitate with immune sera in dilutions up to several millions, elicit anaphylactic shock in passively sensitized guinea pigs,[2] and give immediate skin reactions in convalescent pneumonia patients. Owing to their capacity for combining with antibodies, the specific substances may, according to some experiments, counteract the defensive mechanism of the infected animal and in this respect exert an effect comparable to that of so-called aggressins.[3]

For the tests, since generally immunization with isolated polysaccharides is not a suitable method, the necessary immune sera are prepared by injection of whole bacteria.

The chemical diversity of the specific substances was clearly demonstrated in the first studies which included the representative types I, II, and III (Table 32).

From pneumococci of type II the polysaccharide was obtained as a nitrogen-free substance, weakly acidic and containing a large proportion of glucose;[4] in the specific substance of type I, having 5% total nitrogen and 2.5% amino nitrogen,[5] the presence of an amino sugar and a uronic acid (galacturonic acid) was determined; and the nitrogen-free polysaccharide of type III was found to be composed of glucuronic acid units. Most completely cleared up by subsequent investigations is the constitution of the polysaccharide III.[6] It is built up from an aldobionic acid (glucose-4-β-glucuronide or cellobiuronic acid, $C_{11}H_{19}O_{10}COOH$) which can be prepared from the parent substance by hydrolysis.[7] Upon complete methylation, catalytic reduction and hydrolysis of the acid, 2-3-6-trimethyl glucose and 2-4-dimethyl glucose were obtained in equimolecular amounts; 2-4-dimethyl glucose was identified by comparison of the dimethyl-β-methyl glucoside with the synthetic compound.[8] From these and

[2] Tomcsik (7), Avery and Tillet (8), Morgan (9).

[3] v. (10); Pettersson (11).

[4] Uronic acid was later demonstrated by Goebel (private communication).

[5] The serological activity is destroyed by nitrous acid, or treatment with diazonium compounds (11a).

[6] This carbohydrate is obtainable in relatively large quantity, the yield amounting to about 1 gr. per 5 liters glucose broth culture. On the method of preparation v. Goebel (12).

Superseding previous lower estimates (13), a minimum molecular weight of 62000 is calculated by Heidelberger et al. (14) from the combining ratio with antibodies; a still higher value obtained with the diffusion method of Northrop and Anson has been reported by Babers and Goebel (15); v. (15a) (polysaccharide I).

[7] Heidelberger and Goebel (16).

[8] Hotchkiss and Goebel (17), Reeves and Goebel (18), Adams et al. (251).

TABLE 32

[after Heidelberger and Kendall (23)]

Polysaccharide of pneumococci	$(\alpha)_D$	Acid Equivalent	Total N %	Amino N %	Acetyl %	Sugar after hydrolysis calculated as glucose %	Hydrolysis products
Type I	$+300°$	310	5.0	2.5	0	28	(Galacturonic acid.) (Amino sugar derivative.)
Type II	$+74°$	1250	0.0			70	Glucose.
Type III	$-33°$	340	0.0			75	Aldobionic acid, glucose.[+]
Type IV	$+30°$	1550	5.5	0.1	5.8	71	(Amino sugar derivative), acetic acid.
Type VIII[++]	$+125°$	750	0.2			76	Aldobionic acid, glucose.[+++]
Species specific substance	$+42°$	1050	6.1	0.9	3.7	36	(Amino sugar derivative), phosphoric acid, acetic acid.
Inactive substance	$+10°$	4540	5.9	0.0	5.6	55	(Glucosamine), acetic acid.

The substances in parenthesis have not been definitely identified.
[+] Probably formed by hydrolysis of aldobionic acid. Ratio of glucose to glucuronic acid 1:1.
[++] R. Brown; Goebel (121).
[+++] Ratio of glucose to glucuronic acid about 7:2.

previous results it could be concluded that in the aldobionic acid units glucose is linked to the third carbon atom of the glucuronic acid which again is connected with the fourth carbon atom of the second adjacent glucose molecule.

Intermediate hydrolysis products were found still to be precipitable by horse, not by rabbit antisera (19), (19a).

A survey of the specific substances of types I to XXXIII and some subtypes has been presented by Brown (20). The data gathered comprise the analytical values for nitrogen, amino nitrogen, phosphorus, acetyl, figures for optical rotation and viscosity, tests for uronic acid and amino sugar, and the behavior towards precipitants (tannic acid, phosphotungstic acid, ammonium sulfate, barium chloride, salts of heavy metals). Some of the polysaccharides contained a considerable amount of phosphorus; only that of type I gave a significant figure for amino nitrogen (before hydrolysis); nine preparations showed positive reactions for amino sugar; twenty-one gave the test for uronic acids.

Following the demonstration by Lancefield (21) of a carbohydrate peculiar to streptococci, both the R and the S forms of different pneumococcus types were shown to have a species specific polysaccharide (called the somatic C polysaccharide).[9] Characteristic is the content of phosphorus, in organic combination, of about 4.5% which certainly accounts for small amounts of phosphorus found in some type-specific carbohydrate preparations; among the hydrolysis products amino sugar (more than twenty per cent) was identified (v. p. 225). The preparations may in addition contain another carbohydrate which is derived from the "peptone" of the culture media.[10]

Once the serological significance of protein-free bacterial carbohydrates was established, attention was turned to devising methods of preparation which would yield the polysaccharides unaltered as far as possible. Thus, the acetylated polysaccharide which Avery and Goebel (27) isolated, when degradation by alkali was prevented, was distinguished from the original deacetylated product by its antigenicity (p. 108) and the behaviour towards immune sera. Both substances are precipitated to the same titre, but exhaustion with the acetylated product removes the precipitins completely whereas after

---

[9] Tillett, Goebel and Avery (22), Heidelberger and Kendall (23), Goebel, Shedlovsky, Lavin and Adams (24); v. Wadsworth and Brown (25), Sevag (30).
[10] Goebel (26).

absorption with the deacetylated polysaccharide there still remain precipitins for the acetylated substance.[11] Among other procedures employed for preparing the carbohydrates in or near their native state [12] are adsorption to calcium phosphate [13] and ultra-filtration [14] (prior to the usual fractional precipitation with alcohol). Heidelberger and his colleagues (34), using methods in which heat, alkali and mineral acid were avoided, isolated the polysaccharides as neutral sodium salts characterized by high viscosity of the solutions. Heating diminished the viscosity of these preparations owing probably to depolymerization of the "long, thread-like chains of the native polysaccharide" and rendered the polysaccharide type III dialyzable, to some degree, in the presence of much salt.

The discovery of specific carbohydrates in pneumococci could hardly fail to stimulate the search for similar substances in other bacteria, and many chemical data have been assembled. Heidelberger, Goebel and Avery extended their work to Friedländer's bacillus,[15] a bacterium encapsulated like the pneumococcus. From three types of this bacterium as many polysaccharides were obtained of which those from type B and type C have very similar chemical properties. The three polysaccharides are decomposed by acid hydrolysis into glucose and sugar acids and give a positive colour test with naphthoresorcinol. The substance of type A is probably built up of units consisting of one molecule each of glucose, an aldobionic acid and a second, unidentified sugar acid. The aldobionic acid contains glucose and glucuronic acid and is isomeric with the acid of Pneumococcus III, probably because of a different position of the linkage between the two components. Table 33 summarizes the chemical results on the polysaccharides of Bact. pneumoniae.[16]

In distinction to the polysaccharides considered thus far which reside in the bacterial capsules, the specific capsular substance of

---

[11] Cases of lability to alkali were observed among the antigens of Salmonella bacilli [Landsteiner and Levine (28)]. In the light of the above findings it would be desirable to resume the investigation. In experiments with extracts of cholera vibrios which contained protein, it was found that the extracts, like proteins, lost their antigenic power when treated with alkali, while the reactions in vitro were unaffected.

[12] Enders et al. (29), Sevag (30) (removal of proteins by shaking with chloroform and amyl alcohol); Chow (31); v. Wong et al. (31a) (antigenic preparations).

[13] Felton (32).

[14] Brown, Wadsworth (33), (20).   [15] (35–37).

[16] On the cross reactions of capsulated bacteria v. Julianelle (38).

anthrax bacilli is free of carbohydrate (p. 226) but the bacilli contain a somatic polysaccharide hapten consisting largely of glucosamine and galactose in equimolecular amounts.[17]

From non-capsulated gram-negative bacteria carbohydrates have been obtained which are associated with the O antigens. For the

TABLE 33
[after Goebel (36)]

Polysac- charide of Bac. Friedländer	$(a)_D$	Acid equiva- lent	C	H	N	Sugar after hydrolysis calculated as glucose %	Hydrolysis products
Type A.....	−100°	430	43.95	6.0	0.0	65	Glucose, aldobionic acid
Type B.....	+100°	680	44.6	6.1	0.0	70	Glucose, aldobionic acid
Type C.....	+100°	680			0.0	70	Glucose, aldobionic acid

Salmonella group data on the chemical constitution of the specific substances in the various types are still lacking;[18] at any rate, correspondence was demonstrated between the precipitin reactions of the polysaccharides and the O types as determined by agglutination of the bacteria,[19] and differences in the somatic carbohydrates were found between the S and R forms.[20] Not seldom the O polysaccharides possess the Forssman property.[21] Freeman [22] prepared from B. typhosus an O-specific polysaccharide which, when examined by fractional precipitation and in the ultracentrifuge, proved to be practically homogeneous. After treatment with acetanhydride and saponification of the resulting (chloroform and acetone soluble) acetyl derivative, the substance was recovered in unaltered state. Upon hydrolysis there were isolated — as the respective hydrazones — 40% glucose, 21% mannose, and 17% galactose. The preparation contained 0.2–0.4% nitrogen, 0.4–0.9% P and about 3% acetyl. Examination in

[17] Ivanovics (39). This carbohydrate is chiefly responsible for the thermo-precipitin reaction of Ascoli in which aqueous extracts of infected organs, freed of protein by heat coagulation, are used (40).

On precipitin reactions of split products see (19a).

[18] On differences in the resistance to acid and alkali v. (41), (41a), (41b).

[19] White (42), Furth and Landsteiner (41), Morgan and Beckwith (43).

[20] (41), (44). The presence of small amounts of R polysaccharides in S forms of vibrios is discussed by White (45). Cf. on the polysaccharides of mucoid strains (41b), of ρ and R variants (45a), (252).

[21] v. (28).   [22] (46); v. (47), (48).

the ultracentrifuge indicated a minimum molecular weight of the order of 10000. It did not precipitate with Vi-antiserum.

The polysaccharide of smooth strains of B. dysenteriae Shiga was isolated by Morgan [23] in apparently pure condition, as evidenced by recovery of the substance, constant in its properties, upon employing various purification and fractionation methods and regeneration from acetyl and benzoyl derivatives. From the analytical results it is assumed that the polysaccharide contains units consisting of one acetylated amino sugar and four hexose molecules. Among the hydrolytic products d-galactose and l-rhamnose were identified. An acid equivalent of 9000 was found, but uronic acids were not detected; phosphorus in organic combination was demonstrated. Iodometric titration indicated the presence of one free aldehyde group and a minimal molecular weight of about 5000. Interesting results, to be discussed below, were obtained in investigations on the full bacterial antigen of which the polysaccharide is a constituent.

Cholera vibrios and some related organisms were classified by Linton [24] into three types (v. p. 35), each sort of vibrios having one of three different polysaccharides,[25] containing: (1) aldobionic acid and galactose, (2) aldobionic acid and arabinose, (3) glucose, a nitrogenous compound, and phosphorus. The aldobionic acid was shown to be composed of galactose and glucuronic acid.

A number of serologically different polysaccharide fractions, and others which are inactive, have been separated from tubercle and other acid-fast bacilli.[26] To be noted are the low values obtained in molecular weight determinations on some of the specifically precipitable polysaccharides (55b). The carbohydrates contain d-mannose and d-arabinose in varying proportions. These sugars appear to be the chief constituents; other sugars, inosite and unidentified acids have been recorded as cleavage products of the polysaccharides and phosphatides, still others in the immunologically inert waxes.

Noticeable on account of the varied and sometimes peculiar chemical composition are polysaccharides — not examined serologically — which were isolated from moulds.[27] For instance, one of the sub-

[23] (49); v. (3).
[24] (50) (review); v. White (51).
[25] A characteristic carbohydrate containing hapten was found by White (52) in the gelatinous intercellular substance of rugose variants; for R and ρ forms v. (52a).
[26] Heidelberger et al. (53); (54) (bovine strain); (55) (avian strain); Seibert et al. (55a, 56–60).   [27] Raystrick (61).

stances, that present in cultures of Penicillium luteum, is composed of glucose and malonic acid; another gave on hydrolysis a mixture of d-glucose, d-galactose and either l-altrose or d-idose.

A carbohydrate produced by Trichophyton, which contains glucosamine, may be responsible for the cutaneous trichophytin reactions.[28]

Immunologically reactive carbohydrates have now been found in a great number of bacteria. In many instances little is known about the chemical composition of the preparations aside from the demonstration of reducing sugars after hydrolysis and other qualitative reactions. A partial list of studies on bacterial and some other specific polysaccharides not already referred to is given here: streptococci (63–67), staphylococcus (68–69), gonococcus (68, 70, 71), meningococcus (68, 72, 73), leuconostoc (74), Salmonella bacilli (75–77, 45), Bact. proteus (78–81), Bact. lactis aerogenes (82, 38), Bact. rhinoscleromatis (83–85, 3), Brucella group (86), Phytomonas and Pasteurella groups (87), H. influenzae (88, 89), Bact. mallei (90, 3), fusobacteria (91), C. diphtheriae (92), Cl. welchii (93), spirochaetes (94), Asterococcus (95), Rickettsiae (80), yeasts and fungi (96–100, 62), Helminths (101).[29] [For trypanosomes see (254)].

Besides specific bacterial polysaccharides serologically inactive ones have been encountered.[30]

A polysaccharide isolated by Kendall and coworkers from mucoid strains of streptococcus was found to contain N-acetyl glucosamine and glucuronic acid in equimolecular amounts and, curiously, appeared to be identical with an acid carbohydrate (hyaluronic acid) occurring in human umbilical cord and other tissues, and in bovine vitreous humor.[31]

A source of error in investigations on bacterial polysaccharides was pointed out by Sordelli and Mayer (104) and Morgan (102) who observed that antibacterial immune sera may contain antibodies for the carbohydrates of the agar used in preparing the culture media (p. 98). Zozaya's report (105) on numerous cross reactions of bacterial carbohydrates was called in question on this account

---

[28] Bloch et al. (62).

[29] Campbell recorded highly specific precipitin tests with carbohydrates from several species of worms. Such a preparation from Ascaris lumbricoides was found by Baldwin and King (253) to consist largely of glycogen, and the authors are led to the conclusion that the specific properties must be due to some substance present in small proportion or, conceivably, to groups attached to glycogen.

[30] Oerskov (101a), Morgan (102).

[31] Kendall, Heidelberger and Dawson (103). Loewenthal (67) found the carbohydrate to have serological activity.

(106).[32] Intensive treatment was found by Morgan to be necessary for the production of agar antibodies of high titre.

Investigations on vegetable gums,[33] which should be mentioned here, have serological significance, first because of the fact that (degraded) gums are acted upon by antipneumococcus sera,[34] and secondly through the development of a method for preparing antisera against these carbohydrates (p. 227). Generally, gums consist of pentoses, which are completely or largely split off by mild treatment with acid, and an acidic nucleus, relatively resistant to hydrolysis. In chemical composition these partially hydrolyzed substances resemble bacterial polysaccharides, and it is these acid residues that precipitate strongly with immune sera for pneumococci. Several of the gums have been thoroughly examined. The findings on gum arabic (gum acacia) may be given as example.[35] Gentle hydrolysis removes l-arabinose, l-rhamnose, and a d-galactosido-l-arabinose. The remaining part consists of galactose and an aldobionic acid (galactopyranose-6-β-glucuronopyranoside) in the proportion of two to one. Synthesis of the acid by Hotchkiss and Goebel (112) brought confirmatory proof of its constitution. From the products of methylation and hydrolysis of the resistant nucleus it is probable that the aldobionic acid residues form branches, joined to a main galactose chain through the galactose residues in the former. Further results concern the structure and mode of linkage of the pentoses in the unchanged gum.

Sera have so far been prepared for gum arabic and cherry gum showing in precipitin reactions good specificity with a slight degree of crossing.[36] Inhibition tests with uronic acids gave no indication that these constituents take a decisive part in the reactions, and the antisera failed to precipitate the polysaccharide of type I and II pneumococci. The possibility that the easily detachable pentoses are of serological significance has been suggested.

IMMUNOLOGICAL SPECIFICITY OF POLYSACCHARIDES. — The noteworthy discovery of Heidelberger and Avery that carbohydrates are indeed of no less significance than proteins for the immunological

---

[32] Contamination with agar may be detected by colour tests [Pirie (255)].

[33] For literature on gums, agar, uronic and aldobionic acids v. (107), (108); (109) (constitution of agar).

[34] Heidelberger, Avery and Goebel (110).

[35] Cretcher and Butler; Heidelberger and Kendall (111); Challinor et al.; Jackson and Smith; v. Norman (107).

[36] Partridge and Morgan (113).

specificity of bacteria and of bacterial types came as a surprise, but the existence of a great number of specific polysaccharides is, after all, as intelligible as the enormous multiplicity of proteins. As Heidelberger remarked, very many compounds can arise from the asymmetry of carbon atoms in sugars (pentoses, hexoses) and sugar acids, the position of the oxygen bridges, the α and β-glycoside linkages and the various modes of union of sugars and sugar acids. Great diversity in the make-up of polysaccharides has actually been demonstrated in recent chemical work, and that the differences in constitution suffice to explain the immunological specificity [37] is borne out by the serological reactions of synthetic compound antigens, such as those prepared from stereoisomeric tartaric acids, sugars and sugar acids.

The problem of establishing relationships in carbohydrates between serological specificity and their chemical constitution entails difficulties similar in kind to those encountered with proteins, though less in practice. Already, instructive results have been obtained through the use of artificial compound antigens and of inhibition reactions. The serological relationship of pneumococcus polysaccharides and antigens made from glucuronic or aldobionic acids [38] and the partial inhibition by glucuronic acid and glucuronides in the precipitation of the pneumococcus polysaccharide type II,[39] leave no doubt about the significance of the sugar acid components in the specific structures. Leading in the same direction are the prominent influence of acid groups on specificity in general and experiments of Chow and Goebel (116) in which esterification of the polysaccharide of type I pneumococci by means of diazomethane was shown to abolish its reactivity.[40] Though a promising lead, these observations do not provide a full explanation of the specificities. There are specific polysaccharides without uronic acids, and as for the others, these acids alone cannot constitute the determinant groups, in view of the marked serological differences between polysaccharides containing the same acidic components. Furthermore, after precipitation of

[37] v. Weidenhagen (114) (specificity of carbohydrases).
[38] Goebel (115), Marrack and Carpenter (108).
[39] Woolf (259); (108).
[40] The activity was restituted by treatment with dilute alkali, the resulting substance still containing methyl attached to the primary amino group and to hydroxyl. In experiments of Heidelberger and Kendall (117) the polysaccharide of Pneumococci III reacted after methylation only with a part of the antibodies in type III antisera.

pneumococcus sera with glucuronic acid azoprotein the supernatants still precipitate with the homologous polysaccharide since the antibodies are only in part capable of combining with the artificial antigen.

Illustrative of the synthetic method are other cases, as the pronounced specificity of artificial antigens made from either glucuronic or galacturonic acid in agreement with the difference in polysaccharides containing these respective sugar acids, and the parallelism in serological relationships between the acetylated and deacetylated polysaccharides of type I pneumococci (p. 213), and between aminophenol-β-glucoside and its acetyl derivative (p. 174).

Whereas in general specificity can be correlated to chemical differences in the carbohydrates, this is not always the case, as is exemplified in the polysaccharides of the types B and C of Friedländer's bacilli which appear to be chemically very much alike but entirely different in immunological reactions.

In contrast to the serological diversity of chemically similar substances stand relationships of quite different polysaccharides (v. p. 97). In so far as chemical data are available, these show that the cross reactions depend upon the presence of related or identical constituents. Sugar acids, which occur so frequently, were found to be responsible in a number of instances. This is seen in the inhibiting effect of glucuronic acid and glucuronides [41] on the precipitation of (partly hydrolyzed) vegetable gums with antisera against Pneumococcus II (and III); [42] the reactions take place even though the aldobionic acids in the carbohydrates are not identical. The cross reactivity of the haptens of types III and VIII pneumococci is a similar case. [43] Both substances contain the same aldobionic acid; the first is made up of this acid alone, the second contains in addition glucose, approximately in the ratio of 2 molecules of glucose to 1 molecule of aldobionic acid. [44] Again, the immunological relationship of type B Friedländer bacilli and type II pneumococci is most prob-

[41] Marrack (118), (108); (259). The reactions are not inhibited by galacturonic or mannuronic acid. The expectation, not unreasonable in view of the precipitation of the gums by pneumococcus antisera for types II and III, that the latter polysaccharides would cross react, is not realized [v. (4)].

[42] Heidelberger et al. (110); (108). The amount of precipitate formed varies with different gums.

[43] Sugg et al.; Cooper et al.; Brown (119), Heidelberger et al. (120), (256).

[44] Goebel (121). On the basis of a quantitative study of the two cross reacting systems, and from other evidence, structural formulae for the constitution of the type VIII polysaccharide were proposed by Heidelberger et al. (120).

ably caused by the aldobionic acid constituents of the respective polysaccharides [45] — compare likewise the cross reactions of vegetable gums with antipneumococcal sera (p. 218) — and strikingly corroborating the significance of aldobionic acids is the precipitation of oxidized cotton, which can be assumed to contain cellobiuronic acid units, by antisera for Pneumococci VIII and III.[46] In another group of substances, comprising those from anthrax bacilli, type XIV pneumococci, and from human blood cells, the overlapping is attributable to the characteristic components, acetylglucosamine and galactose, common to these haptens although present in different proportions.

Additional instances, not analyzed chemically, are the reactions of the polysaccharides of gonococci and meningococci with anti-pneumococcus sera for type III,[47] the precipitation of type II pneumococcus carbohydrate by immune sera for a strain of B. lepisepticus,[48] of leuconostoc dextrans [49] by pneumococcus sera, the serological relationship between the carbohydrates of B. proteus X 19 and rickettsiae,[50] and cross reactions of pneumococcus types (124a).

The frequent overlapping reactions of salmonella bacilli (and their polysaccharides) which gave rise to the serological concept of antigenic components and led to assigning "antigenic formulae" to the various types [51] have not been the object of chemical investigation. One may expect that such studies will provide information on the apparent mosaic structure of antigens, a problem referred to in other sections. Evidence for the existence of separate chemical entities underlying the serological reactions may be gained by isolating several specific substances from one bacterium, or, if possible, by demonstrating several specific groupings in homogeneous, well purified polysaccharides by means of cross reactions with substances of known constitution.

Actual separation of different components has been attained in

[45] v. Beeson and Goebel (122); (123); Perlman et al. (124) (comparison of the cross reaction of rabbit and horse antisera, the latter being, as has been noticed in several other instances, considerably more intense).

[46] Heidelberger et al. (256).

[47] Miller and Boor (68).

[48] Dingle (87) reports weak reactions of yeast polysaccharides with plant seed antisera.

[49] Sugg et al. (124a). Owing to the presence of these substances precipitation of commercial sugar samples with pneumococcus sera has been observed.

[50] Castaneda (125), (80), Otto (81).

[51] v. Table 15. A similar description of antigenic relations has lately been proposed for the pneumococcus group [Vammen; Kauffmann (126)].

some instances. Tubercle bacilli were shown to contain a mixture of reactive polysaccharides, and from pneumococci and meningococci type specific as well as species specific polysaccharides could be isolated. Attempts at fractional precipitation of polysaccharides with the aid of precipitating antibodies each corresponding to a single component of the apparent antigen mosaic, in analogy to the partial absorption of antibodies from an immune serum, have not thus far led to the separation of fractions of different specificity; [52] on the contrary, these experiments seemed to prove the homogeneity of the substances tested.

The formation of more than one specific antibody in response to a single polysaccharide has been shown by partial absorption of antisera with heterologous precipitants in practically every case of overlapping reactions; typical examples are the pneumococcus types III and VIII, or the antibodies in hemolytic antibacterial immune sera of which only a fraction combines with blood cells. However, frequently upon absorption of immune sera with heterologous materials antibodies are left behind for the homologous substance; as pointed out before, such results while suggestive cannot be regarded as an equivalent substitute for chemical evidence, in demonstrating the presence of distinct groupings in a molecule.

COMPLEX BACTERIAL ANTIGENS.[53] — It has been mentioned previously that a new and fruitful line of inquiry was opened up by Boivin and his coworkers, and Raistrick and Topley, through the separation from gram-negative bacteria of potent antigenic materials, free from protein and composed of phosphatides, polysaccharides and nitrogenous substances in loose combination. Extraction with trichloroacetic acid or tryptic digestion was first made use of in the preparation; in later methods urea (128a), diethyleneglycol, and formamide have been employed. The substances were found to be toxic and to represent so-called endotoxins;[54] upon injection into animals O-agglutinins (and weak anti-endotoxins) are produced.[55]

---

[52] Furth and Landsteiner (41), Burnet (127), K. Meyer (128), Marrack and Carpenter (108). Burnet writes: ". . . while a given antiserum corresponding to a bacterial polysaccharide antigen can be readily fractionated by immunological methods into dissimilar parts, no such components can be demonstrated for the antigen."

[53] v. (p. 109).

[54] Cf. Topley and Wilson (10), Haas (128b).

[55] Sera made with whole bacteria may, in addition, contain flagellar- and protein-antibodies.

The polysaccharides being the main specific components, great diversity in composition has naturally been encountered among different bacteria; various simple sugars, amino sugars, and uronic acids have been recognized as cleavage products.

The somewhat variable composition of the complex antigens may be instanced by analyses of the substance from Shiga bacilli. Elementary analysis of a typical preparation gave: C 45.5%, H 7.6%, N 3.8%, P 1.3%.

The results obtained with various bacteria (and their variants) are discussed in papers by Boivin and others (129–133). The failure to extract complete antigens of this sort from Gram-positive microorganisms proves again the marked biochemical difference paralleling the distinction by the Gram stain. It may be noted that the antigens underlying the Weil-Felix reaction (134) and Felix's labile Vi antigen,[56] responsible for the virulence of typhoid bacilli, belong to the class of antigens under discussion. The latter could be differentiated from the O antigen. Ultracentrifugal separation of (different) fractions is described by Boivin [v. (134a)].

The constitution of the complex antigens and the immunological significance of the components have been elucidated by further studies. In the work of Morgan and Partridge (135) on Bact. dysenteriae Shiga, the primary product, extracted with diethyleneglycol and fractionated with acetone, was subjected to stepwise dissociation by formamide. In the process first phospholipid was split off, and an antigenic complex of polysaccharide and a substance originally described as "polypeptide like" but later recognized as a protein, was obtained, capable of eliciting agglutinins for Shiga bacilli, precipitins for the polysaccharide and heterogenetic sheep hemolysins. On repeated treatment with formamide this combination was partially broken up and the specific, non-antigenic polysaccharide hapten set free. The protein, itself antigenic, is therefore essential for the immunizing properties,[57] and this is corroborated by the loss of antigenicity upon tryptic digestion of the complex antigen, which goes along with destruction of the protein. Dissociation of the antigenic complex takes place also in strong phenol solution and in aqueous alkaline solution.

According to K. Meyer (258) the heterophile reactivity of the complex antigen is weaker than that of the polysaccharide resulting from its cleavage; the same relation was found for the precipitative capacity (134a).

[56] (129), (48), (252), (257), (41b).　　　　　　　[57] v. (136), (137).

The three constituents were found to amount to: phospholipid 9–12%, polysaccharide 50–55%, protein 17–20%; the presence of other components in small amounts could not be excluded with certainty. In the lipid oleic, palmitic and glycerophosphoric acids were identified, and in the protein 8% tyrosine, 5.5% arginine, and subsequently 6% tryptophane and at least 12% glutamic acid were determined.

In continuing their investigation, Morgan and Partridge showed that the protein is probably conjugated and dissociates in strong phenol solution to yield, presumably by removal of a phosphorus containing prosthetic group, a different protein, characteristically soluble at pH 2.0–2.5, with about 14.5% nitrogen, and free of phosphorus. This protein was obtained also from rough strains which do not contain the complex antigen.

Much the same situation has been encountered in the case of Bact. typhi and Bact. typhi murium. Freeman and Anderson (136) found that the O antigen can be dissociated by mild hydrolysis into a polysaccharide component (v. p. 215), an insoluble "polypeptide," a small amount of lipid, and a soluble nitrogenous component which, however, Morgan and Partridge (48) think belongs to another antigen (Vi). Unlike the Shiga antigen, that from typhoid bacilli is relatively refractory to tryptic digestion.

The antigens of organisms of the Brucella group have repeatedly been explored [58] (with rather inconsistent results), most carefully by Miles and Pirie (86). These workers isolated, by centrifugal fractionation at 14,000 r.p.m., the antigen of Br. melitensis, apparently in its native state, as a material which is birefringent under certain conditions and resembles plant viruses in particle size [59] and anisotropy of flow. It is made up of phospholipids, a protein or peptide, and a formylated amino-polyhydroxy compound, most likely a polysaccharide. From the complex the protein-like substance and lipid can be split off, and the resulting product contains the amino-polyhydroxy compound and phospholipid. This substance has a molecular weight of about one million, is precipitable in high dilution and probably antigenic to some extent. By treatment with sodium dodecyl sulphate it is disaggregated — like the original complex, having then a molecular weight of 1 to 2×10^5. Further degradation by gentle

---

[58] Bibliography in (86).

[59] The demonstration of particulate, serologically characterized antigens in animal tissues may be recalled (pp. 81, 130, see also p. 226).

hydrolysis removes phospholipid and yields the amino carbol drate which is no longer precipitable but still inhibits specifically the agglutination of Br. melitensis.

Results having an important bearing upon the problem of antigenicity [60] were gained in the chemical examination of another bacterial antigen, namely the Forssman antigen of an R variant derived from type I pneumococci which was obtained in a yield of only about 700 mg. from five hundred liters of bacterial culture. The active substance proved to be, like the Forssman substance of animal origin (p. 229), a lipocarbohydrate, differing from the latter in that it contains only about 6% lipoidal material and as much as 4–5% phosphorus, and is a potent antigen without the aid of added protein. It consists of a carbohydrate, seemingly identical with the C polysaccharide of pneumococci, and of firmly bound fatty acids which appear to confer upon the polysaccharide antigenicity and the cross reactivity with sheep blood. In both substances — C polysaccharide and Forssman antigen — a small percentage of the total nitrogen (6%) is not accounted for by amino sugar, indicative of the presence of an additional nitrogenous component. This as well as a sugar constituent, other than amino sugar (probably acetylglucosamine), have not been identified. The most significant finding is the absence, notwithstanding the high antigenicity of the substance, of amino acids in the molecule, as determined by van Slyke's ninhydrin method, a point to which special attention had not always been given in the studies on complex bacterial antigens.

It is quite possible that other bacterial substances belong to the category of amino acid-free antigens, such as the phosphatides recovered from acid-fast bacilli. Specifically reacting,[61] antigenic phosphatide fractions prepared by Anderson [62] from human tubercle bacilli, differing from ordinary phosphatides by the absence of bases and the very small nitrogen content, yielded on hydrolysis several fatty acids, organic phosphoric acids, inosite, mannose (and another hexose).[63] Among the fatty acids in the phosphatides from human

[60] Goebel, Shedlovsky, Lavin and Adams (24).

[61] Boquet and Nègre; Pinner (138), Pedersen-Bjergaard (139), Chargaff and Schaefer (140); v. Klopstock et al. (141), Sandor et al. (142), Witebsky et al. (143).

[62] (144). Lipids and fats, in a surprisingly complex mixture, varying from strain to strain and including substances of peculiar composition, constitute a large part — up to 20% (or more) — of the bodies of tubercle and other acid-fast bacilli [Chargaff et al. (145)]. The literature is reviewed in (146), (147).

[63] Cf. Chargaff and Schaefer (59) (Calmette-Guérin bacilli).

strains of tubercle bacilli an optically active, branched, saturated acid, phthioic acid, was detected which, like Anderson's phosphatides, produced histological changes resembling tuberculous tissue reactions [64] (Sabin et al.). Whether or to what extent the fatty constituents influence the specificity of the phosphatides is undecided.

Macheboeuf,[65] who questions the purity of other preparations found to have immunological activity, prepared complement-fixing, non-antigenic, nitrogen-free lipids of tubercle bacilli which he describes as a mixture, in the form of salts, of glycerophosphoric and inositophosphoric acids in ester linkage with fatty acids, related in composition to substances found in higher plants.

Supplementary references on specific bacterial substances extracted by organic solvents are: (151, 151a) (diphtheria bacilli); (152–159); (160–162) (reviews).

Antigenic particles containing protein, nucleic acid, lipid, carbohydrate and pigment, with a diameter estimated at around 40 mμ, were separated in the high speed centrifuge from streptococci, disintegrated by grinding or by sonic vibration.[66]

A POLYPEPTIDE-LIKE HAPTEN. — In view of the parallelism existing between pneumococci and anthrax bacilli with regard to the correlation between virulence and the ability to form capsules in the animal body, it is of great interest that the specifically precipitable capsular substance of B. anthracis proved to be not a polysaccharide but, as shown by Ivánovics and Bruckner (163), a polypeptide-like substance of unique composition. The colloidal, strongly acid substance is made up of only one amino acid, namely d-glutamic acid, the optical isomer of the "natural" l (+) — glutamic acid. Since in the bacterial capsules there is a considerable amount of this hapten and most enzymes of animal origin so far examined attack only peptides consisting of the natural amino acids, a relationship between virulence and the peculiar chemical character of the hapten has been suggested.

Under the supposition of a simple peptide chain, and from the ratio amino N:total N, it would follow that the substance contains 40–50 glutamic acid residues and has a molecular weight of about 6000, but the possibility of a more complicated structure is left open, especially

---

[64] The histological effects of various of the bacillary fractions are summarized by Sabin (148).

[65] (149), (147); v. Bloch (150), Bailey (150a).

[66] Sevag et al. (162a).

because of the somewhat too high acid equivalent. To be noted is the occurrence of substances indistinguishable from the anthrax polypeptide in the organisms of the mesentericus and subtilis groups; cultures of these bacteria have been recommended as a suitable source material for the preparation of d-glutamic acid.

PREPARATION OF ARTIFICIAL ANTIGENS USING BACTERIAL PROTEINS.[67] — In the study of the complex antigen of dysentery bacilli, Morgan and Partridge observed that its dissociation by formamide is reversible, and that the (conjugated) protein can be recombined with the "undegraded" polysaccharide to form again an antigenic complex. This resynthesis took place not only in formamide but also in slightly alkaline aqueous solution, and an active Shiga antigen could likewise be prepared by combining the protein of typhoid bacilli with the Shiga polysaccharide. The method, reminding one of the immunization with organ extracts and foreign proteins, proved to be generally applicable and enabled, by combination with the protein in formamide solution, potent antigens to be made from agar and plant gums. Combination of the Shiga protein with the group substance A contained in commercial preparations of pepsin or gastric mucin (p. 232) also yielded an antigen capable of eliciting anti-A immune sera of high titre.

In alkaline solution Shiga polysaccharide forms complexes with the bacterial protein after degradation by phenol, and with other proteins,[68] but these combinations differ from the ones made with the native conjugated protein in that they disintegrate on heating and are non-antigenic. Hence one may infer that a special group in the conjugated protein is responsible for the formation of a relatively stable complex. The complexes with ordinary proteins have some importance for the investigation of antigens, inasmuch as these materials are not precipitated by trichloracetic acid, which shows that negative reactions with protein precipitants, as with bacterial products containing polysaccharides, should not be taken as valid proof of the absence of protein.

The claims that adsorption to collodion particles [or flocculation by immune sera (p. 108)] imparts antigenic properties to polysaccharides could not be confirmed with the Shiga polysaccharide. Positive immunization

---

[67] (135), (164), (113).
[68] On the formation of complexes between carbohydrates and proteins v. Przylecki et al. (165).

results obtained with agar or impure gums may well be connected with the findings of Partridge and Morgan.

SPECIFIC SUBSTANCES OF ANIMAL ORIGIN. — The assumption of serologically reactive lipids,[69] suggested by the primary solubility of specific substances in organic solvents, has received some confirmation from reports on antibody reactions of chemically defined lipids (p. 111). In so far as chemical evidence has been procured, however, one may suspect that the specificity of the complex substances in question is carried entirely or in greater measure by carbohydrates than by the fatty components, and haptens not containing lipids have been obtained from animal sources.

Lipids were seen to participate in phenomena that, according to material and technique, fell within the scope of serological work.[70] Among these topics are the neutralization of toxins [71] and lysins, e.g. the inactivation of tetanolysin and other hemolysins (also saponin) by minute amounts of cholesterol,[72] which substance possibly contributes to the antilytic properties exhibited by sera. From studies of Abderhalden (170) and Walbum (171) the hydroxyl group of cholesterol is of importance for the neutralization of saponin while saturation of the double bond by addition of bromine is indifferent; in the series of aliphatic alcohols the highest members (cetyl, myricyl alcohol) showed marked antilytic properties. With saponin, cholesterol forms a non-hemolytic crystalline molecular compound (Windaus).

Further subjects, followed up because of the outward similarity [73] and possible relationship to other forms of cytolysis, are the hemolytic (and bacteriolytic) effects of organ extracts, attributable probably to soaps and fatty acids, and of cobra and other snake venoms. Venom hemolysis is activated or strengthened by lecithin or serum and could be traced to the action of a lecithinase, which results in splitting off oleic acid and converting lecithin into a strongly hemolytic substance — desoleolecithin (lysolecithin).[74]

The subsidiary role of lipids, noted in specific precipitation (p. 246), is prominent in certain complement fixation and flocculation tests, in par-

[69] Cf. reviews (160), (4b). The term lipids (or lipins) while currently applied to phosphatides and cerebrosides has been used, vaguely, in serological papers for substances extractable by organic solvents; in the text it designates substances containing fatty acids, and sterols as well.

[70] The extensive older literature is dealt with in (162).

[71] (166) (neutralization of tetanus toxin by protagon); (167), (168), (261).

[72] Noguchi (169).

[73] An example is the inactivation of the hemolytic action of fatty acids (oleic acid) upon heating in protein solutions.

[74] Flexner and Noguchi (172), Calmette (173), Kyes (174), Lüdecke (175), Manwaring (176), Delezenne and Fourneau (177).

ticular those of syphilis sera where lecithin seems to be requisite and cholesterol is added to the reagent for intensifying the visible reaction.[75]

Isolation in a pure state of the specific substances, tentatively classed as lipids, is a difficult task since it involves the separation of small amounts of active material from a bulk of lipids which have the property of modifying solubilities. Indeed, it has been observed that in the presence of lecithin various substances such as sugar, proteoses, cobra venom, ferrous hydroxide, may pass into organic solvents.

In the case of the Forssman hapten, the circumstance that starting material is available in quantity offered opportunity for chemical study. The Forssman hapten, recovered from horse kidney first in fractions consisting largely of cerebrosides,[76] was on further fractionation shown to possess distinctive properties (181). Active protein-free preparations were separated, dissolving in water or dilute alkali and in pyridine but not soluble, or barely so, in other common organic solvents; in comparison to the known cerebrosides and phosphatides their carbon content was much lower, and a greater amount of reducing sugar was found after hydrolysis, along with a considerable proportion of fatty acids; the preparations contained 2–4% nitrogen, no sulphur or phosphorus.

Even though it could not be claimed that the substances were homogeneous, the observations suggested the carbohydrate nature of the specific groupings and, by presumption, the significance of carbohydrates for the specificity of haptens of animal origin, in analogy to bacterial polysaccharides.[77]

An extensive investigation was conducted by Brunius; [78] several preparations were obtained, one of which appeared to be of greater purity than those examined previously, as estimated by serological tests (inhibition of hemolysis [79] of sheep cells by Forssman antisera). Among the hydrolytic products glucosamine was identified, and the presence of aldo-hexoses was inferred. The amount of glucosamine

[75] v. Hazato (178); (179) (influence of lipids on stability of bacterial suspensions).

[76] Sordelli et al. (180).

[77] (181). Immunologically there is a difference between bacterial polysaccharides and the primarily alcohol-soluble haptens of animal origin in that the latter exhibit the phenomenon of high antigenic activity upon addition of proteins, a feature which may be connected with the fatty acid content.

[78] (182); v. (183).

[79] For the method see Landsteiner and van der Scheer (183a), Brahn and Schiff (199).

increased with the potency of the substances and accounted, in the most active preparations, for one third of the reducing sugar as well as of the total nitrogen. The serological activity was destroyed upon treatment with strong alkali or diazomethane but was not affected by proteolytic enzymes, cobra venom,[80] and by treatment with phenylisocyanate or nitrous acid, the latter result indicating that glucosamine may be present in acylated form. The demonstration of glucosamine in the Forssman hapten is of interest in view of the occurrence of acetyl glucosamine in the serologically related polysaccharides of Shiga bacilli and certain pneumococci, and in the group A substance (see below).

Separation of a water-soluble active material, after hydrogenation, from the acetone-insoluble lipid fraction of horse kidney was reported by Fujimura (185). His results do not justify conclusions as to the chemical nature of the active substance.

The substance which serves as reagent in the Wassermann reaction for the diagnosis of syphilis is, like the "lecithin" hapten, exceptional among serologically active materials on account of its widespread distribution, surpassing in this respect even the Forssman antigens. Originally devised as an application of Bordet-Gengou's complement fixation to a special microbe and carried out with tissues containing syphilis spirochaetes, the reaction was soon found to take place with alcoholic extracts of normal organs; of these, heart muscle proved to be particularly reactive.[81] This tissue is routinely employed for the preparation of diagnostic extracts and has also been used as the material of choice for chemical investigation. Efforts to isolate the active principle gave mostly indefinite and in part discrepant results. Consistently the substance was found to be precipitable by acetone and cadmium chloride, which directed attention to the role of lecithin. In fact, an optically active β-lecithin was reported to be the effective component of the heart extract (188), but this is not consonant with conclusions of other workers.[82] In recent papers by Pangborn (192) evidence is brought forward to show that a non-nitrogenous phospholipid (cardiolipin) is essential for the reaction.[83] Since in the actual

[80] This is in conflict with a report by Mizuhara (184).
[81] Landsteiner, Müller and Pötzl (186); v. Marie and Levaditi (187).
[82] v. Rudy; Oe. Fischer (189), Weil et al. (190), O. Fischer et al. (191). Several studies on the purification of the substance are abstracted in (160).
[83] Her assumption that the water-soluble fraction obtained on hydrolysis of the substance is a polysaccharide could not be confirmed in later experiments (per-

test addition of lecithin and cholesterol is required, Pangborn is inclined to believe that the reactivity of the heart extracts is not due to a single substance but to three components. From the consideration, however, that the union with antibody probably takes place through a special grouping, the chances are that the part played by lecithin and cholesterol is subsidiary. With the purified, certainly specific Forssman substance (181), too, and some other haptens (151), nonspecific lipids are added as adjuvants.

The nature of the specific brain hapten, extractable by cold alcohol, still awaits elucidation. Rudy (194) separated an active fraction, apparently water-soluble, relatively stable against alkali, free from phosphorus, and not related to cerebrosides.

In the investigations on this and other animal haptens, in addition to separation by solvents, adsorption procedures have been employed.[84] It has been possible in this way to separate artificial mixtures of specific substances and, in organ extracts, the "lecithin" or the Wassermann haptens from the Forssman substance. Progress in this field may be anticipated from the use of the methods of adsorption analysis developed by Twsett and others.[85]

BLOOD GROUP SUBSTANCES. — Apart from the findings just outlined, the extraction of species specific substances from erythrocytes by alcohol and the agglutination of bloods (also other than sheep blood) by antipolysaccharide sera (p. 98), little has been learned about the chemical nature of the haptens in animal cells. Some advance has been made with the substances that are associated with the reactions of blood groups. Alcoholic extracts of moderate activity have been obtained from human red cells and tissues,[86] and serologically reactive aqueous extracts [87] from the erythrocytes; but neither

---

sonal communication). From analyses of precipitates formed in the flocculation tests with syphilis sera Brown and Kolmer (193) were led to conclude that a non-nitrogenous substance is responsible for the reaction.

[84] (160), (195–197), (190), (4b).

[85] v. (198).

[86] (199–203). The serological behavior of alcoholic extracts from A cells has been examined by Stuart and Wheeler with several sorts of A-agglutinins (203a).

[87] Brahn and Schiff (199); cf. Lattes (203), Hallauer (204), Ottensooser (205); (206). In trying to reproduce Hallauer's results with extraction by dilute alcohol, the writer has been unable as yet to secure yields comparable to those of this author. In some extracts prepared by other workers the presence of suspended stromata, difficult to remove by centrifugation, has probably been responsible for positive reactions.

The properties M and N of human blood have not been demonstrated in aqueous or alcoholic extracts, or in secretions (207).

of these preparations has served to clear up the chemical nature of the substances. The possibility that the solubility in alcohol may be due to a combination of the specific substances with lipids was considered by Schiff (208).

New materials for chemical studies became accessible when it was found that water-soluble group-specific substances, demonstrable by inhibition of isoagglutination or hemolysis (and giving precipitin reactions with selected immune sera) [88] are present in saliva, gastric juice, urine, and organs of human beings and certain animals. Brahn and Schiff [89] described active preparations from commercial pepsin in which, after hydrolysis, reducing sugar could be demonstrated; galactose was identified, and the presence of amino sugar was surmised. Freudenberg and his coworkers (211) separated from human urine of groups A, B, and O preparations containing galactose and N-acetylglucosamine. The activity of these preparations was weak [90] and later, in an apparently purer preparation, no acetylglucosamine could be detected. Yet the two constituents, galactose and glucosamine, were subsequently again identified in highly active group A-substances [91] from horse saliva and in commercial pepsin,[92] mucin and peptones.

In our preparations from hog stomach the presence of amino acids [v. (214)] was shown by the Sakaguchi reaction for arginine, and by weak biuret and xanthoprotein reactions, and by a faint test with ninhydrin; the reaction with diazobenzenesulfonic acid was positive; sulphur and phosphorus were not demonstrable. Quantitative determinations by Van Slyke's ninhydrin method [93] on preparations from hog stomach, commercial mucin, and pepsin showed that the amino acid N amounted to about 2%, that is roughly 35% of the

[88] v. Poulsen; Boyd (208a) (immunization with saliva).

[89] (209), (210).

[90] v. (212). This indicates that the group substances in urine are either weakly reactive, or that the preparations contained much inactive material, possibly similar in chemical composition to the group substances.

[91] Landsteiner, Chase, Harte (212), (213); Freudenberg (214). (The latter papers should be consulted for detailed descriptions of purification methods). Further analytical results are given by Goebel (215), K. Meyer et al. (216); Witebsky et al. (217) (gastric juice of group B and O individuals). Analysis of substances from hog stomach gave about 55% reducing sugar (as glucose), about 30% hexosamine (213).

[92] A sample of crystallized pepsin (Northrop) contained only traces of group substance.

[93] Because of contrary statements it should be mentioned that glucosamine does not interfere with the method (218).

total N; and about 2.5% amino acid N was found in substances from human saliva. It may be concluded that amino acids, presumably in peptide linkages, are a part of the active molecule; for various methods of fractionation, heating with formamide, tryptic digestion, or acetylation followed by deacetylation of the chloroform-soluble product, according to the method prescribed by Freudenberg, did not succeed in reducing significantly the amino acid content. If our conclusion is correct, the group substances represent a new type, related to glucoproteins but peculiar because of the very high carbohydrate content.

Jorpes and Norlin (219) found that the anti-agglutinating activity of a group substance of rather low potency from urine was greatly diminished by tryptic digestion and described two fractions, one inhibiting isoagglutination, believed to be a protein, the second antilytic and probably a polysaccharide. In our experiments treatment with formamide or tryptic digestion was also seen to reduce greatly the anti-agglutinating power whereas the antilytic activity was even increased by formamide treatment. Both methods, however, did not change substantially the amino acid content.

As for the serological importance of the constituents, there is no doubt that the carbohydrate moiety plays a part. A cogent argument is the fact that antisera [94] specific for the polysaccharides of Pneumococcus XIV [95] and of anthrax bacilli [96] give, in the cold, rather weak but unmistakable precipitation with the blood group substances, directly or — with anthrax serum — after slight hydrolysis; furthermore, the A substance was found to inhibit the agglutination of human blood cells by immune sera for type XIV (p. 98). Secondly, A haptens are destroyed [97] by microorganisms which attack certain bacterial polysaccharides — Morgan's Myxobacterium, Saccharobacterium ovale described by Sickles and Shaw, and a bacterium, exacting in its nutritive requirements, cultivated from leaf mold by Chase (225) — and the urine A substance could be inactivated, simultaneously with liberation of reducing sugar, by an enzyme from

[94] v. (p. 221).

[95] Beeson and Goebel (220). The reaction of polysaccharides of other pneumococcus types with anti-A sera may have to do with the peptone content of the culture media and, particularly in view of Morgan's observations on artificial antigens, it is comprehensible that this sort of contamination can account for false cross reactions of bacteria [Goebel (26), Ottensooser (221), (260)].

[96] Ivanovics (222).

[97] Landsteiner and Chase (223). On decomposition by an enzyme from Cl. welchii v. Schiff (224).

the digestive organs of snails [Freudenberg (211)]. Lastly, the significance of acetyl glucosamine is supported by Freudenberg's observation that group substances are inactivated by partial deacetylation, and the activity restituted upon reacetylation with ketene.

Whether or not the presumptive polypeptide structure is involved in the specific reactions cannot be stated from the evidence at hand. Were the answer in the affirmative, it would lead one to surmise, considering the extraordinary diversity of specific cell properties, the existence of numerous substances, similar in carbohydrate composition, yet serologically different by virtue of peptide components. However that may be, it has not been possible, so far, to distinguish chemically the substances that characterize the several blood groups. Analyses carried out with saliva preparations for N, amino acid N, hexosamine and reducing sugar revealed no significant differences between the substances from groups A, B and O (226), and similar analyses on a saliva preparation from an A non-secreter were likewise not definitely characteristic. (Instances of serological diversity of chemically indistinguishable substances, already noticed, were also encountered in the domain of bacterial polysaccharides).

As in the case of the Forssman hapten, the group substances from various sources although related are not or need not be identical. Thus, serological distinctions could be made between the group substances in human blood and secretions [v. (227)]. The analytical differences, as between the substances of comparable activity from horse saliva and pig stomach, cannot be taken as decisive on account of uncertainty concerning the purity of the preparations. Dialysis experiments seem to indicate that the A substance can exist in various states of dispersion (213).

In view of the negative results the statements concerning differences between the human blood group substances, found in adsorption experiments and by other means,[98] would call for closer examination.

ENZYMES FOR BACTERIAL POLYSACCHARIDES.[99] — The discovery made by Dubos and Avery (231) of an enzyme specific for the carbohydrate of Pneumococcus III affords evidence for the similarity in specificity of immunological and enzyme reactions.[100] After assiduous search this enzyme was detected in mixed cultures of soil bacteria, and with the aid of special culture media containing the specific sub-

---

[98] (228), (229); v. (229a).
[99] v. Dubos (230).
[100] Other examples are bacterial enzymes examined by Karström (232) and dopaoxydase specific for 1-3-4-dihydroxyphenylalanine (233, 234).

stratum, a bacillus was isolated whose capacity to produce the enzyme could be enhanced by continued cultivation in the presence of the substance. While splitting the polysaccharide of type III by depolymerization, this bacillus was without action upon those of the pneumococcus types I and II; still more strikingly, bacterial enzymes for the serologically related polysaccharides of Pneumococci III and VIII attacked only the corresponding, not the other polysaccharide.[101] When the enzyme for type III was allowed to act upon live cocci their capsules were destroyed, which serves to explain that the enzyme exerts protective and curative effects in animals although it does not hinder the growth in culture media.

A Myxobacterium decomposing various bacterial polysaccharides was described by Morgan and Thaysen (236), and Sickles and Shaw (237) found several bacterial strains which produce enzymes attacking pneumococcus polysaccharides.

The high bactericidal potency towards certain bacteria of an enzyme — Fleming's lysozyme — present in egg white, tears and other secretions, is probably due to the decomposition of a mucoid bacterial polysaccharide.[102] Hydrolysis of the acid polysaccharide from mucoid forms of streptococci (p. 217) is effected by enzymes of bacterial as well as animal derivation.[103]

Enzymatic synthesis of polysaccharides has been described by Beyerinck and others, and recently a serologically reactive carbohydrate, the dextran of leuconostoc, has been obtained by the action of an enzyme contained in culture filtrates of the organism.[104]

The relation of bacteriophage activity to the serological specificity of bacterial polysaccharides was noticed by Burnet, and in experiments of Levine and Frisch (242) specific inhibition of the phage action by bacterial extracts, depending upon polysaccharides, could be demonstrated. Confirmatory results were then described by Gough and Burnet (243). White (244) described inhibition of the activity of certain phages by lipoid constituents of cholera vibrios.

TRANSFORMATION OF BACTERIAL TYPES. — A discovery of far reaching importance was made by Griffith (245) who found that one type of pneumococcus can be converted permanently into another under the influence of substances contained in the latter. In the experiments of Griffith the change was brought about in the animal

[101] Sickles and Shaw (235), Dubos (230).
[102] K. Meyer et al. (238), Epstein et al. (239).
[103] K. Meyer et al. (240).
[104] Hehre et al. (241).

body, but Dawson and Sia (246) succeeded in reproducing the phenomenon in vitro by growing R forms in the presence of a very small quantity of killed pneumococci of another type. A further step was made by Alloway (247) when he showed that the transformation can be induced with extracts of the cocci and that these can be purified by adsorption of inactive material to charcoal and precipitation with alcohol or acetone. The activity of the substance withstands heating for 10 minutes at 80°, and sometimes 90°, but is destroyed by boiling and by enzymes which are liberated in autolysis of the cocci. These properties distinguish the specifically active substance from the type-specific polysaccharides which, moreover, are not capable of causing the transformation.

The virulence for rabbits of a Pneumococcus III strain was not lost by undergoing the sequence of transformations: S to avirulent R to S, the last change effected under the influence of killed, even avirulent, smooth type III strains; it follows that factors connected with virulence exist independently of the capsule in the smooth form.[105]

An apparently analogous transformation of a virus, namely the change of rabbit fibroma virus into that of myxomatosis has been described by Berry and Dedrick.[106]

The biological significance of Griffith's phenomenon lies in the initiation by certain substances of inheritable changes in unicellular organisms so that there is reproduced indefinitely in subsequent generations the agent which induces the change, along with the type-specific polysaccharide, not previously present. This calls to mind the effects of plant viruses, bacteriophages and filtrable tumor agents. All these active principles are — under the control of living cells — endowed like genes with the capacity for causing the reproduction of their own kind, and have raised fundamental issues regarding the boundary between living and lifeless matter.

BIBLIOGRAPHY

(1) Heidelberger: Ph 7, 107, 1927; ARB I, II, IV; ChR 3, 403, 1927. — (2) Morgan: JC 55, 284, 1936. — (3) Mikulaszek: EH 17, 415, 1935. — (4) Marrack: CAA. — (4a) Velluz: VI Congr. Chim. Biol., 1937. — (4b) Rudy: KZ 65, 356, 1933. — (5) Kimmig: K 1940, p. 858. — (5a) Levinthal: CB, Beih. 110,

[105] Shaffer, Enders and Wu (248).
[106] (249); v. (250).

30, 1929. — **(6)** Heidelberger et al: JEM *38*, 73, 1923, *40*, 301, 1924, *42*, 727, 1925. — **(7)** Tomcsik: Pr *24*, 812, 1927. — **(8)** Avery et al: JEM *49*, 251, 1929. — **(9)** Morgan: Br *13*, 342, 1932. — **(10)** Topley et al: PB. — **(11)** Pettersson: ZI *99*, 142, 1940. — **(11a)** Eagle et al: JEM *63*, 617, 1936. — **(12)** Goebel: JBC *89*, 395, 1930. — **(13)** Heidelberger: JBC *96*, 541, 1932. — **(14)** Heidelberger et al: JEM *75*, 35, 1942. — **(15)** Babers et al: JBC *89*, 387, 1930. — **(15a)** J. Franklin Inst. *230*, 727, 1940. — **(16)** Heidelberger et al: JBC *74*, 613, 1927. — **(17)** Hotchkiss et al: JBC *121*, 195, 1937. — **(18)** Reeves et al: JBC *139*, 511, 1941. — **(19)** Heidelberger et al: JEM *57*, 373, 1933. — **(19a)** Ivanovics: ZI *98*, 420, 1940. — **(20)** Brown: JI *37*, 445, 1939; Ann. Rep. Division of Lab., New York State Dept. of Health, 1941, p. 22. — **(21)** Lancefield: JEM *47*, 481, 1928. — **(22)** Tillett et al: JEM *52*, 895, 1930. — **(23)** Heidelberger et al: JEM *53*, 625, 1931. — **(24)** Goebel et al: JBC *148*, 1, 1943; JEM *77*, 435, 1943. — **(25)** Wadsworth et al: JI *24*, 349, 1933. — **(26)** Goebel: JEM *68*, 221, 1938. — **(27)** Avery et al: JEM *58*, 731, 1933. — **(28)** Landsteiner et al: JI *22*, 75, 1932. — **(29)** Enders et al: Pr *31*, 37, 1933, *34*, 102, 1936; JEM *60*, 127, 1934. — **(30)** Sevag: BZ *273*, 419, 1934; S *87*, 304, 1938. — **(31)** Chow: JEM *64*, 843, 1936; CJP *11*, 223, 1937. — **(31a)** Wong et al: CA *36*, 2616, 1942. — **(32)** Felton et al: JBa *29*, 149, 1935; JID *56*, 101, 1935; Pub. Health Repts. *53*, 1855, 1938 (B). — **(33)** Brown et al: Pr *34*, 832, 1936; JI *37*, 445, 1939 (B). —**(34)** Heidelberger et al: JEM *64*, 559, 1936; Pr *33*, 445, 1935. — **(35)** Heidelberger et al: JEM *42*, 701, 709, 1925, *46*, 601, 1927. — **(36)** Goebel: JBC *74*, 619, 1927. — **(37)** Mueller et al: Pr *22*, 373, 1925. — **(38)** Julianelle: JI *32*, 21, 1937. — **(39)** Ivanovics: ZI *97*, 402, 1940 (B). — **(40)** Ivanovics: Arch. wiss. Tierheilk. *74*, 75, 1939. — **(41)** Furth et al: JEM *49*, 727, 1929. — **(41a)** Meyer et al: Br *16*, 476, 1935. — **(41b)** Kauffmann: ZH *117*, 778, 1936. — **(42)** White: JP *31*, 423, 1928. — **(43)** Morgan et al: JBa *37*, 389, 1939. — **(44)** Meisel et al: ZI *73*, 448, 1932. — **(45)** White: JP *34*, 325, 1931, *41*, 567, 1935. — **(45a)** White: JP *36*, 65, 1933. — **(46)** Freeman: BJ *36*, 340, 1942. — **(47)** Malek: SB *126*, 127, 1937. — **(48)** Morgan et al: Br *23*, 151, 1942. — **(49)** Morgan: Br *12*, 62, 1931, *16*, 476, 1935; BJ *30*, 909, 1936; H *21*, 469, 1938. — **(50)** Linton: BaR *4*, 261, 1940. — **(51)** White: JP *44*, 706, 1937 (B). — **(52)** White: JP *50*, 160, 1940. — **(52a)** White: JP *51*, 447, 1940. — **(53)** Heidelberger et al: JBC *118*, 79, 1937 (B). — **(54)** Menzel et al: JBC *127*, 221, 1939. — **(55)** Karjala et al: JBC *137*, 189, 1941. — **(55a)** Seibert et al: JBC *140*, 55, 1941. — **(55b)** Tennent et al: JI *45*, 179, 1942. — **(56)** Maxim: BZ *223*, 404, 1930. — **(57)** Masucci et al: ART *22*, 678, 682, 1930, *24*, 737, 1931. — **(58)** Du Mont et al: ZPC *211*, 97, 1932. — **(59)** Chargaff et al: JBC *112*, 393, 1935. — **(60)** Takeda et al: ZPC *262*, 171, 1939. — **(61)** Raistrick: EE *7*, 343, 1938. — **(62)** Bloch et al: ADS *148*, 413, 1925. — **(63)** Lancefield: JEM *42*, 377, 1925, *59*, 441, 1934 (B); HL 1940/41, p. 251. — **(64)** Krestownikowa et al: ZI *78*, 414, 1933. — **(65)** Heidelberger et al: JI *30*, 267, 1936 (B); RI *4*, 299, 1938 (B). — **(66)** Seastone: JEM *70*, 361, 1939. — **(67)** Loewenthal: Br *19*, 164, 1938. — **(68)** Miller et al: JEM *59*, 75, 1934 (B). — **(69)** Julianelle et al: JEM *62*, 11, 23, 31, 1935. — **(70)** Mutermilch et al: SB *120*, 587, 1935. — **(71)** Casper: JI *32*, 421, 1937 (B). — **(72)** Krestownikowa et al: ZI *83*, 164, 1934. — **(73)** Scherp et al: JEM *61*, 753, 1935 (B); JI *37*, 469, 1939. —**(74)** Hassid et al: JBC *134*, 163, 1940 (B). —**(75)** Branham: Pr *25*, 25, 1927. — **(76)** Casper: ZH *109*, 170, 1928. — **(77)** Combiesco et al: Arch. Roum. Path. Exp. Microb. *3*, 189, 1930. — **(78)** Przesmycki: SB *95*, 744, 1926. — **(79)** Meisel et al: SB *114*, 364, 1933. — **(80)** Castaneda: JEM *62*, 289, 1935 (B). — **(81)** Otto: MK *31*, 333, 1935. — **(82)** Tomcsik: Pr *24*, 810, 1927. — **(83)** Prasek et al: CB *128*, 381, 1933. — **(84)** Kurylowicz: ZI *93*, 457, 1938 (B). — **(85)** Wong et al: Pr *39*, 161, 1938, *40*, 357, 1939 (B). —

(86) Miles et al: Br *20*, 83, 109, 278, 1939 (B); BJ *33*, 1709, 1716, 1939.— (87) Dingle: AJH *20*, 148, 1934.— (88) Pittman et al: JI *29*, 239, 1935.— (89) Dingle et al: JI *37*, 53, 1939.— (90) Sakamoto: JI *18*, 331, 1930.— (91) Weiss et al: JEM *67*, 49, 1938.— (92) Wong et al: Pr *41*, 160, 40, 356, 1939.— (93) Meisel: ZI *92*, 79, 1938.— (94) Hindle et al: RS, *B*, *114*, 523, 1934.— (95) Kurotchkin: Pr *37*, 21, 1937.— (96) Tomcsik: ZI *66*, 8, 1930 (B).— (97) Kesten et al: JEM *53*, 803, 1931; JID *50*, 459, 1932.—(98) Sevag et al: AC *519*, 111, 1935.— (99) Klopstock et al: ZI *88*, 446, 1936.— (100) T'ung et al: Pr *41*, 155, 1939.— (101) Campbell: JID *65*, 12, 1939; S *96*, 431, 1942.— (101a) Oerskov: CB *119*, 88, 1930.—(102) Morgan: BJ *30*, 909, 1936.—(103) Kendall et al: JBC *118*, 61, 1937.—(104) Sordelli et al: SB *107*, 736, 1931, *108*, 675, 1931; Fol. Biol. *1*, 97, 1932.— (105) Zozaya et al: JEM *55*, 353, 1932, v. *57*, 41, 1933.— (106) Heidelberger: ARB *1*, 662, 1932.— (107) Norman: ARB *10*, 65, 1941.— (108) Marrack et al: Br *19*, 53, 1938.— (109) Jones et al: JCS 1942, p. 225.— (110) Heidelberger et al: JEM *49*, 847, 1929.— (111) Heidelberger et al: JBC *84*, 639, 1929 (B).— (112) Hotchkiss et al: JBC *115*, 285, 1936.— (113) Partridge et al: Br *23*, 84, 1942.— (114) Weidenhagen: Angew. Chem. *47*, 451, 1934 (B). — (115) Goebel: JEM *68*, 469, 1938.— (116) Chow et al: JEM *62*, 179, 1935. — (117) Heidelberger et al: JEM *61*, 563, 1935.—(118) Marrack: 2d Int. Congr. Microbiol. London 1936, p. 425.—(119) Brown: Pr *32*, 859, 1935.— (120) Heidelberger et al: JEM *75*, 35, 1942 (B).— (121) Goebel: JBC *110*, 391, 1935.— (122) Beeson et al: JI *38*, 231, 1940.— (123) Goebel et al: JEM *46*, 601, 1927.— (124) Perlman et al: JI *43*, 99, 1942.— (124a) Sugg et al: JI *43*, 119, 1942.— (125) Castaneda: JEM *60*, 119, 1934.— (126) Kauffmann: JI *39*, 397, 1940.— (127) Burnet: Br *15*, 354, 1934.— (128) K. Meyer: ZI *69*, 499, 1931.— (128a) Walker: BJ *34*, 325, 1940. — (128b) Haas: ZI *92, 94, 97, 99*.— (129) Boivin: AP *61*, 426, 1938 (B).— (130) Mesrobeanu: Arch. Roum. Path. Exp. *9*, 1, 1936.— (131) Chase et al: ARB *8*, 579, 1939.— (132) Marrack: ARB *11*, 629, 1942.— (133) Mackenzie et al: JBa *40*, 197, 1940.— (134) Ciuca et al: SB *127*, 1414, 1938.— (134a) Hórnus et al: AP *66*, 136, 1941.— (135) Morgan et al: BJ *34*, 169, 1940 (B), *35*, 1140, 1941; JC *60*, 722, 1941.— (136) Freeman et al: BJ *35*, 564, 1941.— (137) Soru et al: SB *133*, 498, 1940. — (138) Pinner: ART *18*, 497, 1928.— (139) Pedersen-Bjergaard: ZI *82*, 258, 1934.— (140) Chargaff et al: AP *54*, 708, 1935.— (141) Klopstock et al: ZI *84*, 34, 1934.— (142) Sandor et al: AP *55*, 38, 163, 1935.— (143) Witebsky et al: Erg. Tuberkulosef. *5*, 125, 1933.— (144) Anderson et al: JBC *136*, 211, 1940 (B).— (145) Chargaff et al: BZ *255*, 319, 1932.— (146) Anderson: Ph *12*, 166, 1932; Fortschr. Chem. org. Naturstoffe *3*, 145, 1939; HL 1939/40, p. 271.— (147) Macheboeuf: VI Congr. chim. biol. 1937, p. 261.—(148) Sabin: Ph *12*, 141, 1932.— (149) Macheboeuf et al: CR *204*, 1843, 1937; AP *55*, 547, 1935.— (150) Bloch: ZPC *244*, 1, 1936.— (150a) Bailey: Year Book Path. a. Immunol., 1940, p. 616.— (151) Freund: JI *13*, 161, 1927.— (151a) Hoyle: JH *42*, 416, 1942.— (152) Boquet et al: AP *37*, 787, 1924.— (152a) Eisler et al: ZI *53*, 151, 1927 (B).— (153) Wells: CAI.— (153a) Przesmycki: ZI *51*, 408, 1927.— (154) Dienes: JI *17*, 85, 157, 1929.— (155) Gundel et al: ZI *66*, 45, 59, 78, 1930; v. ZI *69*, 244, 1930.— (156) Sachs: ZI *69*, 221, 1930.— (157) Nussbaum: ZH *113*, 305, 1932.— (158) Annell et al: ZI *61*, 336, 1929.— (159) Wadsworth et al: JI *21*, 255, 1931.— (160) Weil: BaR *5*, 293, 1941.— (161) Sachs: EH *9*, 29, 1928.— (162) Landsteiner: HPM *1*, 1069, 1929.— (162a) Sevag et al: JBC *139*, 925, 1941.— (163) Ivánovics et al: ZI *90*, 304, 1937, *91*, 175, 1937, *97*, 443, 1940.— (164) Partridge et al: Br *21*, 180, 1940, *23*, 84, 1942; JC *60*, 722, 1941.— (165) Przylecki et al: BZ *286*, 360, 1936 (B).— (166) Landsteiner et al: CB *42*, 562, 1906.— (167) Takaki: BCP *11*, 288, 1908.—

(168) Loewe: BZ *33*, 225, 1911, *34*, 495, 1911.— **(169)** Noguchi: Uni. Penn. Med. Bull. *15*, 327, 1902.— **(170)** Abderhalden et al: Z. Exp. Path. *2*, 199, 1905. — **(171)** Walbum: ZI *7*, 544, 1910.— **(172)** Flexner et al: JEM *6*, 277, 1902.— **(173)** Calmette: CR *134*, 1446, 1902.— **(174)** Kyes: BZ *4*, 99, 1907.— **(175)** Lüdecke: Ing. Diss. München 1905.— **(176)** Manwaring: ZI *6*, 513, 1910.— **(177)** Delezenne et al: Bull. Soc. Chim. *15*, 421, 1914.— **(178)** Hazato: ZI *89*, 1, 1936.— **(179)** White: JP *31*, 423, 1928.— **(180)** Sordelli et al: SB *92*, 898, 1925, *84*, 173, 1921.— **(181)** Landsteiner et al: JI *10*, 731, 1925, *14*, 81, 1927.— **(182)** Brunius: Chemical Studies, etc., Fahlcrantz, Stockholm, 1936.— **(183)** ARB *6*, 621, 1937.— **(183a)** Landsteiner et al: JEM *41*, 427, 1925.— **(184)** Mizuhara: ZI *40*, 84, 1924.— **(185)** Fujimura: JB *32*, 329, 1940.— **(186)** Landsteiner et al: WK 1907, p. 1565.— **(187)** Marie et al: AP *21*, 138, 1907.— **(188)** Sakakibara: JB *24*, 31, 1936.— **(189)** Oe. Fischer: ZI *79*, 391, 1933, *89*, 133, 139, 1936.— **(190)** Weil et al: ZI *78*, 316, 308, 1933 (B).— **(191)** O. Fischer et al: ZI *87*, 400, 1936 (B), *85*, 233, 1935.— **(192)** Pangborn: JBC *143*, 247, 1942 (B).— **(193)** Brown et al: JBC *137*, 525, 1941 (B).— **(194)** Rudy: BZ *267*, 77, 1933 (B).— **(195)** Balbi: ZI *79*, 372, 1933.— **(196)** Rudy: BZ *253*, 204, 1932; K 1933, p. 1279.— **(197)** Klopstock et al: ZI *79*, 53, v. 39, 1933.— **(198)** Zechmeister et al: Principles and Practice of Chromatography, Chapman, London, 1941.— **(199)** Brahn et al: K 1926, p. 1455.— **(200)** Landsteiner et al: JEM *42*, 123, 1925; Pr *22*, 289, 1925.— **(201)** Doelter: ZI *43*, 95, 1925.— **(202)** Witebsky: ZI *48*, 369, 1926, *49*, 1, 1926.— **(203)** Lattes et al: WK 1928, p. 1038.— **(203a)** Wheeler et al: JI *33*, 393, 1937.— **(204)** Hallauer: ZI *83*, 114, 1934.— **(205)** Ottensooser: ZI *77*, 140, 1932.— **(206)** Kossjakow et al: ZI *98*, 261, 1940 (B), *99*, 221, 1941.— **(207)** Boyd: JI *27*, 485, 1934.— **(208)** Schiff: Ueber die gruppenspezifischen Substanzen, etc., Fischer, Jena, 1931.— **(208a)** Boyd: JI *37*, 65, 1939.— **(209)** Brahn et al: K 1932, p. 1592.— **(210)** Schiff et al: BZ *235*, 454, 1931.— **(211)** Freudenberg et al: AC *510*, 240, 1934, *518*, 97, 1935; Nw 1936, p. 522.— **(212)** Landsteiner: JEM *63*, 185, 1936.— **(213)** Landsteiner et al: JEM *63*, 813, 1936, *71*, 551, 1940 (B).— **(214)** Freudenberg: Sitzungsber. Heidelberg Akad. d. Wiss. 1938–40.— **(215)** Goebel: JEM *68*, 221, 1938.— **(216)** Meyer et al: JBC *119*, 73, 1937.— **(217)** Witebsky et al: JEM *73*, 655, 1941 (B).— **(218)** Van Slyke et al: JBC *141*, 627, 1941.— **(219)** Jorpes et al: ZI *81*, 152, 1933; APS *11*, 91, 99, 1934.— **(220)** Goebel: JBC *129*, 455, 1939; JEM *70*, 239, 1939.— **(221)** Ottensooser et al: Schweiz. Z. allg. Path. *1*, 421, 1938.— **(222)** Ivanovics: ZI *98*, 373, 1940.— **(223)** Landsteiner et al: Pr *32*, 713, 1208, 1935.— **(224)** Schiff: JID *65*, 127, 1939.— **(225)** Chase: JBa *36*, 383, 1938.— **(226)** Landsteiner et al: JBC *140*, 673, 1941.— **(227)** Wiener: Blood Groups and Blood Transfusion, Thomas, Springfield, 3d ed., 1943.— **(228)** Schroeder: ZI *75*, 77, 1932.— **(229)** Dujarric et al: AP *55*, 331, 1935.— **(229a)** Weltner: BZ *297*, 142, 1938.— **(230)** Dubos: HL 1939/40, p. 223; BaR *4*, 1, 1940.— **(231)** Dubos et al: JEM *54*, 51, 1931.— **(232)** Karström: EE *7*, 350, 1938.— **(233)** Bloch: ZPC *98*, 226, 1916; K 1932, p. 10.— **(234)** Peck et al: K 1932, p. 14.— **(235)** Sickles et al: Pr *32*, 857, 1935.— **(236)** Morgan et al: N *132*, 604, 1933; v. Br *16*, 476, 1935.— **(237)** Sickles et al: Pr *31*, 443, 1934, *32*, 857, 1935 (B); JID *53*, 38, 1933; JBa *28*, 415, 1934.— **(238)** Meyer et al: JBC *113*, 479, 1936.— **(239)** Epstein et al: Br *21*, 339, 1940.— **(240)** Meyer et al: JEM *71*, 137, 1940.— **(241)** Hehre et al: JEM *75*, 339, 1942.— **(242)** Levine et al: JEM *59*, 213, 1934 (B).— **(243)** Gough et al: JP *38*, 301, 1934; AJE *15*, 227, 1937. — **(244)** White: JP *43*, 591, 1936.— **(245)** Griffith: JH *27*, 113, 1928.— **(246)** Dawson et al: JEM *54*, 681, 701, 1931.— **(247)** Alloway: JEM *55*, 91, 1932, *57*, 265, 1933.— **(248)** Shaffer et al: JEM *64*, 281, 1936.— **(249)** Berry et al: JBa *31*, 50, 1936, IC, p. 343.— **(250)** Gardner et al: JID *71*,

47, 1942. — **(251)** Adams et al: JBC *140*, 653, 1941. — **(252)** Henderson: Br *20*, 11, 1939. — **(253)** Baldwin et al: BJ *36*, 37, 1942. — **(254)** Poindexter: JEM *60*, 575, 1934. — **(255)** Pirie: Br *17*, 269, 1936. — **(256)** Heidelberger et al: PNA *28*, 516, 1942, (B). — **(257)** Felix: IC, p. 798. — **(258)** K. Meyer: SB *129*, 825, 1938 (B). — **(259)** Woolf: RS *130*, 70, 1941. — **(260)** Bliss: JI *34*, 337, 1938.

VII

ANTIGEN-ANTIBODY REACTIONS [1]

Toxin neutralization. — The study, inaugurated by Ehrlich, of the quantitative relations in the neutralization of toxin by antitoxin was years ago the source of much controversy concerning the nature of antibody reactions. His concept that toxin and antitoxin combine in effect irreversibly and in a fixed ratio met with difficulties. For on this assumption, if the quantity of diphtheria toxin which on sub-cutaneous injection kills a guinea pig of 250 gm. within 4 days is called a minimum lethal dose (M.L.D.): if, further, an L_0 dose (L=limes) is defined as the greatest amount of toxin which is neutralized by one (arbitrary) unit of antitoxin, and an L_+ dose is defined as the smallest quantity which, mixed with one unit of antitoxin, kills a guinea pig: then the difference between L_+ and L_0 should be equal to one minimal lethal dose. Actually, quite a number of M.L.D. must be added to the neutral mixture of toxin and one antitoxin unit to get the lethal effect. In general, equal fractions of a unit of antitoxin, added to an L_0 dose, neutralize successively diminishing quantities of toxin, and when the hemolytic power of mixtures, for instance of tetanus hemo-toxin and antitoxin, is plotted against the amount of antitoxin, smooth curves, resembling adsorption isotherms, are obtained (similar in shape to that to be drawn from the values for agglutination in Table 35). Combination in definite ratios would, however, yield straight line graphs, as does the reaction between strong acids and bases, and so Ehrlich was forced by his hypothesis to the supposition, now generally abandoned, that the neutralization curve comes from the pres-

---

[1] For detailed information on physico-chemical aspects see the comprehensive reviews by Marrack (1), Heidelberger (2), Boyd (3) which have been made use of in the presentation. See, also, Haurowitz (4), Marrack and Smith (5), (145); Glenny (22) (toxin neutralization), von Dungern (6) (precipitin reactions).

ence in the toxins of constituents differing in toxicity and affinity for antitoxin. The competing "adsorption" theory of Bordet (7), embracing the whole of serology, set the quantitative relations down to combination in continuously varying proportions and by this criterion placed the serological reactions together with adsorption phenomena in a class distinct from ordinary chemical reactions. However, not only has the chemical character of serological specificity become more and more manifest, but failure to sort out definite proportions from the reactions of a complex system does not disprove the actual existence of stoichiometric relations or provide evidence for the nature of the forces involved; and many instances of so-called adsorption have been recognized as due to chemical affinity, a point touched upon by Bordet himself. A third theory advocated by Arrhenius and Madsen (8) approaches current ideas, but their assumptions are at variance with experimental results subsequently obtained. It conceives of toxin neutralization as a chemical reaction which, comparable to the neutralization of a weak acid by a weak base, leads to an equilibrium in accordance with the law of mass action. The equation derived by the authors fits observed numerical data but other equations can be offered that do about equally well.[2] At present the neutralization curve is explained like other antiprotein reactions (v. p. 256 and Table 34b) on the basis of combination in multiple proportions.[3]

For the relation of the values obtained in antitoxin titration with the flocculation method of Ramon (p. 32) and by neutralization tests in animals (general toxicity or skin reactions), the reader is referred to papers by Glenny, and others;[4] often the flocculation and neutralization values are not far apart. Discrepancies in the combination with antitoxin, between toxin and toxoid are discussed by Glenny (15).

The flocculation point has been found to correspond to the ratio of two molecules of antitoxin to one toxin molecule, i.e. less than the ratio of toxin to antitoxin at the antibody excess end of the equivalence zone.[5] The question is thereby raised how neutralization is effected and why such incompletely saturated toxin should be no longer toxic; possibly by combination of toxin with antitoxin the toxic groups are blocked even if they are not really the binding sites (Boyd).

[2] v. Boyd (3), Ghosh (9).

[3] (10), (11); v. (12), (2). A dissenting opinion reviving Ehrlich's theory was offered by O'Meara (13) to the effect that the phenomena are accounted for by the presence of two components in diphtheria toxin.

[4] (14–17), (10), (75).

[5] Pappenheimer et al. (18) (ultracentrifuge experiments); Kekwick and Record (19), Boyd (20).

On aging, toxoid is formed from toxin which leads to increasing discrepancy between the L_0 and L_f doses, the latter representing the amount of toxin equivalent by flocculation to one unit of antitoxin.

Technical prescriptions for the standardization of antitoxin are given by Wadsworth (21), Glenny (22), and Boyd (3).

Antitoxins produced in the rabbit are generally of much lower titer than horse antitoxins [v. Freund (22a)].

QUANTITATIVE STUDIES ON PRECIPITIN REACTIONS. — When serial tubes are set up with a constant amount of antiserum and increasing quantities of antigen, the amount of precipitate rises to a maximum, then, as a rule, falls off and finally vanishes (inhibition zone), as soluble compounds of antibody with much antigen are formed.[6] The excess of antigen necessary for complete inhibition is different with various antigens and sera. Often the amount of precipitate is still increasing in the zone (equivalence zone) in which practically all antibody has combined and at best only traces are detectable in the supernatant solution, owing to the binding of more antigen to the precipitate. With certain immune sera a marked inhibition zone occurs also with an antibody excess, particularly noticeable in Ramon's flocculation of diphtheria toxin by horse antitoxin, which takes place only within a very narrow range. This was taken to be an exceptional case but Pappenheimer showed that a horse immune serum for egg-white behaves in the same manner as the antitoxin,[7] and diphtheria rabbit antitoxin reacts like other antiprotein rabbit sera, i.e. without inhibition by antibody excess. The peculiarity therefore appertains to antiprotein immune sera produced in the horse rather than to the special antigen.[8]

Why the compounds containing much antigen are more often soluble than those rich in antibody — even when serum globulin, a protein so close to antibodies, is used as antigen — is an interesting question not definitely answered as yet. It has been attributed to the greater number of binding groups in antigens as compared with antibodies [9]

[6] Ultracentrifugal determinations on precipitins, dissolved in an excess of antigen, were carried out by Heidelberger et al. (20a) and Pappenheimer et al. (18).

[7] (23). Horse antiserum for hemocyanin, likewise, gave no precipitation with a great excess of antibody but a wider precipitation zone than diphtheria antitoxin [Hooker and Boyd (24)]. To a lesser degree such zones were found with other sera, as certain horse antipneumococcal sera (v. 25).

[8] Cf. (1). It may be that in the horse the route of injection influences the character of the antibody [Heidelberger et al. (26)].

[9] (1), (4); (pp. 58, 140).

which indeed is an ostensible reason for the unsymmetric shape of the reaction curve. Marrack suggests that when antibody is present in excess, hydrophile polar groups are occluded through the tight packing of the antibody molecules, whereas with an excess of antigen such blocking does not occur because of the smaller number of antigen molecules scattered on the surface of the complex. If this is the solution, one must explain why there are sera that give an inhibition in the zone of antibody excess, which shows that special properties of the sera play a part, possibly the greater or lesser hydrophilic character of the antibodies [10] or the mode of distribution of the binding sites [Pappenheimer et al. (18)].

As seen from Table 34, a and b, the composition of the precipitates varies with the proportions of the reactants. Molecular ratios are given which show that antigens are multivalent with regard to antibodies and that the antibody-antigen ratio is high with the large molecules of thyroglobulin and hemocyanin. It is in most cases about twice as high in the region of antibody excess as in the equivalence zone. From determinations of Malkiel and Boyd (28) the ratio in the zone of antigen excess is proportional to the amount of antiserum.

TABLE 34a. — RATIO OF ANTIGEN TO ANTIBODY IN PRECIPITATES FORMED IN EQUIVALENCE ZONE
[after Marrack (1)]

	Approximate molecular weight	1:	1:
Egg albumin	44000	8.6	−15
Pseudoglobulin horse	167000	3.2	− 4.5
Hemocyanin (Busycon)	6680000	0.59−	0.75

On prolonged immunization higher antibody-antigen ratios and broadening of the equivalence zone were observed (v.p. 144), [Heidelberger et al. (27), Malkiel and Boyd (28)].

With heterologous systems lower antibody-antigen ratios were found than in homologous reactions and the combination may be markedly dissociable.[11] As an example, curves are given in the following figure of the reactions of an azoovalbumin antiserum with the homologous antigen (I), the unchanged ovalbumin (II), and of anti-ovalbumin serum with ovalbumin (III). The relations, however, differ with various cross reacting systems. Infrequently heterologous an-

[10] Brown; Boyd (20).
[11] Heidelberger et al. (30), Hooker and Boyd (31).

TABLE 34b. — MOLECULAR COMPOSITION OF SPECIFIC PRECIPITATES FROM RABBIT ANTISERA [after Heidelberger (29), Pappenheimer (18), and Boyd (3)]

Antigen or Hapten	At extreme antibody excess	At antibody excess end of equivalence zone	At antigen excess end of equivalence zone	In inhibition zone	Soluble compounds
Cryst. egg albumin (Ea)	EaA_5	EaA_3	Ea_2A_5	EaA_2	(EaA)[‡]
Cryst. serum albumin (Sa)	SaA_6	SaA_4	SaA_3	SaA_2	(SaA)
Thyroglobulin (Tg)	TgA_{40}	TgA_{14}	TgA_{10}	TgA_2	(TgA)
Type-III pneumococcus polysaccharide (S)	SA	S_3A_2	S_2A	S_5A	(S_5A)
Viviparus Hemocyanin (H)		HA_{120}	HA_{88}	HA_{86}	
Diphtheria toxin (T)	TA_8	TA_4	TA_2[*] T_2A_3	TA and T_2A	

A = antibody; S = minimum polysaccharide chain weight reacting.
The molecular weight of rabbit-antibody is taken as 150,000.
* Flocculation point [cf. (19)].
‡ Data in parentheses are uncertain.

tigens are found that give more precipitate than the homologous,[12] presumably because of greater precipitability of the former.

In the case of azoproteins the number of antibody molecules bound increases with the number of haptenic groups, as would be expected, each group combining with one antibody molecule; but above a limit which is probably imposed by steric hindrance, progressively less of the added hapten groups are occupied [Haurowitz (34)]. An anal-

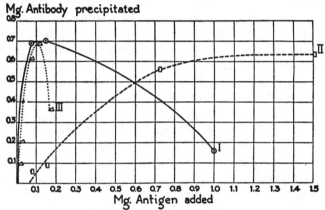

Quantitative relations between an R-salt-azo-benzidine-azoovalbumin immune serum and (I) the corresponding azoovalbumin, (II) unaltered ovalbumin, and between anti-ovalbumin and ovalbumin (III). The special type of heterologous reaction (II) may be attributable to high dissociation of the antigen-antibody complex [cf. (30)].

ogous condition was seen in the precipitation of iodoproteins, except that only a few of the iodized groups, in Haurowitz' opinion those on the surface of the molecule, reacted with antibodies (35).

The ratio of antigen and antibody in specific precipitates is easily determined in the case of antigens with "tracers" such as iodoprotein, hemoglobin, hemocyanin, or arsenic-containing azoproteins which can be quantitatively estimated in the presence of other proteins, and simply by nitrogen analysis with precipitates of nitrogen-free polysaccharides. With ordinary protein antigens, if an excess of antigen is used, the amount of antigen remaining in the supernatant fluid can be estimated by precipitin tests with a serum of known potency and, on subtracting this from the total added, the amount present in the precipitate is found.[13] For the analyses the precipitates are washed with cold saline, which is feasible on account of their

[12] Kleczkowski (32), Heidelberger et al. (33).
[13] Heidelberger and Kendall (36); v. (28). For microdetermination of polysaccharides by precipitin reactions see (36a). A quantitative study of phage-antiphage reactions was made by Hershey, Kalmanson and Bronfenbrenner (36b).

mostly negligible solubility (or dissociation) amounting, for example, at low temperature to about 0.005 mg. in 1 ml. saline for an ovalbumin precipitate.[14] Naturally, the solubility or dissociation will vary with the nature and composition of the precipitate and the temperature; [15] certain precipitates were seen to go into solution on warming, or were formed only at a temperature near 0°.[16]

Other proteins than those involved in the specific combination may influence the course of serum reactions but are not carried down in precipitates to an appreciable extent, as shown repeatedly.[17] The values for the lipid content [18] of precipitates were found to vary greatly with different precipitating systems, from less than 1% in diphtheria toxin precipitates and 2–8% in those of hemoglobin to large percentages in precipitates obtained with anti-pneumococcus sera if but small quantities of antigen are used (47). Lipids play a part in the flocculation since, as Hartley (48) demonstrated, protein antigens and antibodies do combine (and the mixture may become opalescent) but no precipitate is formed when both have been extracted with ether; the immunizing power of proteins is not impaired by the extraction.[19] The observation of Horsfall and Goodner, that the precipitating and agglutinating activity of pneumococcus sera after extraction with organic solvents could be restituted by small quantities of lipids, is noteworthy — particularly, that horse sera were restored by "lecithin," rabbit sera by "cephalin." Antibodies from other animals fell into the one or the other class, and various immunological differences were found between the two categories (concerning the behavior in complement fixation, passive sensitization, protective capacity, prozones, molecular weight, etc.).[20]

AGGLUTINATION. — The union of agglutinins and lysins with cells was examined quantitatively by estimating the amount of free anti-

---

[14] (36), v. (1), (24), (37).

[15] High salt concentrations and major changes in pH interfere with the antibody reaction (38), (39). Detailed data on the influence of pH and salt concentration on the combination of antigen and antibody and on flocculation under various conditions are given in (1), (3), (39a). According to Duncan (39b) the binding of bacterial agglutinins is lessened below a certain salt optimum. For inhibition of precipitation by "lyotropic" anions (salicylate) v. (39c).

[16] (40), (41), (42).

[17] Marrack and Smith (43), Haurowitz and Breinl (44), Heidelberger and Kendall (45); v. (1). At low temperature antibodies were seen to be non-specifically adsorbed to precipitates [Goodner et al. (45a)].

[18] (46), (1).

[19] v. (34).

[20] Horsfall, Goodner, MacLeod (49), (47); v. (50).

body in the supernatant fluid after removing the cells by centrifugation, in the early work of Eisenberg and Volk, and by other investigators.[21] In this way it was established that a very much greater amount of agglutinin or lysin is more or less firmly fixed by cells than that necessary for agglutination. Furthermore, it can be calculated that quantities of agglutinins much too small to cover the surface of the cells suffice for agglutination,[22] and in an experiment by Jones (54) collodion particles which were successively treated with five different antigens could be agglutinated by any of the corresponding antisera although evidently each antigen occupied but a part of the surface.[23] Conformable inference was drawn from experiments on the agglutination by an anti-azoprotein serum of erythrocytes coupled with diazonium compounds,[24] and from the quantitative relations obtaining in the flocculation reactions of lipid particles.[25]

By nitrogen determination Heidelberger and Kabat (58) found that pneumococci bound up to more than 1 mg. antibody N per mg N in the bacteria. An increase in the volume of bacteria upon agglutination of 30 to 65% on the average, was reported by Jones and Little (59); of this only a fraction can be attributed to attached antibody.

The ratio of absorbed to free agglutinins at varied antibody concentrations (Table 35) fits at least within a certain range Freundlich's adsorption equation ($\frac{x}{m} = k\,c^{1/n}$) [26] which, however, does not warrant conclusions to be drawn on the nature of the reaction. The "equilibrium" reached is generally shifted by an increase in temperature in the direction of diminished absorption.

Some sera show an inhibition zone ("prozone"), and frequently agglutination is slowed down at high agglutinin concentrations. This phenomenon was found to be well-marked with some heated immune sera and was attributed to modified agglutinins — agglutinoids — that are capable of combining but do not cause clumping, or to coating of the cells with non-specific proteins or other substances.[27]

[21] Eisenberg and Volk (51), Cromwell (52), Dreyer and Douglas (53).
[22] v. (1); (53a) (study of hemolysins).
[23] v. de Kruif and Northrop; Eagle (55).
[24] Pressman, Campbell and Pauling (56).
[25] Eagle (57); v. (57a).
[26] x=the amount absorbed, c=the amount in the supernatant solution, m=the quantity of adsorbent, k and n constants.
[27] v. (1), (51), (60–64), (52).

RATE OF REACTION.[28] — Binding of antibodies, as a rule, proceeds fast [29] and in several experiments was found to be accomplished, for the most part, in less than a minute, or within not more than 15 minutes at the outside; [30] from experiments of Mayer and Heidelberger [31] with polysaccharide antisera it is concluded that the reactions went

TABLE 35. — ABSORPTION OF ANTITYPHOID HORSE SERUM BY A CONSTANT AMOUNT OF BACILLI

Serum dilution	Agglutinin units	Units absorbed	Percentage absorption
10000	2	2	100
1000	20	20	100
500	40	40	100
300	67	67	100
100	200	180	90
50	400	340	85
10	2000	1500	75
2	10000	6500	65
1	20000	11000	55

One unit of agglutinin is here taken as the smallest amount of serum still giving definite agglutination.

to 90% completion in less than three seconds at 0°. The second stage, the clumping of cells or formation of precipitates, is a slower process — markedly so in the Ramon flocculation — and may require hours to reach a maximum although with sera of good potency precipitation starts immediately after mixing. By centrifuging, agglutination is greatly hastened and this procedure, when applied to red cells (a routine often used in blood group determination), proves again the rapidity of the primary combination since the whole manipulation takes but a few minutes.

The time required for the formation of precipitates, and for agglutination and lysis, has been studied under varied experimental conditions.[32] The rate is influenced by the concentration of reacting substances and by the environment, such as salt content, pH, presence of other proteins and non-electrolytes. Clearly, higher concentrations will increase the number of collisions between molecules and particles

[28] v. (1). For theoretical deductions see Hershey (65).
[29] This is easily demonstrated with particulate antigens which can be speedily separated by centrifuging.
[30] (52), (27); v. (66), (67).
[31] (68); v. (27).
[32] v. (70), (71).

capable of combining, and thus in general also the velocity of flocculation. As Hooker and Boyd found (69), the rate of agglomeration, in a zone of considerable antibody excess, is proportional to the concentration of antigen, in accordance with Smoluchowski's theory of the flocculation of colloids. The velocity of flocculation is increased by stirring and by a rise in temperature up to the limit where the active substances are deteriorating. However, higher temperature generally favours dissociation of the antibody combination and in cases in which dissociation (or solubility) is greatly enhanced the acceleration by warming may be overbalanced.

Mention has been made, in connection with the titration method of Dean and Webb (p. 15), of the dependence of the flocculation rate on the proportions of the reactants. Theoretical considerations have centered around this topic and especially the unanticipated fact that the proportion at which flocculation proceeds most rapidly is not the same when antibody is kept constant and antigen is added in different quantities, as it is when one uses varying amounts of antibody and a constant amount of antigen (so-called α and β procedures where, of course, not the same tubes are compared).[33] The β (constant-antigen) optimum is always shifted in the direction of increasing antibody. Referring for a detailed discussion to the writings of Topley, Marrack, Boyd,[34] it may here be said that the discrepancy between the α- and β-series seems to be related to differences in solubility of the compounds containing a large proportion of antibodies. In fact, in the Ramon titration of diphtheria-antitoxin — performed with the β-procedure — precipitation is prevented also by an antibody excess (p. 242) and, concurrently, there is generally approximate correspondence between the α- and β-optima.

Boyd, in a thorough reinvestigation of the question of flocculation optima, with twenty antisera, arrived at a distinction of roughly two types: H sera, represented by diphtheria antitoxin and other horse antiprotein sera, and R sera — most of the rabbit antisera. The former show optima with both the α and β technique; the latter give an α optimum but no consistent β optimum with varied antigen concentrations. The formation of soluble complexes with sera of the H anti-protein type in the zone of antibody excess is reasonably attributed to the greater solubility of the H antibodies (v. p. 242).

[33] v. Ramon (72), Miles; Taylor, Adair and Adair (17), Brown (73). On optima in agglutination v. Duncan (41), Miles (74).

[34] (75), (1), (3).

FIRMNESS OF COMBINATION. — Total or partial reversibility of serological reactions is evidenced in the recovery of antibodies (p. 137) from precipitates and agglutinated cells by various means, as warming or high salt concentrations — most readily from complexes rich in antibody — and by the possibility of liberating unchanged antigen (e.g. diphtheria toxin) from the combination with antibody by dilution, heating, acids, etc.[35] A conspicuous demonstration of reversibility is the following. Red cells are charged with hemolysin in excess and separated from the solution; when new erythrocytes are added to a suspension of the sensitized cells and later, after incubation, complement, all the cells will be hemolyzed.[36] In the degree of dissociation of the antigen-antibody complex there are wide differences, as is seen in the behaviour of hemagglutinins. Thus red cells agglutinated by normal sera commonly release much agglutinin upon being warmed to 50–60° while agglutinins of immune sera are much more firmly bound; [37] and when red cells are agglutinated, in the cold, by serum of the same animal (p. 85) the clumps disperse even at room temperature and the agglutinin is set free.[38] Differences depending on greater or lesser affinity (avidity) of antibodies were further noted in the speed of flocculation by precipitins, and diphtheria antitoxins were found to vary in reversibility, therapeutic effect and the ratio of flocculation and neutralization titer.[39] Of interest in this connection are observations by Woolf (86) on the reactions, with a pneumococcus antiserum, of the homologous pneumococcus polysaccharide and of cherry gum. There was a parallelism, depending on the difference in affinity, between the amount of precipitate, the speed of its formation and the concentration of hapten (sodium euxanthate) required for inhibition, which was much higher in the case of the homologous pneumococcus polysaccharide.[40]

[35] Madsen et al. (76), Morgenroth (77), Otto and Sachs (78), Glenny et al. (79), Morris (80), Hartley (81).

[36] Morgenroth; Philosophow (82).

[37] Landsteiner et al.; Prášek (83). In experiments of Follensby and Hooker the equilibrium attained in the combination of antitoxin or antiovalbumin with the corresponding antigens was practically not affected by changes in temperature (71).

[38] Landsteiner (84). The characteristic property of combining at low temperature only, is shared by the autolysin responsible for the disintegration of red cells in paroxysmal hemoglobinuria.

[39] Glenny et al. (14, 15), Ramon (85).

[40] On account of the relative magnitude of this effect Woolf considers inhibition reactions as a serviceable measure of antibody affinity (v. p. 186).

The presence of both antigen and antibody in the supernatant fluids of precipitates in the equivalence zone, detected at times by adding more of the reactants, may be caused, apart from dissociation or solubility of precipitates, by imperfect homogeneity of antigen and antibody.[41]

Although from the facts above one must, theoretically, suppose dissociation to occur always to some extent, it is often inappreciable under ordinary conditions.[42] A recognized instance is the long known Danysz effect which is seen when to a certain amount of diphtheria or tetanus antitoxin a suitable (equivalent) amount of toxin is added at once and, in a parallel test, in fractions at intervals of say 15 minutes; rather unexpectedly the two mixtures behave differently, the first being neutral, the second toxic.[43] Apparently the reaction is not easily reversible, and in the second experiment a complex is formed initially with a higher proportion of antitoxin than necessary for neutralization which, perhaps after a secondary change (maybe agglomeration),[44] does not react readily with a new portion of toxin. That there is some degree of reversibility may be indicated by experiments of Healey and Pinfield (12).

Evidence for a firm combination is the considerable heat evolution, determined calorimetrically by Boyd et al.,[45] in the reaction of an antihemocyanin serum. In order to exclude any heat produced by the formation of a precipitate, it was measured in the zone of antigen excess and amounted to 40000 calories per mol antibody.

Secondary changes of antibodies which tend to reduce the reversibility of antigen-antibody compounds were described by Otto and Sachs, Krogh, and others;[46] however, inamuch as both components may be recoverable in active form from the combination, this cannot be an intrinsic feature of the antigen-antibody reaction.[47] As regards the solubility of precipitates in an antigen excess, no reduction of reversibility with time was observed by Boyd who found that (ovalbumin) precipitates were not less soluble, after storage for 10 months

---

[41] v. (87), (1).

[42] v. Burnet (88). Martin and Cherry (89) were unable to recover active snake venom from the mixture with antitoxin even by heating to a temperature at which antitoxin but not toxin is destroyed.

[43] On similar phenomena with other serological systems v. (82), (90). A "Danysz effect" was observed by Bordet also in the absorption of dyes by blotting paper (91).

[44] Pappenheimer and Robinson (10); v. (1).

[45] (92); v. Smith and Marrack (93).

[46] (78), (94); v. (1); Enders and Shaffer (95), Pope; Hershey (128).

[47] On X-ray evidence v. (1).

at low temperature, than originally [48] and, conversely, soluble mixtures of antibody with an excess of antigen give precipitates on addition of more antibody even after a long time. Also the solubility of precipitates was found by Woolf not to be appreciably diminished on standing for 24 hours.[49]

STAGES OF ANTIBODY REACTION. — In serological reactions two stages have been distinguished since Bordet (99) found that after fixation of lysins to cells addition of complement is necessary to bring about dissolution and that, although agglutinins combine with bacteria in salt-free solutions, clumping does not occur unless electrolytes are present. The first stage, then, is the specific union between antibodies and the substrate, and this is followed by secondary, visible changes, such as flocculation or lysis — except in the case of many haptens where no second stage ensues and the union with antibodies is demonstrable by inhibition tests only.

The effect of electrolytes (cations) on bacteria impregnated with agglutinins [50] (polyvalent cations having greater effect) at once brought to mind the precipitation of colloidal solutions and fine suspensions by salts and, with Bordet, most investigators looked upon the second stage of agglutination and precipitation as analogous to this familiar phenomenon. A physico-chemical interpretation was given by Northrop and de Kruif who from studies on bacterial agglutination concluded that the stability of bacterial suspensions is the resultant of opposing forces, the cohesion of the bacteria, measurable by the force needed to pull apart cover slips coated with bacteria and, on the other hand, the attraction for water and a repulsive force depending on the charge of the cells. Upon treatment with agglutinins, the surface potential of the (strongly electronegative) bacteria is mostly decreased [51] to about the level of serum globulins, and the cohesive force is enhanced, the bacteria getting less hydrophilic and susceptible to agglomeration by small concentrations of electrolytes.[52]

More recently the theory has been put forward that specific forces intervene in the second stage of precipitation and agglutination and,

[48] Boyd (98).

[49] (86). Some observations on diminished solubility are difficult to judge because of the coarseness of the precipitates which impedes the interaction.

[50] For details v. (1).

[51] A weak effect of agglutinins may reveal itself under the microscope by unevenness in the distribution of red cells. Cf. (99a) (observation of bacterial agglutination in cinematographic pictures).

[52] Northrop et al. (100), (101); v. (1), (102–104), (75).

in the opinion of some authors, these are sufficient to account for the formation of large insoluble aggregates. As Topley and his colleagues showed, when a suspension made of two sorts of bacteria was acted upon by a mixture of the corresponding agglutinins, two sorts of clumps were formed, each composed of one bacterial species only (105). This result supports Marrack's theory (1) which is upheld by Heidelberger (2) that owing to the specific forces during the second stage of agglutination and precipitation a lattice is formed consisting, in the words of Topley, of "masses of antigen and antibody molecules bound together by specific linkages, any one antigen molecule within the mass being united to two or more molecules of antibody, any one molecule of antibody to two or more molecules of antigen. . . ."

In similar experiments [53] mixed clumps, contrary to Topley's result, were obtained by Abramson, and by Hooker and Boyd but, again, homogeneous ones were seen by Wiener and Herman who used a suspension containing pneumococci of type I and human red cells and a mixture of the corresponding immune sera (or erythrocytes and a "Wassermann antigen"). Remarkably, pneumococci of type XIV and human red cells gave mixed aggregates upon addition of immune serum for Pneumococcus type XIV which is in good accord with the lattice [54] idea since this serum gives cross agglutination with human erythrocytes.[55] This experiment would eliminate the objection that the formation of homogeneous aggregates is due only to a specific attraction between like bacilli. The case for the older theory, experimental and theoretical, was stated by Hooker and Boyd [56] and, as Marrack (116) remarks, "the observations that flocculation does not take place in the absence of salt, although antibody combines with antigen" are obvious arguments for the intervention of non-specific factors in the flocculation phase.[57] Topley's result,

[53] (106), (107), (108).
[54] Pauling suggests instead the name "framework theory" on account of the special meaning of "lattice" in crystallography. Also, the words "alternation" and "mutual multivalence" theory have been proposed.
[55] With ricin and two sorts of readily agglutinable red cells the author observed "mixed" agglutination.
[56] (107), (109–111), (3); v. the reply offered by Heidelberger (2) and by Duncan (112), and Marrack's discussion (1).
[57] Cf. Heidelberger (2) (discussion of the salt effect).
The fact that agglutinated cells and precipitates may stick to the walls of test tubes, and that sensitized cells tend to adhere to leucocytes or to platelets (113), (114), may possibly be caused, in keeping with the lattice theory, by antibody adsorbed to the glass surface, and perhaps on cells.

on the other hand, is strong evidence for the lattice theory. An experiment with pneumococci combined with a large amount of antibody was given as proof by Heidelberger: Reagglutination into larger clumps occurred promptly when either fresh pneumococci or the corresponding polysaccharide were added to the washed and evenly resuspended cells; and as further evidence the agglutination of cells by quantities of antibody that can cover only a very small part of their surface and cannot change appreciably the surface properties has been offered.[58] Putting all arguments together, it may well be that both specific and non-specific forces are operative in the second stage of agglutination and precipitation.[59]

If the presence in a substance of at least two binding groups were a sufficient condition for precipitation by antibodies, as has been postulated for the lattice theory, then haptens even of small molecular size should invariably precipitate with immune sera. Experiments to test this proposition were made with numerous compounds, chiefly azodyes. Pauling and his coworkers [60] examined compounds containing arsonic acid groups and found that all those possessing two or more such haptenic groups were precipitable by immune sera for antigens containing the phenylarsonic acid group. Hooker and Boyd,[61] however, found a number of compounds, including substances other than azodyes (biphenyl-p-p'-dicarboxylic acid, p-p'-dicarboxylazobenzene), that were not precipitable although possessing two or more reactive groups; [62] in particular none of those containing carboxyl gave (with weak sera) a positive result. And we may recall that azoproteins possessing only few combining sites were found not to be precipitable (p. 160). Other probably pertinent instances are conjugated deuteroproteoses (119) that were not precipitated under the same conditions that azoproteins were, and unaltered proteoses that do not give precipitin reactions with the corresponding anti-protein sera. Hence, from the available evidence, the presence of two or more binding groups alone appears not to be sufficient for precipitability; that it is a necessary condition is suggested by the

[58] Pressman et al. (56) (experiments with azo-erythrocytes).
[59] v. (115), (116).
[60] Pauling, Campbell and Pressman (117).
[61] (118), (111).
[62] Some of the results with arsenic compounds were contradicted by Pauling and his collaborators (117).

Hooker and Boyd (109) were unable to demonstrate birefringence of solutions containing divalent haptens and the corresponding antibodies, as might be expected from the lattice theory.

fact that no precipitation took place, in the tests of either group of workers, with any substance having only a single binding group.

It must be considered, in judging the above experiments, that the precipitin reactions of azodyes vary considerably in strength, and in the cases examined are in general much weaker than those of proteins and polysaccharides, or of the particular azodyes with which the phenomenon was originally observed. In the author's study precipitation with azodyes made from resorcinol and tartranilic or arsanilic acids took place very slowly and was feeble in comparison to the strongly and immediately precipitating dyes. Evidently, the degree of precipitability depends on peculiarities in chemical structure; what these are is not well enough known to predict the outcome in every instance. It is supposed that precipitability is influenced by several factors, the greater or lesser correspondence in structure of the hapten to the antibody-producing groups, the blocking of solubilizing groups through mutual neutralization of the reactants, favouring precipitation, and the spacing of the reacting groups in the hapten which controls the extent of steric hindrance "exerted by one antibody molecule on another" (Boyd, Pauling). From everyday serological experience — i.e. the contrast between proteins or colloidal polysaccharides, and their split products [63] — one may think that the molecular or particle size is a material factor, and it would be desirable to determine whether or not precipitable azodyes are aggregated in solution.

To explain the hydrophobic property of specific precipitates there has been suggested an alteration of antibodies, similar to denaturation, after union with antigen,[64] yet this view is not sustained by actual evidence (1). One has assumed instead a decrease in affinity for water, due to occlusion of polar groups and dehydration of the antigen (Reiner, Marrack). This concept was proposed by Reiner from model experiments with tannic acid (102).

In considering the conditions for precipitation, the dependence on lipids associated with the antibody globulin and on electrolytes should also be remembered.

In line with the lattice theory, the existence of immune sera that react with antigens (and inhibit the precipitation by more potent antisera) but do not precipitate them has been hypothetically ex-

---

[63] v. Heidelberger et al. (120), Ivánovics (121).
[64] (122), (96); v. (97).

plained on the assumption that the sera contain antibodies with only one combining group.[65] This supposition may be right but it cannot account for other cases, for example the observation that rabbit immune sera while precipitating macromolecular polysaccharides do not, in distinction to horse antisera, precipitate degradation products thereof.[66] It seems adequate to explain these results by greater solubility of the rabbit antibody-globulins. More to the point are findings like those of Hooker and Boyd (24) who showed that anti-hemocyanin sera, modified by heating, were still reactive, but failed to precipitate except, apparently, in conjunction with unheated antisera. These observations (pp. 141, 257) would seem to be attributable to weakened affinity rather than to univalence of antibodies through selective loss of binding groups.

QUANTITATIVE THEORIES. — The first attempt, by Arrhenius, at a theoretical treatment of precipitin and agglutinin reactions was rationally based on the mass law but included a number of quite improbable propositions. A new theory was developed by Heidelberger and Kendall [67] who took advantage, in their experimental approach, of the specific precipitin reactions of N-free polysaccharides which by microanalyses for nitrogen [68] allow one to determine directly the quantity of antibody in precipitates. The theory assumes that both antigen and antibody are "multivalent," i.e. contain two or more combining groups and that the reactions follow the mass law; it is further said that "for convenience of calculation the . . . antibody reaction is considered as a series of successive bimolecular reactions which take place before precipitation occurs" and that "no evidence of dissociation could be found over a large part of the reaction range." The lattice concept [69] of specific forces operative in flocculation is embodied in the statement that "the final precipitate, then, would in each case consist of antibody molecules held together in three dimensions" by antigen molecules, for example:

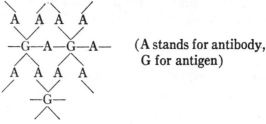

(A stands for antibody,
G for antigen)

From their assumptions the authors derived, for the region of antibody surplus, the formula $y = 2 R x - \dfrac{R^2 x^2}{A}$ where y is the antibody in the precipitate in mg., x the antigen in mg., A the total antibody and R the ratio of antibody to antigen at the equivalence point. A large body of accurate quantitative data was obtained on the precipitation of polysaccharides and protein antigens (also the agglutination of bacteria), and in the majority of the examined instances the equation gave satisfactory agreement between observed and calculated values in the range considered; it accounts for the variable composition of the precipitates and the approximate independence, in the authors' experiments, of changes in volume.[70]

It may be noted that in experiments of Goodner (125) with antipneumococcus sera, when the amount of antibody precipitated was plotted against the quantity of antigen, the graph, instead of being a smooth curve, was a figure of linear segments each of which seemed to be related to a certain grouping in the antigen. With a cross reacting precipitation system similar results were obtained by Heidelberger et al. (126).

A priori it would seem that difficulties in the way of a rigorous physico-chemical treatment would arise from the presence of antibodies of different reactivity in the same immune serum and of groups unequal in specificity and binding capacity in antigen molecules. With regard to the unhomogeneity of immune sera Heidelberger remarks that "although the anticarbohydrate is known to be a mixture of antibodies of different reactivities it may be treated mathematically as if its average behavior were that of a single substance."

In addition to other variations, "low-grade" antibody fractions [70a] have been demonstrated by Heidelberger and Kendall that are not capable of producing precipitation but are included in precipitates formed by more potent antibodies; and there exist antibodies that precipitate only at low temperature (p. 246).

---

[65] Pappenheimer (23), Heidelberger et al. (27), Tyler (123).
[66] (120), (121).   [67] (87), v. (2), (124).
[68] The lipid content of precipitates could be neglected.
[69] Concerning the explanation of the flocculation effect of electrolytes, v. (2).
[70] For criticisms and the controversy elicited by the theory which cannot be presented in brief, the reviews by Marrack (1), Heidelberger (2) and Boyd (3) should be consulted. Boyd submits, along with other things, that several equations can be formulated which fit approximately the numerical data over parts of the reaction range. [See also (109)].
[70a] See pp. 194, 256, 146.

In view of the similarity in quantitative relations between specific and non-specific reactions (e.g., precipitation of serum proteins by isamine blue),[71] it is suggested by Marrack (116) that the investigation of relatively simple non-specific systems may be a means for elucidating the problem of antigen-antibody reactions; and, apart from questions of specificity, there is much to favor the idea.

A theory put forward by Kendall (127) through different reasoning and avoidance of some assumptions of the original theory of Heidelberger and Kendall — which the authors felt to be oversimplified — arrived, for bivalent antibodies, at the same equation as above. The following of his postulates should be stated. "Both antigen and antibody may be multivalent with respect to each other . . . reactive groups upon a given molecule may all have the same specificity or they may be different." "The maximum number of molecules of antibody bound by one antigen molecule may be determined by the number of reacting groups upon the antigen, or it may be limited by steric factors." "The reactivity of an antigen or antibody group upon the surface of a precipitate is the same as the reactivity of the same group on the surface of a molecule in solution." The equilibria are established "between the free and combined antigen and antibody groups in the system" and "between reactive groups upon the surface of precipitates and the molecules in solution." Kendall concludes that the equation is valid for homogeneous antibodies.

Another formulation derived by Hershey (128), based upon kinetic considerations, expresses the equilibrium in the precipitin reaction between multivalent antigens and antibodies in terms of the valence, the initial molar concentration of the reactants, and the dissociation constant of the "specific valence," regarded as the main characteristic of individual immune sera. The theory was found to be compatible with experimental results and presumptive evidence for the valence of antibodies.

The specific precipitation of azodyes (and some other simple compounds) and the inhibition of these reactions by univalent haptens has received theoretical treatment by Pauling et al. (129). The experiments dealing also with the influence of dilution and of hydrogen ion concentration showed that with precipitable haptens the molar antibody-antigen ratio (somewhat less than unity) remained constant

[71] Dean (124a).

throughout the reaction range, in compatibility with assumptions on the effective valence of the dyes and bivalence of antibodies.

In work with arsanilic acid immune serum and an heterologous arsanilic acid azoprotein, to avoid reactions depending on the protein moiety of the antigen, Haurowitz (34, 4) obtained precipitates of constant composition which led him to the opinion that the varying composition of precipitates is in general an effect produced by antibodies of different combining properties. In similar experiments of the same author on iodoproteins, however, the outcome was different, and it is questionable whether his interpretation is generally applicable (35).

It was brought out by Hooker and Boyd [72] that the ratio, by weight, of antigen to antibody in precipitates in the equivalence zone diminishes with increasing molecular weight of the antigen. As an interpretation, the authors assumed that in this range the antigen may be just covered by spherical antibody units of molecular weight of 35,000 corresponding to "Svedberg units," [73] or by long, flexible ellipsoids, and, applying spherical trigonometry, deduced an equation expressing the antibody-antigen ratio as a function of the molecular weight of the antigen. While the postulate that in the equivalence zone the antigen molecules are always completely covered may not be accurate [74] and the flexibility of antibody molecules may be less than supposed, the ratios obtained with various precipitating systems obey the relation fairly well in the majority of cases [75] and, at any rate, show unquestionably the influence of the molecular weight of the antigen. Haurowitz, and Marrack,[76] who used immune sera for arsanilic acid azoprotein, found that the ratio in question depends also on the number of combining groups in the antigen, the amount of antibody equivalent to a certain quantity of antigen being greater the higher the As-content.

From their model Hooker and Boyd calculate a thickness of about 40 Å for the antibody layer coating an antigen molecule, and

[72] (130); v. (20), (3); (131) (differences in the ratios observed with spherical bushy stunt virus and rod-like tobacco mosaic virus); (132).

[73] The molecular weights of proteins were found by Svedberg with the method of ultracentrifugation to be approximately multiples of 17500 or 35000.

[74] v. Hooker and Boyd (109), Pappenheimer et al.; Boyd (20), (3).
As is easily understood, the number of antibody molecules bound may because of steric hindrance be smaller than the number of sites which can combine.

[75] A somewhat better fitting formula was given by How (133). Calculations by Pauling (134) on the assumption of bivalent spherical antibody molecules yielded similar results.

[76] (34), (1).

this would imply that antibodies do not form a thin extended film on the antigen, as has been suggested. The latter assumption is contradicted by the experimental fact that precipitates may contain a quantity of antibody too great to form simply an extended film (about 10 Å) on the antigen.[77] By electron microscopy of precipitates of tobacco mosaic virus, Anderson and Stanley (135) arrived at the figure of about 225 Å for the thickness of the coating antibody layer [78] which is in fair agreement with the dimensions of antibodies calculated by Neurath. Anderson and Stanley suggest from their measurements that the antibody molecules are arranged around the antigen molecule with the long axis perpendicular to its surface, as was deduced by Boyd and Hooker (137), and Marrack (116), from the number of antibody molecules that, at antibody excess, can be placed without deformation about an antigen molecule. Observations with the electron microscope on antibacterial sera yielded much lower values.[79] The discrepancy may find an explanation if the antibody molecules are oriented in different ways, according to the concentration applied — either patchy, or as a complete layer with the long axis parallel to the surface, or endwise at high concentrations.

Values of the order computed by Hooker and Boyd for the equivalence zone were obtained with antibody layers fixed to expanded antigen films.[80]

THE MODE OF COMBINATION OF ANTIGENS WITH ANTIBODIES. — While the investigations on the serological behaviour of substances (or groupings) of known constitution have provided data which were a prerequisite for correlating serological specificity and chemical structure, no finished theory of antibody reactions has yet been attained that is comparable to those that cover and make it possible to formulate the reactions of organic chemistry. This is not too surprising in view of the fact that even the theory of organic molecular compounds is not perfected, and there is the additional difficulty with immunological reactions that the chemical structures responsible for

---

[77] v. (1).

[78] With regard to a comment by Marrack (115), Stanley points out (136) that the number of elongated antibody molecules, having the dimensions 4 x 27 mμ, that can be attached radially to a cylindric virus molecule of 15 mμ diameter and 280 mμ length, approximates satisfactorily the value of about 1000 found by Kleczkowski for the ratio of antibody to antigen at antibody excess.

[79] Mudd and Anderson (138).

[80] Rothen and Landsteiner (139), (p. 57); v. Harkins et al. (140), Bateman et al. (141); (t: 6).

the specificity of one of the reactants — the antibodies — are still unknown.

Although the highly specific adjustment that antibodies exhibit is, likely enough, due to the peculiar chemical nature and complex organization of the protein molecule with its variety of reactive structures, it was natural, as an indirect approach to the problem, to look for phenomena that to some extent resemble in specificity those of serology. In experiments along this line by Schulman and Rideal (142), chiefly with monolayers, the association of different molecules was seen to depend on attraction, selective in some measure, between the polar and between the non-polar parts of the substances. The specificity of the forces that hold molecules in a crystal lattice, and cases of specific adsorption on the surface of crystals, were adduced by Marrack (1). Methylene blue is adsorbed by diamond and not by graphite, succinic acid in the opposite way, although the two adsorbents are chemically alike and are different solely as regards the spacing of the carbon atoms; and likewise dependent on spatial configuration is the adsorption by potash alum of only one of two isomeric azodyes differing in the position of the substituents in the naphthalene ring. A number of similar observations have been recorded in papers by France in which it is shown that the adsorption of azodyes on to inorganic crystals is affected, for example, by replacement of OH by NH_2 or by the position of substituents in the benzene ring.[81] In the formation of molecular compounds [82] specific relationships are not seldom encountered, salient cases being the reactions of Cu, Fe, Tl, Ni, Co, with certain groups ($OH - \overset{|}{C} - \overset{|}{C} = NOH$ with Cu, $-CO - CH_2 - CO -$ with Tl, etc.) in organic substances.[83] Of interest for our question, because they deal with amino acids, are observations of Bergmann (150) and others, e.g., the formation of insoluble salts of glycine with potassium trioxalatochromiate or nitranilic acid, the precipitation of proline and some peptides by rhodanilic acid.

These and other examples of reactions, specific to a certain degree,

---

[81] (143). On the significance of the corresponding spacing of NH_2 groups and Cl ions for the adsorption of urea on crystals of ammonium halides v. (144), (145).

[82] On selective reactions between organic molecules cf. Pfeiffer (146). Evidence concerning the influence of polar groups on the specificity of intermolecular reactions is offered in studies of Labes (147).   [83] v. (148), (149).

would indicate that the reactions underlying the combination of antibodies are not altogether unique, which seems probable of itself.
However, the analogies available at present do not suffice to form
more than a vague picture of the determinant structures in antibodies
that could account for the most striking experience in the field of
serology — the virtual existence of specific reagents for, one might
almost say, any organic substance, whether simple or highly complex — or to suggest the possibility of discovering synthetic reagents
which, like precipitins, would generally distinguish closely related
proteins.

As regards the affinities involved, one may say that in velocity and
reversibility antibody reactions for the most part differ from those
due to primary valences, and a decisive argument against the assumption of covalent bonds is the fact which emerged from the serological
study of azoproteins and simple compounds, that quite different substances are capable of reacting with antibodies all in a like manner,
regardless of their chemical character. These investigations afforded
further definite information on the specificity and nature of antibody
reactions by establishing the significance of polar groups, and the
effect of spatial structure as shown in the reactions of substituted
benzene compounds and unequivocally by the differentiation of
stereoisomeric substances. Evidently, since compounds identical in
chemical constitution, except for the interchange of two atoms or
groups on one asymmetric carbon atom (phenylaminoacetic acid,
tartaric acid, sugars), are serologically distinct, antibodies must be
adjusted to the steric configuration of the determinant structures in
the antigen.

The precise nature of the forces involved in serum reactions is still
open to discussion, but it is supposed [84] that the union between antigen and antibody is brought about chiefly through electrically
charged acid or basic groups, van der Waals forces and interaction
between polar groups (or polar groups and ions) in which, as Pauling emphasizes, hydrogen bonding presumably plays an important
part. The assumption is further made that the reacting groups in the
antigen or hapten have counterparts in complementary groups on the
surface of the antibody, and that firm compounds can be formed,
even if the single bonds are weak, through the participation of a

[84] Marrack (1), Haurowitz (4), Pauling (134).
For general information on the nature of intermolecular forces v. (Chapter
VIII); Hammett (151), Evans (152), Briegleb (153).

number of combining groups. Hence with haptens of low molecular weight without strongly active groups the forces may be insufficient to bring about demonstrable combination despite the presence of specific determinants, as in the case of simple sugars and peptides.

In the field of enzyme reactions, which to all appearances are related to serum reactions with regard to the affinities concerned, Bergmann et al. (154) came to the conclusion that the proper spatial distribution of at least three binding groups in peptides, COOH, NH, NH_2, is essential for the specific union with peptidases.[85]

That there may be interaction of acid and basic groups, e.g. COOH (or phenolic groups) in the antigen and NH_2 in the antibody is suggested by the predominant influence of acid groups in conjugated antigens and the change in specificity through esterification [86] of proteins, or of hapten groups in conjugated antigens. Of equal evidential weight are the cross reactions of pneumococcus immune sera with azoproteins containing quite different acid determinants (p. 175). And, while experiments by Michaelis and Davidsohn, and de Kruif and Northrop (158) contradict the assumption that the combination of antibody and antigen (both having a negative charge under common conditions) is caused by opposite electric charges of the molecules, they do not preclude that among the combining structures involved there are acid and basic groups reacting with each other.[87]

Chow and Goebel offered as evidence the fact that the precipitating capacity of antibodies for a strongly acid polysaccharide is (reversibly) inactivated when their amino groups are blocked by treatment with formaldehyde or ketene (159). This argument is questionable, however, because of the observation that the activity of antibodies for an azoprotein made from aminoantipyrine, which does not possess any acid binding groups, was likewise destroyed by formaldehyde (164).

Positive evidence for the participation of acid groups is afforded by the simplest instances which have so far been examined, namely the specific inhibition of azoprotein antisera by compounds like

[85] On the specificity of enzyme reactions v. (155), (155a); Bergmann and Fruton (156), Marrack (156a) (compares the specificity of carbohydrates and antibodies to carbohydrates).

[86] The combining capacity of blood stromata was seen to be abolished or greatly diminished by treatment with acid alcohol in tests with serum agglutinins (157).

[87] (159); v. (160). Data on the influence of hydrogen ion concentration on antibody union and flocculation are provided in (161–163), (73), (34), (1).

benzoic, benzenesulfonic, phenylarsenic, succinic acids (p. 186). It is difficult to arrive at any other conclusion than that in these reactions acid groups, COOH, SO_3H, AsO_3H_2, are of chief importance for the union since in the inhibition of aminobenzoic acid sera benzoic acid can be replaced to some extent by certain fatty acids, and the sera for aminophenylarsenic acid react not only with various substituted phenylarsenic acids but also, weakly, with arsenic acid. For all these reactions only small combining groups in the antibodies would seem to be required, yet these are in all probability commonly parts of larger determinant structures that are able to react with groupings in the immunizing protein.

The best examples of bonding through non-ionic polar groups are the reactions of tartaric acid and of simple sugars in which the specificities are conditioned by the steric arrangement of hydroxyls which therefore would supply points of linkage; in such cases, as Marrack points out, the spatial arrangement, on account of the short range of the forces, will be of prime importance. Other suggestive instances are the conspicuous influence of the acetylamino group CH_3CONH as substituent in the benzene ring, in comparison to halogen or CH_3, in the reactions of azoproteins made from substituted anilines, and the increase in specificity of fatty acid chains produced by the NHCO linkage.

A difference in the serological significance of COOH and the strongly polar $CONH_2$ group is indicated by the observation that through conversion of a pentapeptide into an amide the determinant influence of the terminal groups was abolished or greatly reduced (pp. 178; 208 (ref. 82)).

Changes in specificity can be produced also by inert groupings, as shown in the marked differences between antigens containing benzene and naphthalene or ethyl- and butylbenzene. Possibly, the effect of weakly polar or non-polar groups is mainly steric, due to modification of shape of the determinants (Pauling), or the specificity of neighboring structures may be affected; [88] at least there is no conclusive evidence to show autonomous serological reactivity of aliphatic or aromatic hydrocarbons.[89] Further information on this point, how-

[88] v. (191). Cases in which reactivity was increased by introduction of groups unrelated to the antibody are the inhibition reactions with dipeptide sera and acylated amino acids, and observations by Pressman, Brown and Pauling (164a).

[89] Attraction between long hydrocarbon chains attributable to van der Waals forces were reported in the studies of Schulman and Rideal (142).

ever, would be desirable and could be obtained by inhibition tests with soluble compounds.[90] The data collected hitherto give no indication of cross reactions referable to substituents like Cl, NO_2, CH_3. Instances of changes in specificity by such substituents are presented in Table 18.

The significance of electronic structure was considered by Erlenmeyer and Berger (165) and a broad parallelism drawn between similar or identical serological reactivity and the formation of mixed crystals (v. p. 170). Their view that immunological similarity follows from the presence of an equal number of valence electrons in atoms or small groups of atoms ("pseudoatoms") was based upon tests with conjugated antigens in which CH_3 and halogen, and the groups CH_2, NH and O appeared to be serologically equivalent. Furthermore, correspondence was found in the reactions of azoproteins prepared from the isomorphic aminophenylarsenic and aminophenylphosphinic acids, in distinction to aminophenylstibinic acid [v. (44)]; and benzene sulfonic and benzene selenonic acids were serologically similar, and different from benzene sulfinic acid. The presumption of a sharp difference between O or CH_2, and CO was not fully substantiated in experiments of Jacobs (166) with antigens made from amino diphenylmethane or aminodiphenylether and aminobenzophenone (and with aminodiphenyl). More significant is his result that in inhibition tests 4-bromoaniline-2-sulfonic acid and 4-bromoaniline-3-sulfonic acid proved to be serologically distinct; therefore the expectation that Br and NH_2 are serologically equivalent appears not to be fulfilled.[91]

A conjecture of Lettré (168) that the specific combination of antibodies with antigen consists of the formation of partial racemates owing to stereoisomeric groupings in the two reactants does not allow for the reactions of many haptens having no asymmetric structure.

Another hypothesis by Jordan (169) holds that antigen and antibody attract each other through resonance, by virtue of groupings, identical or nearly so, which are present in both, a supposition which will have been seen to be incompatible with known facts about antibodies.[92] From the viewpoint of the theory of resonance the hypothesis has been criticized by Pauling and Delbrück (172).

Attempts to elucidate the nature of serum reactions by physicochemical studies, chiefly on colloids, at one time in the foreground of discussion,[93] were prompted by several considerations: the colloidal

[90] v. Creech and Franks (164b).
[91] v. also (1), (167).
[92] v. (170), (171).
[93] These investigations are aptly reviewed in Wells (173), including endeavours

character of antigens and antibodies, which are all macromolecular substances, their combination in seemingly variable, indefinite proportions, the incomplete reversibility of the reactions, inhibition zones as in the mutual precipitation of positive and negative colloids, the flocculation of sensitized bacteria by low concentrations of electrolytes, the imitation of serological hemolysis and hemagglutination by means of inorganic colloids (p. 6). Apart from analogies, serological reactions certainly take place on the surface of the antibody molecules, these being impenetrable by any but compounds of small molecular size, and the degree of dispersion of antigens (particle size) is a factor which actually modifies the appearance of the reactions and largely controls the quantitative relations [94] (p. 259). However, the "colloid theory," stressing the relationship to surface and colloidal reactions, failed to give a clue to the specificity of serum reactions and, indeed, as we have seen, very simple non-colloidal substances react specifically with antibodies.

REMARKS ON SEROLOGICAL SPECIFICITY. — The high specificity of many serum reactions led Ehrlich to the view that each antibody is sharply adjusted to one particular structure (receptor), and accordingly that overlapping reactions of antigens must depend upon the presence in each of them of identical substances or chemical groupings. The idea was widely accepted and is still taken as a matter of course in some recent writings. From numerous observations on artificial conjugated antigens, however, this notion is seen to be inadmissible, and it is certain that antibodies react most strongly upon the homologous antigen, but also regularly, with graded affinity, on chemically related substances. Or, as Haldane (176) wrote regarding an enzymic reaction, "the key does not fit the lock quite perfectly, but exercises a certain strain upon it."

The specificity of a lower order, shown by many normal antibodies and plant agglutinins, differs from those of immune antibodies in not being directed towards a particular antigen; but there is no good reason to believe that the specificity of normal antibodies is different in principle. As a matter of fact, it is not always possible to decide to which of the two kinds an antibody belongs.

Another class of reactions often showing only a moderate degree of specificity is the combination of enzymes with their substrates. An example

---

to demonstrate antigen-antibody reactions through physical measurements, especially of changes in surface tension. For literature see also (173a), (173b).

[94] An illustration is the difference seen in complement fixation reactions with

taken from recent studies is the selectivity of peptidases which require for their action the presence of certain amino acids in the peptides while wide changes in other amino acids do not abolish the enzymatic effect [Bergmann (156)].

That no strict correlation exists between antigens and antibodies, as the assumption of absolute specificity and a rigid concept of receptors would demand, is shown clearly by the occurrence of non-reciprocal reactions [95] in which antisera for an antigen A react also with another antigen B, but those for B only slightly or not at all with A. This relation, for example, obtains between the antibodies elicited by sheep blood and by certain bacteria, or between immune sera for native protein and peptic metaprotein. Whereas with natural antigens the cause of the phenomenon remains uncertain and has been attributed to complexity of the antigens,[96] the state of affairs can be judged with greater assurance in the reactions of synthetic conjugated antigens. Thus, when immune sera for o-aminobenzoic acid react strongly with o-aminobenzene sulfonic acid antigen, while o-sulfonic sera do not precipitate o-aminobenzoic acid azoprotein,[97] or do so but weakly, this must be due to the difference in specificity between the two sera, which proves once more that even a small determinant structure can combine with quite different antibodies.

The combining part of an antigen or hapten one may conceive of as a structure — an area on the surface in the case of large molecules — that is brought into play in diverse ways, depending upon the antibody involved in the reactions. Hence, determinant structures are defined only in relation to a certain antibody. For example, the affinity of meso-tartaric acid is presumably determined by the carboxyl groups and, in addition, according to whether it is reacting with the homologous or with d- or l-tartaric acid sera, by either both or only one of the asymmetric configurations, while in the case of larger molecules the situation will be still more complicated. Indeed,

---

Forssman hapten or the Wassermann substance according to whether the alcoholic extracts are diluted with saline solution slowly or quickly. [Sachs and Bock (174)]; similarly, the alcohol concentration of the emulsion is of influence [Browning et al. (175)].

[95] v. Doerr (177).

[96] (178), (179).

[97] Additional examples will be found among the reactions of disaccharide glucosides [v. Goebel, Avery and Babers (180)].

Both reactions of o-aminobenzoic acid sera are inhibited by o-aminobenzoic acid and by o-aminobenzene sulfonic acid, but the sulfonic acid sera markedly only by o-aminobenzene sulfonic acid.

it would hardly be possible to interpret consistently on the receptor principle all the cross reactions of disaccharide glycosides, described by Goebel, Avery and Babers (180), even though the chemical and steric structure of the substances is well known.[98] For example, immune sera for lactose reacted with cellobiose antigens although the terminal hexose, which in other instances has dominant influence on the specificity, is different; the similarity between the two sugars consists of the β-configuration of the terminal hexose and the linkage, through the fourth carbon atom, between the two monosaccharides. But neither of these structural features causes a cross reaction in other cases examined. Summarizing, one may say that the antibodies for disaccharide glycosides seem to reflect the pattern of the homologous hapten as a whole [99] while being capable of interacting with different integral patterns. A similar situation was encountered in the study of immune sera to peptides. There was no evidence for the presence of antibodies against amino acid residues, and sera for a pentapeptide (diglycylleucylglycylglycine) contained antibodies which were highly specific for the homologous peptide in its entirety (p. 179). These observations would be accounted for if the compounds, under the influence of intramolecular forces, should assume a characteristic spatial pattern.

Accepted physico-chemical theories offer no basis for the assumption of forces dependent upon entire, large structures. In the opinion of London (181) the existence of long-range forces which are due to "properties of the molecule as a whole" cannot be excluded to a certainty.[100] However, his theory is developed with reference mainly to the special case of conjugated double bonds.

The view here submitted bears upon the question previously discussed (p. 114) as to the necessity of discrete chemical structures for the production of various antibodies. That in this way multiple antibodies can be produced is unambiguously evidenced by the complexity of the immune sera that are produced with azoproteins (p. 159) or iodoproteins (35). In order to investigate this issue with small structures of known constitution, reactions of azoprotein sera were examined by partial absorption and by inhibition tests. In one sort of experiment [101] azoproteins were employed with two different

[98] v. (p. 174), Marrack (1), (156a).
[99] Goebel et al. (180).
[100] v. Langmuir (182).
[101] Landsteiner and van der Scheer (183); v. Haurowitz et al. (184).

active groups in the same molecule, viz. succinic and phenylarsenic acids, or glycine and leucine. The immune sera engendered by these antigens were analyzed by absorption with azoantigens possessing either one or the other of the residues and were found chiefly to contain two kinds of antibodies each directed towards only one of the two groups, possibly along with small amounts of antibody reacting with both. (By experiments of this kind it might be possible to provide a striking proof for the assumed bi- or multivalence of antibodies.) [102]

TABLE 36
[after Landsteiner et al. (183)]

	Repeated Absorptions with G Test antigens			Repeated Absorptions with L Test antigens		
	GIL	G	L	GIL	G	L
I	++±	o	++	++	+	o
	+++±	o	++±	++±	+	o
II	++	o	+±	+±	+	o
	+++±	o	++±	++	+	o
III	++	o	+±	+	+	o
	+++±	o	++±	++	+	o
Unabsorbed immune serum	+++	+	++	+++	+	o
	+++±	+	++±	+++±	+	o

Serum for sym. aminoisophthalyl glycine-leucine (GIL) was absorbed separately with glycine (G) and with leucine (L) azostromata. This procedure was repeated twice; the supernatant fluids were tested with the homologous substance and with azoproteins made from glycine and from leucine.

The above mentioned results prove that separate antibodies can be formed to several groupings within a rather small molecule and are in agreement with prevailing opinion, but a different aspect was revealed by experiments in which determinants of simple structure were put to use. Naturally, absorbing an immune serum with a suitable quantity of an heterologous antigen of weak affinity will remove the reaction for this but not entirely for more reactive antigens; though occasionally a moderate amount of cross reacting heterologous antigen may exhaust a serum completely. The more significant phenomena observed with a number of azoprotein antisera may be illustrated by the behaviour of m-aminobenzene sulfonic acid sera.

[102] v. Pauling (134), Hooker and Boyd (109); (p. 140).

These showed overlapping reactions as presented in Table 21. Absorption with not too large an amount of o-aminobenzene sulfonic acid antigen which gives an intense cross reaction yielded a solution strongly precipitating the homologous meta-, but no longer the ortho-antigen, and control tests with dilutions of the immune sera proved that the effect cannot be attributed to diminution of a single antibody since dilutions of unabsorbed serum were distinctly less specific than the supernatant fluid after absorption. When the experiment was performed with several heterologous antigens, it appeared that often the resulting fluids differed in their specificity, as shown in Table 37, treatment with each antigen impairing principally the corresponding precipitin reaction.

TABLE* 37
[after Landsteiner and van der Scheer (185)]

Immune sera for meta-amino-benzene sulfonic acid after absorption with	Azoproteins made from chicken serum and			
	Ortho-amino-benzene sulfonic acid	Meta-amino-benzene sulfonic acid	Meta-amino-benzene arsenic acid	Meta-amino-benzoic acid
o-Aminobenzene sulfonic acid	o o	++± +++±	± ±	+ +
m-Aminobenzene arsenic acid	+± ++	+++ ++++	o o	+ +±
m-Aminobenzoic acid	+± ++	+++ ++++	± ±	o ≐
Unabsorbed immune serum	++ +++	+++± ++++	+ ++	+± ++±

Reading: first line, after standing 1 hour at room terperature;
second line, after standing overnight in the icebox.
Since the test antigens contained the same proteins, unrelated to the horse serum used for immunization, the protein component could not be responsible for the differential reactions [v. (37)].

These facts can scarcely be interpreted otherwise than by assuming a multiplicity of antibodies, and a glance at the structural formulae of the substances examined shows that the single antibodies cannot be related to special groupings — of which at least four would be required in the example given — present in the homologous substance.[103] Thus to a multiplicity of serological reactions there need

[103] It is essential to note that the sera produced to m-aminobenzene sulfonic acid do not by any means react with all compounds containing sulfonic acid groups.

not correspond a coordinate mosaic of chemical structures in the antigen. One may conclude that the antibodies formed in response to one determinant group are, though related, not entirely identical but, as evident from their cross reactions with heterologous antigens, vary to some extent around a main pattern, and that what ordinarily is spoken of as an antibody is generally a mixture of specifically different components.[104] The existence of diverse antibodies, more or less closely adjusted to an antigenic pattern, would be adequate to explain differences in affinity and cross reactivity of antisera produced in individual animals, or taken from the same animal at different times, without necessarily resorting to the hypothesis of antibodies varying in the number of combining sites (p. 140). Certainly, as there was occasion to mention, these considerations apply also to natural antigens, and the results obtained in absorption tests suggesting discrete antigenic elements will include effects such as were described above, in addition to those attributable to the presence of more than one specific substance or determinant group.

The tendency for the production of multiple antibodies may be greater with more complex determinant structures, and one may assume that there are transitions up to those cases in which sharply separated determinants exist (v. pp. 115, 159).

Instances of several antigenic determinants in one molecule are presumably proteins and, quite possibly, polysaccharides,[105] e.g. those of the salmonella group in which distinct serological properties are characterized by antisera absorbed with heterologous substances.

Experiments on the deterioration of the B. dysenteriae polysaccharide by alkali gave no proof of the presence in the molecule of a special "heterophile" structure, even though in the complete antigen the heterophile antigenic function was more labile than its capacity to engender precipitins.[106] For the cross reactions of plant gum with antipneumococcus serum, Marrack and Carpenter (189) came to the conclusion that these are better explained by antibodies with different affinity against related determinant groups than by antibody fractions corresponding to separate groups in the homologous polysaccharide. In particular, although cherry gum gives a greater amount of precipitate than gum arabic, which could indicate the presence of an additional determinant in the former, both reactions are inhibited by

[104] v. p. 144. Landsteiner et al. (186), (185); Heidelberger et al. (87), (36), (126); Haurowitz (37).
[105] v. Goodner (125), Heidelberger (126), Morgan (191); White (187) (determinant groups in vibrios); (pp. 221, 222).
[106] Morgan (188).

glucuronic acid, and the reaction of the immune serum with the homologous pneumococcus carbohydrate, after exhaustion with cherry gum, is again inhibited by glucuronic acid, to the same degree as the reaction of the unabsorbed serum.

It will be well to mention that also in the case of artificial antigens which have groupings in common, such as disaccharides, containing the same hexose or peptides with the same amino acids, the cross-reacting antibody fractions are specifically adjusted to the homologous antigen. This is demonstrated by the fact that the precipitation of heterologous antigens is mostly better inhibited by the substances homologous to the immune serum than by those corresponding to the antigen tested. Representative experiments are reproduced in Table 38, indicating that the antibodies which react with the leucine resi-

TABLE 38
[after Landsteiner et al. (186)]

Antigens:		L				GL		
Substances tested for inhibition (nitrobenzoyl peptides)	L	GL	LL	C	L	GL	LL	C
		dilution 1:4				dilution 1:16		
Immune	o	+	±	++	o	±	⊹	++
serum	o	+	+	++	o	+	+	++
L	o	+±	+±	++±	o	+±	+±	++±
	±	++	++	++±	±	+±	++	++±
		dilution 1:16				dilution 1:4		
Immune	±	o	o	+±	++	o	+±	++±
serum	±	o	±	++	++	o	+±	++±
GL	±	o	±	++	++±	o	++±	+++
	+	o	⊹	++±	+++±	±	+++	++++

L (Leucine), GL (Glycyl-leucine), LL (Leucyl-leucine), C (Control). Dilutions of the substances tested for inhibition in terms of a 1/10 molar solution.
Readings after 15 min., 1 hour, 3 hours, and the next day.

dues in the (azobenzoyl) leucine- or glycylleucine-antigens are not the same in the two sera, each having a special combining structure. That reactions with the immunizing antigen may be inhibited by the homologous hapten alone,[107] as in experiments by Avery, Goebel and Babers with glucosides, is easy to understand because of the differences in affinity (pp. 186, 250).

[107] v. (190).

BIBLIOGRAPHY

(1) Marrack: CAA; ARB *11*, 629, 1942. — (2) Heidelberger: BaR *3*, 49, 1939. — (3) Boyd: FI. — (4) Haurowitz: Fortschritte der Allergielehre, Karger, Basel, 1939, p. 19. — (5) Marrack et al: Br *12*, 182, 1931 (B). — (6) von Dungern: CB *34*, 355, 1903. — (7) Bordet: AP *17*, 161, 1903. — (8) Arrhenius et al: Immunochemistry, MacMillan, New York, 1907. — (9) Ghosh: Ind. J. Med. Res. *23*, 285, 837, 1935/36. — (10) Pappenheimer et al: JI *32*, 291, 1937 (B). — (11) Eagle: JI *32*, 119, 1937. — (12) Healey et al: Br *16*, 535, 1935 (B). — (13) O'Meara: JP *51*, 317, 1940. — (14) Glenny et al: JP *28*, 279, 317, 1925 (B). — (15) Glenny et al: JP *35*, 91, 1932, *47*, 27, 1938. — (16) Bayne-Jones: NK, p. 759. — (17) Taylor et al: JH *34*, 118, 1934. — (18) Pappenheimer et al: JEM *71*, 247, 1940. — (19) Kekwick et al: Br *22*, 29, 1941. — (20) Boyd: JEM *74*, 369, 1941. — (20a) Heidelberger et al: JEM *65*, 393, 1937. — (21) Wadsworth: SM. — (22) Glenny: SystB *6*, 106, 1931. — (22a) Freund et al: JI *45*, 71, 1942. — (23) Pappenheimer: JEM *71*, 263, 1940. — (24) Hooker et al: ANY *43*, 107, 1942. — (25) Sobotka: JEM *47*, 57, 1928. — (26) Heidelberger et al: FP *1*, 178, 1942, Boston Meeting. — (27) Heidelberger et al: JEM *71*, 271, 1940. — (28) Malkiel et al: JEM *66*, 383, 1937. — (29) Heidelberger: JAC *60*, 242, 1938. — (30) Heidelberger et al: JEM *59*, 519, 1934; *65*, 487, 1937. — (31) Hooker et al: JI *30*, 41, 1936. — (32) Kleczkowski: Br *21*, 1, 1940. — (33) Heidelberger et al: JAC *63*, 498, 1941. — (34) Haurowitz: ZPC *245*, 23, 1936. — (35) Haurowitz et al: ZI *95*, 478, 1939; K 1937, p. 257. — (36) Heidelberger et al: JEM *62*, 697, 1935. — (36a) Heidelberger et al: JEM *55*, 555, 1932. — (36b) Hershey et al: JI *46*, 267, 281, 1943. — (37) Haurowitz: JI *43*, 331, 1942. — (38) Landsteiner et al: ZI *8*, 397, 1911. — (39) Heidelberger et al: JEM *63*, 819, 1936. — (39a) Szent-Györgyi: BZ *113*, 36, 1921. — (39b) Duncan: Br *18*, 108, 1937. — (39c) Coburn et al: JEM *77*, 173, 1943. — (40) Heidelberger et al: JEM *50*, 809, 1929, *61*, 559, 563, 1935. — (41) Duncan: Br *15*, 23, 1934. — (42) Beeson et al: JEM *70*, 239, 1939. — (43) Marrack et al: Br *12*, 30, 182, 1931, *11*, 494, 1930. — (44) Haurowitz et al: ZPC *214*, 111, 1933 (B). — (45) Heidelberger et al: JEM *61*, 559, 1935. — (45a) Goodner et al: JEM *66*, 437, 1937. — (46) Breinl et al: ZPC *192*, 45, 1930. — (47) Horsfall et al: JEM *62*, 485, 1935, *64*, 583, 855, 1936 (B); JI *31*, 135, 1936. — (48) Hartley: Br *6*, 180, 1925. — (49) Horsfall et al: S *84*, 579, 1936 (B). — (50) Ivanovics: ZI *97*, 443, 1940. — (51) Eisenberg et al: ZH *40*, 155, 1902. — (52) Cromwell: JI *7*, 461, 1922 (B). — (53) Dreyer et al: RS B, *82*, 185, 1910. — (53a) Heidelberger et al: JGP *25*, 523, 1942. — (54) Jones: JEM *48*, 183, 1928. — (55) Eagle: JI *29*, 485, 1935 (B). — (56) Pressman et al: JI *44*, 101, 1942. — (57) Eagle: JI *29*, 467, 485, 1935. — (57a) Abramson: JGP *14*, 163, 1930. — (58) Heidelberger et al: JEM *60*, 643, 1934, *74*, 105, 1941 (B). — (59) Jones et al: JEM *57*, 729, 1933. — (60) Streng: ZH *62*, 281, 1909. — (61) Heuer: ZH *95*, 100, 1922. — (62) Shibley: JEM *50*, 825, 1929. — (63) Coca et al: JI *6*, 87, 1921. — (64) Jones et al: JI *27*, 215, 1934; JEM *47*, 245, 1928. — (65) Hershey: JI *41*, 299, 1941. — (66) Dreyer et al: RS B, *82*, 168, 1910. — (67) Madsen et al: SB *102*, 1091, 1929. — (68) Mayer et al: JBC *143*, 567, 1942. — (69) Hooker et al: JGP *19*, 373, 1935. — (70) Eagle: JI *23*, 153, 1932. — (71) Follensby et al: JI *37*, 367, 1939. — (72) Ramon: SB *135*, 296, 1941. — (73) Brown: Br *16*, 554, 1935. — (74) Miles: Br *14*, 43, 1933. — (75) Topley: PB. — (76) Madsen et al: CB *36*, 242, 1904; SB *102*, 1091, 1929. — (77) Morgenroth: BK 1905, p. 1550. — (78) Otto et al: Z. exp. Path. *3*, 19, 1906. — (79) Glenny et al: JP *35*, 91, 142, 1932 (B). — (80) Morris: JI *42*,

219, 1941. — (81) Hartley: SystB 6, 262, 1931. — (82) Philosophow: BZ 20, 292, 1909 (B). — (83) Prášek: ZI 20, 146, 1913. — (84) Landsteiner: MM 1903, p. 1812. — (85) Ramon: SB 104, 31, 1930. — (86) Woolf: RS B, 130, 70, 1941. — (87) Heidelberger et al: JEM 62, 467, 1935. — (88) Burnet: JP 34, 471, 1931. — (89) Martin et al: RS A 63, 420, 1898 (B). — (90) Grassberger et al: "Toxin and Antitoxin," Deuticke, Leipzig, 1904. — (91) Bordet et al: CB 49, 260, 1909. — (92) Boyd et al: JBC 139, 787, 1941. — (93) Smith et al: Br 11, 494, 1930. — (94) von Krogh: ZH 68, 251, 1911. — (95) Enders et al: JI 32, 379, 1937. — (96) Chow et al: Pr 37, 460, 1937. — (97) Heidelberger et al: JEM 67, 181, 1938. — (98) Boyd: JI 38, 143, 1940. — (99) Bordet: AP 13, 225, 273, 1899. — (99a) Pijper: JBa 42, 395, 1941. — (100) Northrop et al: JGP 4, 639, 655, 1922, 6, 603, 1924. — (101) NK. — (102) Reiner et al: ZI 61, 317, 397, 1929. — (103) Mudd et al: JPC 36, 229, 1932; JGP 16, 947, 1933, 18, 615, 1935. — (104) Shibley: JEM 44, 667, 1926. — (105) Topley et al: Br 16, 116, 1935. — (106) Abramson: N 135, 995, 1935. — (107) Hooker et al: JI 33, 337, 1937. — (108) Wiener et al: JI 36, 255, 1939 (B). — (109) Hooker et al: JI 45, 127, 1942; v. JI 33, 57, 1937. — (110) Boyd et al: Pr 39, 491, 1938. — (111) Boyd: JEM 75, 407, 1942. — (112) Duncan: Br 19, 328, 1938. — (113) Raffel: AJH 19, 416, 1934 (B). — (114) Mudd et al: JI 42, 251, 1941. — (115) Marrack: ARB 11, 629, 1942. — (116) Marrack: IC, pp. 87, 821. — (117) Pauling et al: PNA 27, 125, 1941; JAC 64, 2994, 1942, 65, 728, 1943. — (118) Hooker et al: JI 42, 419, 1941. — (119) Landsteiner et al: ZH 113, 1, 1931. — (120) Heidelberger et al: JEM 57, 373, 1933, 75, 135, 1942. — (121) Ivánovics: ZI 97, 402, 1940. — (122) Eagle: JI 18, 393, 1930. — (123) Tyler: FP 2, 102, 1943. — (124) Heidelberger: ChR 24, 323, 1939. — (124a) Dean: JP 45, 745, 1937. — (125) Goodner: IC, p. 818. — (126) Heidelberger et al: JEM 75, 35, 1942. — (127) Kendall: ANY 43, 85, 1942. — (128) Hershey: JI 42, 455, 1941, 45, 39, 1942, 46, 249, 1943. — (129) Pauling et al: JAC 64, 3003, 3010, 1942. — (130) Boyd et al: JGP 17, 341, 1934, 22, 281, 1939. — (131) Kleczkowski: Br 22, 44, 1941. — (132) Zinsser: JI 18, 483, 1930. — (133) How: JI 37, 77, 1939. — (134) Pauling: JAC 62, 2643, 1940. — (135) Anderson et al: JBC 139, 339, 1941. — (136) Stanley: personal communication. — (137) Boyd et al: XV Int. Physiol. Congr., Leningrád, 1935, p. 567. — (138) Mudd et al: JI 42, 251, 1941. — (139) Rothen et al: JEM 76, 437, 1942. — (140) Harkins et al: JBC 132, 111, 1940. — (141) Bateman et al: JI 41, 321, 1941 (B). — (142) Schulman et al: RS B, 122, 29, 46, 1937. — (143) France et al: JPC 45, 395, 1941, 42, 1079, 1938. — (144) Bunn: RS A, 141, 567, 1933. — (145) Marrack: IV congr. internaz. patol. comp. 1, 331, 1939. — (146) Pfeiffer: Organische Molekülverbindungen, Enke, Stuttgart, 1927. — (147) Labes: AEP 190, 421, 1938. — (148) Feigl: Specific and Special Reactions, Elsevier Co., New York, 1940; v. N 128, 987, 1931. — (149) Baudisch: BC 49, 172, 1916. — (150) Bergmann et al: JBC 109, 317, 1935, 143, 121, 1942; HL 1935/36, p. 37. — (151) Hammett: Physical Organic Chemistry, McGraw-Hill, New York, 1940, p. 33. — (152) Evans: An Introduction to Crystal Chemistry, Cambridge, 1939. — (153) Briegleb: Zwischenmolekulare Kräfte, Enke, Stuttgart 1937. — (154) Bergmann et al: JBC 109, 325, 1935 (B). — (155) Nord et al: Handb. der Enzymol., Akad. Verlagsges., Leipzig, 1940. — (155a) Johnson et al: AE 2, 69, 1942 (B). — (156) Bergmann et al: AE 2, 49, 1942 (B). — (156a) Marrack: EE 7, 281, 1938. — (157) Landsteiner et al: ZI 17, 363, 1913. — (158) de Kruif et al: JGP 5, 127, 1922 (B). — (159) Chow et al: JEM 62, 179, 1935, v. 66, 204, 1937. — (160) Heidelberger: Medicine 12, 279, 1933. — (161) Coulter: JGP 3, 309, 513, 1921. — (162) Mason: BJH 33, 116, 1922. — (163) Euler et al: ZI 72, 65, 1931. — (164) Landsteiner et al: JI 33, 265, 1937. — (164a) Pressman et al: JAC 64,

3015, 1942. — **(164b)** Creech et al: AJC *30*, 555, 1937. — **(165)** Erlenmeyer et al: BZ *252*, 22, 1932, *255*, 429, 1932, *262*, 196, 1933; H *16*, 733, 1381, 1933. **(166)** Jacobs: JGP *20*, 353, 1937. — **(167)** Lettré: BC *73*, 1150, 1940. — **(168)** Lettré: Angew. Chemie *50*, 581, 1937. — **(169)** Jordan: ZI *97*, 330, 1940. — **(170)** Hooker: Yearbook of Pathology, 1940, p. 614. — **(171)** Haurowitz et al: JI *43*, 327, 1942. — **(172)** Pauling et al: S *92*, 77, 1940. — **(173)** Wells: CAI. — **(173a)** Gengou: Arch. Internat. Physiol. 7, 1, 1908. — **(173b)** Porges et al: HPM 3d ed. *1*, 1027, 1929. — **(174)** Sachs et al: AIF *21*, 159, 1928; BK 1908, p. 1968. — **(175)** Browning et al: JP *31*, 541, 1928. — **(176)** Haldane: Enzymes, Longmans, Green and Co., New York, 1930, p. 182. — **(177)** Doerr: HPM 3d ed., *1*, 796, 1929. — **(178)** Sachs: EH *9*, 39, 1928. — **(179)** Andrewes: JP *28*, 355, 1925. — **(180)** Goebel et al: JEM *60*, 599, 1934. — **(181)** London: JPC *46*, 305, 1942. — **(182)** Langmuir: Proc. Physic. Soc. *51*, 592, 1939. — **(183)** Landsteiner et al: JEM *67*, 709, 1938; unpublished experiments. — **(184)** Haurowitz et al: Br *23*, 146, 1942. — **(185)** Landsteiner et al: JEM *63*, 325, 1936. — **(186)** Landsteiner et al: Convegno Volta 1933, Reale Accademia d'Italia. — **(187)** White: JP *44*, 706, 1937. — **(188)** Morgan: BJ *31*, 2003, 1937. — **(189)** Marrack et al: Br *19*, 53, 1938. — **(190)** Avery et al: JEM *55*, 769, 1932. — **(191)** Morgan: JH *37*, 372, 1937.

VIII

MOLECULAR STRUCTURE AND INTERMOLECULAR FORCES

By Linus Pauling

As our knowledge of the structure of molecules has greatly increased in recent years it has become clear that the physiological activity of substances is correlated not alone with their ability to take part in reactions in which strong chemical bonds are broken and formed, but also with the relatively weak forces which their molecules exert on other molecules. It is, indeed, probable that the high specificity which often characterizes physiological activity is in most cases specificity of intermolecular interaction rather than primarily of chemical reaction with the rupture and formation of strong bonds.

Atoms interact with other atoms in many ways. In the present discussion we divide interatomic forces into two classes, strong forces and weak forces. The strong forces are those which are responsible for the existence of stable molecules; they are the forces which lead to the formation of strong chemical bonds, with bond energies between 10 and 100 kilocalories per mole. The weak interatomic forces, including van der Waals forces and "hydrogen-bond" forces, have energies of a few kilocalories per mole of interacting atom pairs; these forces are effective in holding molecules together without disrupting their individual structures, and also in operating between different parts of a large "loose-jointed" molecule in such a way as to hold it to a particular configuration.

The selection of the energy value of about 10 kilocalories per mole as the transition value between the two classes of interatomic interactions has its justification in statistical considerations based on thermal equilibrium at room temperature. Under ordinary circumstances at room temperature (that is, with the reacting substances present in reasonable concentrations) interaction of the "strong" class between two atoms will hold them together as a complex which is not significantly dissociated by the disrupting action of thermal agitation, whereas interaction of the "weak" class will not do this. Only by cooperation of the interactions between many atoms of one molecule and many atoms of another molecule can weak interatomic forces give rise to a stable intermolecular bond.

Weak interatomic forces have especial significance in serological phenomena — it is presumably these forces which lead to combination between an antibody molecule and a homologous antigen molecule or cell, and which also, operating between parts of the antibody molecule, are responsible for holding it in a configuration which confers on it the power of specific attraction for that antigen.

The interaction between molecules is determined in large part by the detailed structure of the molecules. In the following sections there are discussed first the nature of the chemical bond and the structure of molecules, and then the nature of intermolecular interactions.

THE STRUCTURE OF MOLECULES

The strong interatomic forces which lead to chemical bond formation are varied in nature. For convenience several extreme types, which are in general not sharply demarcated but show gradual transitions through intermediate types, are recognized; these are called ionic bonds, covalent bonds, metallic bonds, and other bonds of less importance.

Atoms such as sodium and chlorine can by losing or gaining electrons achieve stable electronic configurations, such as those of the noble gases. The resulting electrically charged ions with opposite sign, such as Na^+ and Cl^-, attract each other with the strong inverse-square Coulomb attraction. If the dielectric constant of the medium is small this attraction is very strong; it pulls the ions together until they are in contact. The size of an ion is determined by its electron distribution. Anions range in radius from 1.4 Å (for F^- and $O^=$) to 2.1 Å (for I^-), and cations from very small values (about 0.6 Å for Li^+, Be^{++}, Al^{+++}) to a maximum of 1.6 Å for Cs^+. When two ions are at the internuclear distance equal to the sum of their conventional crystal radii their outer electron shells interpenetrate to such an extent as to give rise to a repulsive force which balances the force of Coulomb attraction.

The energy of such an ionic bond in a medium of low dielectric constant is large—of the order of 100 kilocalories per mole — and the ionic bond is a strong bond. Strong ionic bonds exist between sodium ions and chloride ions in the sodium chloride crystal and in sodium chloride gas molecules (at high temperatures), and also, for example, between iron ions and magnesium ions and the surrounding nitrogen atoms of the porphyrin groups in hemoglobin and chlorophyll.

In water or other medium of high dielectric constant the Coulomb forces between ions are very greatly reduced in magnitude, so that in general they no longer lead to the formation of strong chemical bonds. The magnitude of the bond energy for Coulomb attraction of two electrically charged atoms or groups can be easily calculated; it is the product of the two electrical charges divided by the distance between them and by the dielectric constant of the medium. For example, the Coulomb bond strength for a carboxyl ion group $RCOO^-$ and an ammonium ion group RNH_3^+, which can approach each other until the center of the nitrogen atom comes to within about 3 Å of the center of an oxygen atom, is calculated to be about 4.5 kcal. per mole for a medium with dielectric constant 25 (the effective value for water for two charges at this distance [1]). This bond energy is not great enough to cause the two groups to form a stable aggregate; however, as is mentioned in a later section, hydrogen-bond formation between the groups increases the strength of their interaction greatly, and together with the direct ionic attraction gives rise to a rather strong bond between the groups.

Of by far the greatest importance in determining the structure of organic molecules is the covalent bond or shared-electron-pair bond. This is the ordinary valence bond of the organic chemist. The single

[1] Schwarzenbach (1).

covalent bond results from the sharing of an electron pair (two electrons) between two atoms

$$(\text{H–H or H:H};\quad \underset{\overset{|}{H}}{\overset{\overset{H}{|}}{H-C}}-\underset{\overset{|}{H}}{\overset{\overset{H}{|}}{C}}-H \quad \text{or} \quad H:\overset{\cdot\cdot}{\underset{\cdot\cdot}{C}}:\overset{\cdot\cdot}{\underset{\cdot\cdot}{C}}:H)$$

the double bond from the sharing of two electron pairs ($H_2C=CH_2$ or $H_2C::CH_2$), and the triple bond from the sharing of three electron pairs ($HC\equiv CH$ or $HC:::CH$). The carbon atom usually forms four covalent bonds (counting the double bond as two and the triple bond as three), the nitrogen atom three (as in the amines, $\;:N\overset{\diagup R}{\underset{\diagdown R}{-}}$)

or four, in which case it assumes a positive charge $\;(-CH_2-{}^+N\overset{\diagup H}{\underset{\diagdown H}{-}})$

the oxygen atom two ($(CH_3)_2C=O$ or CH_3-O-CH_3) or one (^-O-H), and halogen atoms one

$$(HC\overset{\diagup Cl}{\underset{\diagdown Cl}{-}}Cl)$$

or zero (chloride ion, Cl^-).

The covalent bond holds atoms compactly together: the $C-C$ single bond distance is 1.54 Å, the $C=C$ double-bond distance 1.33 Å, etc. The values of these and many other interatomic distances have been found by experimental investigations, mainly by the diffraction of x-rays and electrons, and it is now possible to predict with confidence, in most cases to within 1 or 2%, the interatomic distances for any molecule of known chemical structure.

The remaining information needed to define the configuration of a covalently bonded molecule relates to the mutual orientation in space of the bonds. For the atoms which in the main occur in molecules of biological substances single bonds usually are directed toward the corners of a regular tetrahedron, making with one another approximately the tetrahedral angle 109°28'. The picture of the tetrahedral angle of Van't Hoff and Le Bel similarly leads to the following values of angles involving multiple bonds: double bond and single bond, 125°16'; two double bonds, 180°; triple bond and single bond, 180°. These predictions have all been verified as holding to within a few degrees by extensive structural studies (electron diffraction, x-ray diffraction) carried out in recent years.

The rotational configuration about a double bond is highly restricted by the requirement that the system, including the adjacent single bonds, be coplanar. Except for possible uncertainty in identification of *cis* and *trans* configurations, this requirement specifies uniquely the configuration about the double bond.

The orientation about a single bond is less well defined. It has been discovered recently that the "staggered" orientation

$$HH\diagdown_{H\diagup}C-C\diagup^{H}_{\diagdown HH}$$

for ethane and related molecules is more stable by a small amount of energy (about 3 kilocalories per mole) than the "eclipsed" configuration

$$H\diagdown_{HH\diagup}C-C\diagup^{H}_{\diagdown HH}$$

which differs from the "staggered" configuration by rotation of one group through the angle 60° relative to the other. However, the requirement that each single bond have the "staggered" orientation often leaves accessible several alternative configurations for a complex molecule among which the molecule easily changes. For a hydrocarbon chain, for example, there exist alternatives such as the extended configuration

$$C\diagup^{C}\diagdown_{C}\diagup^{H\ H\ H\ H}_{C\diagdown_{C}\diagup^{C}\diagdown_{C}\diagup^{C}\diagdown C}$$
$$\underset{H\ H}{\diagup\diagdown}$$

and the various coiled configurations which result from this by rotating through 120° about any of the carbon-carbon bonds. X-ray and electron-diffraction studies have shown that molecules of the normal hydrocarbons and related substances have the extended configuration in crystals, but also assume other coiled configurations in the vapor state at higher temperatures.

The phenomenon of *resonance among alternative valence-bond structures* is of great importance in determining the configuration of aromatic and conjugated systems. For example, for the benzene molecule the two reasonable valence-bond structures

(the Kekulé structures) can be written. These structures, although they are the most satisfactory single structures proposed for the molecule, are unsatisfactory in several ways. Thus benzene does not show the unsaturation which would be expected for a molecule containing double bonds, and it does not form the pairs of isomeric compounds, such as

which would be expected if the individual Kekulé structures existed independently. The solution of these and many other difficulties has been provided recently by the idea, which is a consequence of the physical theory of quantum mechanics, that a molecule such as benzene, for which two or more alternative reasonable electronic structures can be written, is represented not by any one of these structures but by a combination of them; the molecule is described as resonating among all of the alternative structures, and its properties are determined by all of them. For benzene, which resonates between the two Kekulé structures, each of the carbon-carbon bonds has 50% double-bond character; this gives each carbon-carbon bond the stereochemical properties (coplanarity) of a double bond, which requires the molecule to be a completely coplanar regular hexagon and to have carbon-carbon bond distances (1.39 Å) between the single-bond and double-bond values.

A very important consequence of the theory of resonance is that a molecule which resonates between two or more alternative electronic structures is more stable thermodynamically than it would be if it had any one of the structures alone. This energy of stabilization, called the *resonance energy*, amounts for benzene to 39 kcal. per mole. The benzene molecule is more stable by 39 kcal. per mole than it would be if it had a non-resonating Kekulé structure; and since similar resonance stabilization does not occur for its addition compounds, benzene is much more resistant to hydrogenation and similar reactions than are molecules containing non-resonating double bonds. In the same way the resonance energy of other aromatic molecules stabilizes them and causes them to behave as saturated rather than unsaturated substances.

It might be expected that the OH group in acetic acid or other

carboxylic acid $R-C\underset{\overset{\cdot\cdot}{O}-H}{\overset{\overset{\cdot\cdot}{O}:}{\diagdown}}$

would dissociate to about the same extent as in the alcohols. Experiment shows, however, that the dissociation constants for these acids are very much larger (by factors of about 10^{12}) than those for the alcohols. The explanation of this fact provided by the theory of resonance is that the carboxylic acids have the resonating structure

$$\left\{ R-C\diagup^{\ddot{O}:}_{\ddot{O}-H} \qquad R-C\diagup^{\ddot{O}\bar{:}}_{O^{\pm}H} \right.$$

and the contribution of the second of these structures, which assigns a carbon double bond to the oxygen atom with hydrogen attached, causes the proton to tend to leave the molecule. A similar explanation, involving structures such as

$$\overset{\overset{+}{\ddot{O}}-H}{\Big\|}$$

as well as the normal structures, accounts for the greater acidity of phenols as compared with aliphatic alcohols. In a conjugated system such as benzoic acid

in which the double bond of the carboxyl group is conjugated with
the double bonds of the benzene ring, the connecting single bond has
enough double-bond character to keep the carboxyl group coplanar
with the ring. This double-bond character results from resonance
with structures such as

Steric interactions between atoms in a molecule are often important
in preventing the assumption of an otherwise stable configuration.
Thus the carboxyl group in o-methylbenzoic acid is prevented from
becoming coplanar with the benzene ring by the steric repulsion between the methyl group and the carboxyl oxygen atoms.

The examples given above do no more than to indicate the breadth
of the field of application of the theory of resonance. Many further
applications are discussed in treatises on molecular structure and
theoretical organic chemistry.[2]

[2] Pauling (2), Hammett (3), Branch et al. (4), Rice (5), Remick (13).

INTERMOLECULAR INTERACTIONS

Two molecules may attract each other through interactions classified as electronic van der Waals attraction, Coulomb attraction of groups with opposite electric charges, attraction of electric dipoles or multipoles, hydrogen-bond formation, etc. Whatever the nature of the attraction may be, the forces of attraction increase in general as the molecules approach one another more and more closely, and the bond between the molecules reaches its maximum strength when the molecules are as close together as they can come. The molecular property which determines the distance of closest approach of two molecules is the electronic spatial extension of the atoms in the molecules. It is possible to assign to each atom a *van der Waals radius*, which describes its effective size with respect to intermolecular interactions. These radii vary in value from 1.2 Å for hydrogen through 1.4–1.6 Å for light atoms (fluorine, oxygen, nitrogen, carbon) to 1.8–2.2 Å for heavy atoms (chlorine, sulfur, bromine, iodine, etc.). The shape of a molecule can be predicted by locating the atoms within the molecule with use of bond distances and bond angles and then circumscribing about each atom a spherical surface corresponding to its van der Waals radius. This shape determines the ways in which the molecule can be packed together with other molecules.

The most general force of intermolecular attraction, which operates between every pair of molecules, is *electronic van der Waals attraction*. This type of electronic interaction between molecules was first recognized by London (6). A molecule (of argon or carbon tetrachloride, for example) which has no permanent average electric dipole moment may have an instantaneous electric dipole moment, as the center of charge of the electrons, in their rapid motion in the molecule, swings to one side or the other of the center of charge of the nuclei. This instantaneous dipole moment would produce an instantaneous electric field. Any other molecule in the neighborhood would be polarized by this field, which would cause its electrons to move relative to its nuclei in such a way as to give rise to a force of attraction of the second molecule toward the first. By the time that this electronic polarization had occurred the electrons of the first molecule would have moved some distance, and its instantaneous dipole moment would have changed in value. It is clear that the theoretical discussion of the phenomenon is not simple; it was carried out by London, who found that the effect as calculated by quantum-mechanical methods corresponds closely in magnitude with the observed van der Waals attraction for various simple substances (noble gases, hydrogen, oxygen, methane, etc.), and explains quantitatively the intermolecular attraction which leads to their condensation to form liquids.

This electronic van der Waals attraction operates between every atom in a molecule and every atom in other molecules in the near neighborhood. The force is stronger for heavy atoms than for light atoms. It increases very rapidly with decreasing interatomic distance, being inversely proportional to the seventh power of the interatomic distance. Because of this, the electronic van der Waals attraction between two molecules in contact is due practically entirely to interactions of pairs of atoms (in the two molecules) which are themselves in contact; and the magnitude of the attraction is determined by the number of pairs of atoms which can be brought into contact. In consequence, two molecules which can bring large portions of their surfaces into close-fitting juxtaposition will in general show much stronger mutual attraction than molecules with less extensive complementariness of surface topology.

A measure of the energy of the van der Waals attraction of molecules is given by the heats of sublimation of their crystals. This energy is small for light molecules and larger for heavier molecules. For the hydrogen molecule it is about 0.25 kcal. per mole, for nitrogen and oxygen molecules about 1.9, for argon 2.0, for krypton and methane 2.7, and for xenon 3.8. For large molecules the energy of van der Waals attraction is correspondingly larger, in proportion to the number of atoms in the molecule.

Except for molecules with very large electric dipole moment or capable of forming hydrogen bonds, the van der Waals attraction is responsible for the major part of the interaction with other molecules.

Other types of molecular interactions result from the possession by one or both of the interacting molecules of a permanent electric dipole moment or electric moment of higher order. The effects of these permanent moments have been classified in various ways, as dipole-dipole forces, the forces of electronic polarization of one molecule in the dipole field of another, etc. In general, however, these electric forces are of minor importance, except when an isolated or essentially isolated electric charge is involved. Thus for hydrogen chloride, with a rather large value of the electric dipole moment, about five-sixths of the energy of intermolecular attraction is due to electronic van der Waals attraction and only about one-sixth is the result of dipole interaction.

A type of intermolecular attractive force which ranks in importance with the electronic van der Waals attraction is that associated with the structural feature called the hydrogen bond. The importance and generality of occurrence of the hydrogen bond were first pointed out in 1920 by Latimer and Rodebush (7). A hydrogen bond results from the attraction of a hydrogen atom attached to one electronegative atom for an unshared electron pair of another electronegative atom. The strength of a hydrogen bond depends on the electronegativity of the two atoms which are bonded together by hydrogen; fluorine, oxygen, and nitrogen, the most electronegative of all atoms, are the atoms which form the strongest hydrogen bonds. The energy of a hydrogen bond between two of these atoms is of the order of magnitude of 5 kcal. per mole. This is so large as to have a very important effect on the intermolecular interactions of molecules capable of forming hydrogen bonds and on the properties of the substances consisting of these molecules. It is hydrogen-bond formation between water molecules which gives to water its unusual physical properties — abnormally high melting point, boiling point, heat of fusion, heat of vaporization, dielectric constant, etc. Hydrogen-bond formation is also responsible for the existence of the hydrogen fluoride ion HF_2^-, and for the polymerization of hydrogen fluoride, the carboxylic acids, and other substances.

The hydrogen bond is not specific, inasmuch as a hydrogen atom of a hydroxyl group, for example, will attract the unshared electron pair of any electronegative atom which comes into its neighborhood. But the attracted atom must be able to come to a definite position in space in order that a stable hydrogen bond may be formed. This position is along the line of the OH axis (for a hydroxyl group) and at a determined distance, which is about 2.7 Å for an O-H-O hydrogen bond. This steric restriction and the limitation of the class of atoms capable of forming good hydrogen bonds give to the hydrogen bond somewhat greater stereochemical significance than is shown by the electronic van der Waals attraction.

The third type of attractive force which is of significance in protein structure is the interaction of electrically charged groups, such as the carboxyl ion side chains and ammonium ion side chains of amino acid residues. As stated above, the direct electrostatic interaction of two such groups amounts to about 4.5 kcal. per mole. However, the hydrogen atoms of the ammonium ion group are able to form hydrogen bonds with the oxygen atoms of the carboxyl ion group, increasing very much the energy of the interaction of these two groups, and permitting a stable complex, with bond energy of the order of 10 kcal. per mole, to be formed.

The peptide chain provides illustration of several structural features.

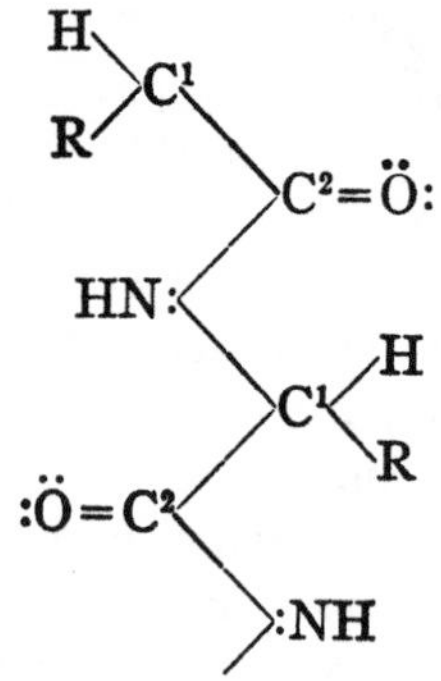

The conventional electronic structure shown above does not provide a complete representation of the chain; each peptide group shows resonance of the type

which gives the C^2–N bond some double-bond character. In consequence of this and the stereochemical property of coplanarity about double bonds, the stable configurations of the chain are those which make coplanar the group of atoms

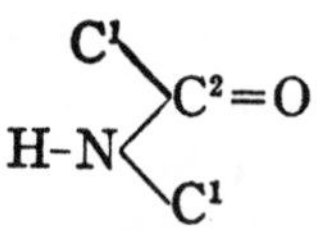

There are two such configurations; the extended one given above, and the contracted one

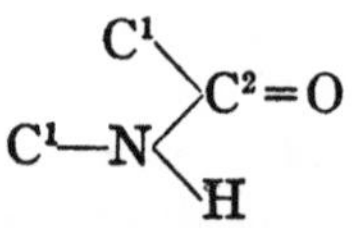

In addition, the chain has considerable freedom of orientation about

the single bonds C^1–C^2 and N–C^1. Further selection among the many corresponding configurations for the peptide chain results from the requirement that no two non-bonded atoms of the chain can come closer together than their limiting contact distance, the sum of their van der Waals radii. Interatomic distances and bond angles for the fully extended peptide chain, as given by Corey (8), are

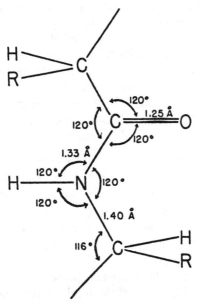

Another result of the resonance of the double bond in the peptide group is that the oxygen atom receives an increased negative charge and the nitrogen atom an increased positive charge. Both of these effects tend to increase the tendency of these atoms to take part in hydrogen-bond formation, and in consequence we would expect stable configurations of protein molecules to be those in which many hydrogen bonds between peptide chains are formed. Such a structure, for extended polypeptide chains, is the following:

A structure of this type was first suggested, for β-keratin, by Astbury, on the basis of his x-ray studies.

Whereas a clear distinction can be made between electronic van der Waals forces and other intermolecular forces, designated as Coulomb attraction of ions, attraction of electric dipoles and multipoles, and hydrogen bonds, these latter types are not clearly demarcated. A neutral molecule containing a carboxyl-ion group $-COO^-$ and an ammonium-ion group $-NH_3^+$ may be described as having an electric dipole moment, and the force of attraction between two such zwitterionic molecules may be said to result from the interaction of two electric dipoles; this, however, is not the best way of describing the system except when the two molecules are very far apart. If the molecules are close together, the carboxyl-ion group of one molecule may be very near to the ammonium-ion group of the other — within 4 Å — with the other charged groups much farther away, perhaps

$$H_3N^+ \underline{\hspace{2cm}} COO^- \qquad H_3N^+ \underline{\hspace{2cm}} COO^-$$
$$\longleftarrow \text{10 Å} \longrightarrow \quad \leftarrow 4\text{ Å} \rightarrow \quad \longleftarrow \text{10 Å} \longrightarrow$$

10 Å or 20 Å away if the molecules are large. The intermolecular attraction for this configuration will be very nearly the same as for a carboxyl ion and an ammonium ion at 4 Å; and it can be expressed still more closely as the sum of the Coulombic interactions of the charged groups in pairs.

Similarly the forces between two protein molecules may be conveniently described in terms of the total electric charges on the molecules and their electric dipole moments and multipole moments so long as the molecules are far apart (compared with their own diameters of say 50 or 100 Å); when the molecules approach one another more closely, and especially when they come into contact, the forces are more conveniently discussed by considering the interactions of small parts of one molecule and small parts of the other.

Since the distances between atoms in contact are about 3 Å or 4 Å, and the dimensions of usually recognized atomic groups are not much greater, electric charges in a molecule which are more than about 5 Å from other charges in the molecule may conveniently be considered separately, whereas pairs of opposite charges less than 5 Å apart in the same group may be described as forming the electric dipole of the group. The dipole moments of groups are usually small — less than 2×10^{-18} e.s.u. This moment corresponds to unit positive and negative charges (equal in magnitude to the electronic charge) separated by less than 0.5 Å. It is seen that the force of attraction of two such dipoles or of a dipole and an ion at a distance of 4 Å or more would be very much less than the Coulomb attraction of two ionic groups, since the effect of a positive charge would be nearly completely neutralized by the opposite effect of the nearby negative charge. It is for this reason that the dipole and multipole moments of groups make little contribution to the forces between molecules, compared with ionic groups. Moreover, as mentioned above for hydrogen chloride, the intermolecular forces due to dipole moments are also much smaller than those due to electronic van der Waals attraction.

The hydrogen bond may be said to arise in large part from the attraction of the dipole moments of the groups involved, such as

$$\text{>N} \xleftarrow{\quad +} \text{H} \qquad \text{and} \qquad \text{:Ö} \xleftarrow{\quad +} \text{C<}$$

(the arrow pointing from the positive charge toward the negative charge of the dipole), with the dipole-dipole attraction unusually large because of the close approximation of the groups permitted by the small size of the hydrogen atom. Other factors (specific interaction of the proton and an electron pair of the electronegative atom of the approaching group) also play a part, however, and it is customary to designate this special sort of interaction of groups as hydrogen-bond formation.

The theoretical discussion of the interaction of molecules in solution is complicated by competition with solvent molecules. A solute molecule may have as strong electronic van der Waals attraction for the solvent molecules as for other solute molecules. The electronic van der Waals attraction of atoms for all other atoms operates in every condensed system to bring all the atoms into as close packing as is permitted by the chemical bonds which determine the shapes of the molecules and by the other intermolecular forces. This is the reason that the molal volumes of most liquids and crystals can be expressed to within a few percent as the sums of definite atomic volumes. Only rarely (as in the case of ice, in which a rather open structure is stabilized by hydrogen-bond formation at tetrahedral angles) do exceptions occur. The effective electronic van der Waals attraction of two solute molecules in aqueous solution may be very small, since the close approach of the two molecules involves the replacement of the water molecules adjacent to each molecule. Similarly the effective hydrogen-bond attraction between two solute molecules is large only if the solute-solute hydrogen bonds plus the corresponding water-water hydrogen bonds are significantly stronger than the ruptured solute-water hydrogen bonds.

Of the great number of well-defined intermolecular compounds which are known, precise structural information is available for only a few; it is only very recently that x-ray and electron-diffraction methods and related techniques have been sufficiently developed to permit them to be applied to complexes containing more than a dozen atoms or so, and their extensive successful application in this field has not yet been accomplished. Reliable information has been obtained about the dimer of formic acid

$$H-C\begin{array}{c} O-H\cdots O \\ \diagdown \qquad \diagup \\ O\cdots H-O \end{array}C-H$$

and about a number of other polymers in which the intermolecular attraction is due to hydrogen bonds. Detailed studies have also been made of Werner coordination complexes, such as the hydrated and ammoniated ions $Zn(NH_3)_4^{++}$, $Ni(H_2O)_6^{++}$, $Co(NH_3)_6^{+++}$, etc. The forces which hold the water or ammonia molecules to the central ion are in part the electrostatic attraction of the ion for the electric dipole molecules; more important is the formation of covalent bonds between the central atom and the attached molecules, with use for each bond of an electron pair provided by the molecule. Most atoms tend to form four bonds at tetrahedral angles (or in special cases four coplanar bonds directed toward the corners of a square) or six bonds directed towards the corners of a regular octahedron. The formation of bonds of this sort may be of general importance in holding the metal-containing prosthetic groups to their protein molecules in the respiratory pigments and related substances. There is strong evidence indicating that in hemoglobin each protoheme is held to globin not only by Coulomb attraction between the propionate ion side chains and positively charged groups in the globin but also by the formation of an octahedral bond between the iron atom in the center of the four nitrogen atoms of the porphyrin and a nitrogen atom of the imidazole ring of a histidine residue in the globin. It seems likely that in some heme pigments, including cytochrome c, each iron atom forms two bonds in this way with two parts of the protein molecule. Similar bonds are probably present also in hemocyanin and other pigments containing metal atoms, aiding in holding the complexes together.

Because of the lack of detailed structural information, there is doubt as to the nature of the forces which lead to compound formation between nitro compounds such as picric acid and various aromatic hydrocarbons. It is probable that the electric dipoles of the nitro groups are important here, and that the high polarizability of aromatic molecules in the plane of the rings gives rise to strong attraction to the nitro molecule when it is suitably oriented.[3] The unusually strong bonds formed between antibody molecules and some nitro haptens [4] may result from a similar effect.

An extraordinary set of molecular compounds is formed by certain sterids, especially desoxycholic acid. In the compounds between desoxycholic acid and various carboxylic acids, dicarboxylic acids, esters, and other molecules it is found [5] that one molecule of the latter combines with 1, 2, 3, 4, 6, or 8 molecules of desoxycholic acid, the number increasing with increase in size of the molecule (1 for acetic acid, 2 (or 3) for propionic acid, and so on to 8 for pentadecylic and larger acids). That hydrogen bonds are involved in the formation of these complexes is indicated by the fact that desoxycholic acid and also α-apocholic acid and β-apocholic acid, which form similar complexes, contain two hydroxyl groups, in essentially the same stereochemical configuration, whereas other related substances without hydroxyl groups or with different configurations are ineffective. It is likely that several molecules of desoxycholic acid pile up in such a way that a cavity is formed through them into which fits the foreign molecule, which thus serves to key them together; but experimental substantiation of this surmise is lacking.

The forces of van der Waals attraction, hydrogen-bond formation, and interaction of electrically charged groups are in themselves not specific; each atom of a molecule attracts every other atom of another molecule by van der Waals attraction, each hydrogen atom attached to an electronegative atom attracts every other electronegative atom with an unshared electron pair which comes near it, and each electrically charged group attracts every other oppositely charged group in its neighborhood. Similarly the van der Waals repulsive forces are non-specific; each atom in a molecule repels

---

[3] Briegleb (9), Pauling (10).
[4] Pressman, Brown and Pauling (11).
[5] For references see Sobotka (12).

every other atom of another molecule, holding it at a distance corresponding to the sum of the pertinent van der Waals radii.

We see, however, that specificity can arise in the interaction of large molecules as a result of the spatial configuration of the molecules. Two large molecules may have such shapes that the surface of one cannot be brought into contact with the surface of the other except at a few isolated points. In such a case the total electronic van der Waals attraction between the two molecules would be small, because only the pairs of atoms near these few isolated points of contact would contribute appreciably to this interaction. Moreover, the distribution of positively and negatively charged groups and of hydrogen-bond forming groups of the two molecules might be such that only a small fraction of these groups could be brought into effective interaction with one another for any position and orientation of one molecule with respect to the other. The energy of attraction of these molecules would then be small. If, on the other hand, the two molecules possessed such mutually complementary configurations that the surface of one conformed closely to the surface of the other, there would be strong electronic van der Waals attraction between all of the atoms on the surface of one of the molecules and the juxtaposed atoms of the complementary surface of the other molecule. And if, moreover, the electrically charged groups of one molecule and those of the other were so located that oppositely charged groups were brought close together as the molecules came into conformation with one another, and if the hydrogen-bond forming groups were also so placed as to form the maximum number of hydrogen bonds, the total energy of interaction would be very great, and the two molecules would attract one another very strongly. It is clear that this strong attraction might be highly specific in the case of large molecules which could bring large areas of their surfaces into contiguity. A molecule would show strong attraction for that molecule which possessed completely complementariness in surface configuration and distribution of active electrically charged and hydrogen-bond forming groups, somewhat weaker attraction for those molecules with approximate but not complete complementariness to it, and only very weak attraction for all other molecules.

This specificity through complementariness of structure of the two interacting molecules would be more or less complete, depending on the greater or smaller surface area of the two molecules involved in the interaction. It may be emphasized that this explanation of specificity, as due to a complementariness in structure which permits non-specific intermolecular forces to come into fuller operation than would be possible for non-complementary structures, is the only explanation which the present knowledge of molecular structure and intermolecular forces provides.

BIBLIOGRAPHY

(1) Schwarzenbach: ZP *A* 176, 133, 1936. — (2) Pauling: The Nature of the Chemical Bond, Cornell University Press, Ithaca, 2d ed., 1940. — (3) Hammett: Physical Organic Chemistry, McGraw-Hill Book Co., New York, 1940. — (4) Branch et al: Theory of Organic Chemistry, Prentice Hall, New York, 1941. — (5) Rice: Electronic Structure and Chemical Binding, McGraw-Hill Book Co., New York, 1940. — (6) London: Z. f. Physik *63*, 245, 1930. — (7) Latimer et al: JAC *42*, 1419, 1920. — (8) Corey: ChR *26*, 227, 1940. — (9) Briegleb: ZP *B 26*, 63, 1934, *B 31*, 58, 1935. — (10) Pauling: PNA *25*, 577, 1939. — (11) Pressman et al: JAC *64*, 3015, 1942. — (12) Sobotka: The Chemistry of the Sterids, Williams and Wilkins Co., Baltimore, 1938. — (13) Remick: Electronic Interpretations of Organic Chemistry, John Wiley and Sons, Inc., New York, 1943.

TEXTBOOKS, REVIEWS, MONOGRAPHS, AND SURVEYS

TEXTBOOKS OF SEROLOGY AND IMMUNOLOGY

Bordet: *Traité de l'immunité dans les maladies infectieuses*. Paris: Masson, 2nd ed., 1939.

Boyd: *Fundamentals of Immunology*. New York: Interscience Publishers 1943.

Hammerschmidt and Müller: *Serologische Untersuchungstechnik*. Jena: Fischer 1926.

Karsner: *The Principles of Immunology*. Philadelphia and London: Lippincott 1921.

Kolmer: *Infection, Immunity and Biologic Therapy* etc. Philadelphia and London: Saunders, 3rd ed., 1925.

Kolmer and Boerner: *Approved Laboratory Technic*. New York: Appleton, 3rd ed., 1941

Metchnikoff: *L'immunité dans les maladies infectieuses*. Paris: Masson 1901.

Müller, P. Th.: *Vorlesungen über Infektion und Immunität*. Jena: Fischer 1909.

Sherwood: *Immunology*. St. Louis: C. V. Mosby Co., 2nd ed., 1941.

Topley and Wilson: *Principles of Bacteriology and Immunity*. Baltimore: William Wood and Co., 2nd ed., 1936.

Zinsser, Enders and Fothergill: *Immunity: Principles and Application in Medicine and Public Health*. New York: The Macmillan Company, 5th ed., 1939.

REVIEWS ON THE SPECIFICITY OF SERUM REACTIONS

Doerr: Allergie und Anaphylaxie. In *Handbuch der pathogenen Mikroorganismen*, vol. 1, p. 785, 790. Jena: Fischer 1929.

Hartley: The Effect of Physical and Chemical Agencies on the Properties of Antigens, and Antibodies. In *System of Bacteriology* 6, p. 224, London, 1931.

Haurowitz: Chemie der Antigene und der Antikörper. In Kallós: *Fortschritte der Allergielehre*, p. 19. Basel: Karger 1939.

Marrack: The Chemistry of Antigens and Antibodies. *Med. Res. Council, Spec. Rep. Ser.* 230, London 1938.

Marrack: Immunochemistry and its Relation to Enzymes. *Ergeb. Enzymforsch.*, vol. 7 (1938) p. 281.

Pick and Silberstein: Biochemie der Antigene und Antikörper. In *Handbuch der pathogenen Mikroorganismen*, vol. 2, p. 317. Jena: Fischer 1929.

Sachs: Antigene und Antikörper. Bethe, *Handbuch der normalen und pathologischen Physiologie*, vol. 13, p. 405. Berlin: Springer 1929.

Thomsen: *Antigens in the Light of Recent Investigations*. Copenhagen: Levin and Munksgaard 1931.

Wells: *The Chemical Aspects of Immunity*. New York: Chemical Catalog Company, 1929.

Westphal: Fermente und Immunchemie. Nord and Weidenhagen, *Handbuch der Enzymologie*, vol. 2 (1940) p. 1129.

Witebsky: Biologische Spezifität. Bethe, *Handbuch der normalen und pathologischen Physiologie*, vol. 13 (1929) p. 473; vol. 18 (1932) 319.

MONOGRAPHS ON SPECIAL CHAPTERS OF SEROLOGY AND IMMUNOLOGY, ETC.

Arrhenius: *Immunchemie.* Leipzig: Akademische Verlagsgesellschaft 1907.

Browning: *Immunochemical Studies.* London: Constable 1925.

Buchbinder: Heterophile Phenomena in Immunology. *Arch. Path.* 19 (1935) 841.

Chargaff: Methoden zur Untersuchung der chem. Zusammensetzung von Bakterien. Abderhalden: *Handb. der biol. Arbeitsmethoden*, Abt. XII, Teil 2, H. 2, Lief. 423.

Coca, Walzer and Thommen: *Asthma and Hay Fever in Theory and Practice.* Springfield, Ill.: Thomas 1931.

Dujarric de la Rivière and Kossovitch: *Antigènes, Hétéroantigènes et Haptènes.* Paris: Baillière et fils 1937.

von Dungern: *Die Antikörper.* Jena: Fischer 1903.

Faust: *Die tierischen Gifte*, Heft 9. Braunschweig: Vieweg 1906.

Gay et al.: *Agents of Disease and Host Resistance.* Springfield, Ill.: Thomas 1935.

Graetz: Ueber Probleme und Tatsachen aus dem Gebiet der biologischen Spezifität der Organantigene etc. *Erg. Hyg.* 6 (1924) 397.

Great Britain: Medical Research Council: *A System of Bacteriology in Relation to Medicine.* London: His Majesty's Stationery Office, vol. 6, 1931.

Handbuch der Biochemie des Menschen und der Tiere, Oppenheimer, vol. 3. Jena: Fischer 1924, Ergänzungsw. 1933.

Handbuch der pathogenen Mikroorganismen, Kolle, Kraus, Uhlenhuth, vols. 1, 2, 3. Jena: Fischer, 3rd ed., 1929–1931.

Handbuch der Technik und Methodik der Immunitätsforschung, Kraus and Levaditi. Jena: Fischer 1907.

Heidelberger: Relation of Proteins to Immunity. In Schmidt: *The Chemistry of the Amino Acids and Proteins*, p. 953. Springfield: C. C. Thomas 1938. And in Schmidt: *Addendum to The Chemistry of the Amino Acids and Proteins*, p. 1258. Springfield, C. C. Thomas 1943.

Klopstock: *Immunität, Medizinische Kolloidlehre.* Dresden: Steinkopff 1932.

Kraus and Werner: *Giftschlangen.* Jena: Fischer 1931.

Levaditi: *Antitoxische Prozesse.* Jena: Fischer 1905.

Mollison: Serodiagnostik etc. Berlin-Wien: Urban and Schwarzenberg 1924. (Abderhalden: *Handb. der biol. Arbeitsmethoden* Abt. IX, Teil 1.)

Oppenheimer: *Toxine und Antitoxine.* Jena: Fischer 1904.

Osborn: *Complement or Alexin.* London: Oxford University Press 1937.

Piettre: *Biochimie des Protéines.* Paris: Baillière et fils 1937.

Sachs and Klopstock: *Methoden der Hämolyseforschung.* Berlin-Wien: Urban and Schwarzenberg 1928. (Abderhalden: *Handb. der biol. Arbeitsmethoden*, Abt. XIII, Teil 2.)

The Newer Knowledge of Bacteriology and Immunology, Jordan and Falk. Chicago: The University of Chicago Press 1928.

Velluz: *Chimie et Immunité.* Le VI Congrès de Chimie Biologique, 1937, p. 173.

Wadsworth: *Standard Methods of the Division of Laboratories and Research of the New York State Department of Health.* Baltimore: Williams and Wilkins, 2nd ed., 1939.

Wiener: *Blood Groups and Blood Transfusion.* Springfield: C. C. Thomas, 3rd ed., 1943.

SURVEYS OF THE LITERATURE

Haurowitz: *Fortschritte der Biochemie.* Dresden: Steinkopff 1932, p. 100; 1938, p. 136.

Immunochemistry, in *Annual Review of Biochemistry*: Stanford University P. O., California: Annual Reviews, Inc.
Heidelberger, vol. 1 (1932) p. 655; vol. 2 (1933) p. 503; vol. 4 (1935) p. 569.
Landsteiner and Chase, vol. 6 (1937) p. 621.
Chase and Landsteiner, vol. 8 (1939) p. 579.
Marrack, vol. 9 (1942) p. 629.

Immunology, in *The Year Book of Pathology and Immunology*: Chicago: The Year Book Publishers, Inc., 1940 and 1941.

Schmidt: *Fortschritte der Serologie.* Leipzig: Steinkopff 1933.

THE BIBLIOGRAPHY OF DR. KARL LANDSTEINER

TOGETHER WITH PAPERS FROM HIS LABORATORY PUBLISHED BY HIS COLLEAGUES
1892-1943

PREPARED BY THE STAFF OF THE LIBRARY OF THE ROCKEFELLER INSTITUTE FOR MEDICAL RESEARCH AND BY MERRILL W. CHASE

Reprinted from THE JOURNAL OF IMMUNOLOGY
Vol. 48, No. 1, January, 1944

The great use of a life is to spend it for something that outlasts it.
—WILLIAM JAMES

YEAR

1892
1. LANDSTEINER, K. Ueber den Einfluss der Nahrung auf die Zusammensetzung der Blutasche. (Z. physiol. Chem. 1892, 16, 13-19)
2. FISCHER, E., AND LANDSTEINER, K. Ueber den Glycolaldehyd. (Ber. deut. chem. Gesell. 1892, 25, 2549-2554)

1893
3. BAMBERGER, E., AND LANDSTEINER, K. Das Verhalten des Diazobenzols gegen Kaliumpermanganat. (Ber. deut. chem. Gesell. 1893, 26, 482-495)

1894
4. LANDSTEINER, K. Ueber Cholsäure. (Z. physiol. Chem. 1894, 19, 285-288)

1895
5. LANDSTEINER, K. Ueber die Farbenreaktion der Eiweisskörper mit salpetriger Säure und Phenolen. (Centr. Physiol. 1894/95, 8, 773-774)
6. LANDSTEINER, K. Ueber die Farbenreaktion der Eiweisskörper mit salpetriger Säure und Phenolen. Nachtrag. (Centr. Physiol. 1895/96, 9, 433-434)

1896
7. SCHOLL, R., AND LANDSTEINER, K. Reduction der Pseudonitrole zu Ketoximen. (Ber. deut. chem Gesell. 1896, 29, 87-90)

YEAR

1897
8. LANDSTEINER, K. Ueber die Folgen der Einverleibung sterilisirter Bakterienculturen. (Wien.klin. Woch. 1897, 10, 430–444)

1898
9. AUSTERLITZ, L., AND LANDSTEINER, K. Ueber die Bakteriendichtigkeit der Darmwand. (Centr. Bakt. Orig. 1898, 23, 286–288)
10. AUSTERLITZ, L., AND LANDSTEINER, K. Ueber die Bakteriendichtigkeit der Darmwand. (Sitzungsber. k. Akad. Wissensch., Math.-naturwissensch. Cl., Wien. Abt. 3, 1898, 107, 33–67)
11. LANDSTEINER, K. Ueber die Wirkung des Choleraserums ausserhalb des Tierkörpers. (Centr. Bakt. Orig. 1898, 23, 847–852)

1899
12. LANDSTEINER, K. Zur Kenntnis der spezifisch auf Blutkörperchen wirkenden Sera. (Centr. Bakt. Orig. 1899, 25, 546–549)

1900
13. LANDSTEINER, K. Zur Kenntnis der antifermentativen, lytischen und agglutinierenden Wirkungen des Blutserums und der Lymphe. (Centr. Bakt. Orig. 1900, 27, 357–362)

1901
14. LANDSTEINER, K. Zur Kenntnis der Mischegeschwülste der Speicheldrüsen. (Z. Heilk. 1901, 22, 1–18)
15. DONATH, J., AND LANDSTEINER, K. Ueber antilytische Sera. (Wien. klin. Woch. 1901, 14, 713–714)
16. LANDSTEINER, K. Ueber degenerative Veränderungen der Nierenepithelien. (Wien. klin. Woch. 1901, 14. 956–958)
17. LANDSTEINER, K. Ueber Agglutinationserscheinungen normalen menschlichen Blutes. (Wien. klin. Woch. 1901, 14, 1132–1134)
18. HALBAN, J., AND LANDSTEINER, K. Ueber Unterschiede des fötalen und mütterlichen Serums und über eine fällungshemmende Wirkung des Normalserums. (Abstract.) (Wien. klin. Woch. 1901, 14, 1270–1271)

1902
19. LANDSTEINER, K., AND STURLI, A. Ueber die Hämagglutinine normaler Sera. (Wien. klin. Woch. 1902, 15, 38–40)
20. HALBAN, J., AND LANDSTEINER, K. Ueber Unterschiede des fötalen und mütterlichen Serums und über eine fällungshemmende Wirkung des Normalserums. (Münch. med. Woch. 1902, 49, 473–476)
21. LANDSTEINER, K. Review of "Arbeiten aus der pathologisch-anatomischen Abteilung des königlichen hygienischen Institutes zu Posen." (Herrn Geh. Rath Dr. R. Virchow zur Feier seines 80. Geburtstages.) Wiesbaden, Bergmann, 1901. (Wien. klin. Rundschau 1902, 16, 460)
22. LANDSTEINER, K., AND CALVO, A. Zur Kenntnis der Reaktionen des normalen Pferdeserums. (Centr. Bakt. Orig. 1902, 31, 781–786)
23. HALBAN, J., AND LANDSTEINER, K. Zur Frage der Präcipitationsvorgänge. (Centr. Bakt. Orig. 1902, 32, 457–458)
24. DONATH, J., AND LANDSTEINER, K. Zur Frage der Makrocytase. (Wien. klin. Rundschau 1902, 16, 773–774)
25. LANDSTEINER, K. Beobachtungen über Hämagglutination. (Wien. klin. Rundschau 1902, 16, 774)
26. LANDSTEINER, K. Ueber Serumagglutinine. (Münch. med. Woch. 1902, 49, 1905–1908)

YEAR

1903

27. LANDSTEINER, K., AND VON EISLER, M. Ueber Präzipitin-Reaktionen des menschlichen Harnes. (Wiener klin. Rundschau 1903, **17**, 10)

28. LANDSTEINER, K., AND RICHTER, M. Ueber die Verwertbarkeit individueller Blutdifferenzen für die forensische Praxis. (Z. Medizinalbeamte 1903, **16**, 85–89)

29. DONATH, J., AND LANDSTEINER, K. Ueber antilytische Sera und die Entstehung der Lysine. (Z. Hyg. 1903, **43**, 552–580)

30. LANDSTEINER, K. Ueber trübe Schwellung. (Beitr. path. Anat. 1903, **33**, 237–280)

31. LANDSTEINER, K., AND JAGIĆ, N. Ueber die Verbindungen und die Entstehung von Immunkörpern. (Münch. med. Woch. 1903, **50**, 764–768)

32. LANDSTEINER, K. Bemerkung zu der Arbeit von K. Glaessner "Ueber antitryptische Wirkung des Blutes." (Beitr. chem. Physiol. u. Path. 1903, **4**, 262)

33. LANDSTEINER, K. Ueber Beziehungen zwischen dem Blutserum und den Körperzellen. (Münch med. Woch. 1903, **50**, 1812–14)

1904

34. LANDSTEINER, K., AND JAGIĆ, N. Ueber Analogien der Wirkungen kolloidaler Kieselsäure mit den Reaktionen der Immunkärper und verwandter Stoffe. (Wien. klin. Woch. 1904, **17**, 63–64)

35. LANDSTEINER, K. Ueber das Sarkom der Gallenblase. (Wien. klin. Woch. 1904, **17**, 163–165)

36. LANDSTEINER, K., AND STOERK, O. Ueber eine eigenartige Form chronischer Cystitis (v. Hansemann's "Malakoplakie"). (Beitr. path. Anat. 1904, **36**, 131–151)

37. LANDSTEINER, K., AND VON EISLER, M. Zur Arbeit von Hans Friedenthal: "Weitere Versuche über die Reaktion auf Blutverwandtschaft." (Wien. klin.-therap. Woch. 1904, **11**, 656–657)

38. LANDSTEINER, K., AND VON EISLER, M. Ueber die Wirkungsweise hämolytischer Sera. (Wien. klin. Woch. 1904, **17**, 776–778)

39. LANDSTEINER, K., AND JAGIĆ, N. Ueber Reaktionen anorganischer Kolloide und Immunkörperreaktionen. (Münch. med. Woch. 1904, **51**, 1185–1189)

40. DONATH, J., AND LANDSTEINER, K. Ueber paroxysmale Hämoglobinurie. (Münch med. Woch. 1904, **51**, 1590–1593)

41. LANDSTEINER, K., AND MUCHA, V. Ueber Fettdegeneration der Nieren. (Centr. allg. Path. 1904, **15**, 752–755)

1905

42. LANDSTEINER, K. Ueber die Unterscheidung von Fermenten mit Hilfe von Serumreaktionen. (Centr. Bakt. Orig. 1905, **38**, 344–346)

43. LANDSTEINER, K., AND VON EISLER, M. Ueber die Wirkung der Hämolysine. (Wien. klin. Rundschau 1905, **19**, 220–221; 421)

44. LANDSTEINER, K., AND LEINER, K. Ueber Isolysine und Isoagglutinine im menschlichen Blut. (Centr. Bakt. Orig. 1905, **38**, 548–555)

45. LANDSTEINER, K. Bemerkung zur Mitteilung von Jean Billitzer: *Theorie der Kolloide. II*. (Z. physikal. Chem. 1905, **51**, 741–742)

YEAR

1905

46. FINGER, E. A first report of work of Finger, E., and Landsteiner, K., on "Untersuchungen über die Syphilis der Affen." (K. Akad. der Naturwissenschaften, Wien. Math.-naturwissensch. Cl., Anzeiger, 1905, 42, 219-224)

47. FINGER, E., AND LANDSTEINER, K. Untersuchungen über Syphilis an Affen. Erste Mitteilung. (Sitzungsber. k. Akad. Wissensch., Cl., Wien.-Abt. 3, 1905, 114, 497-539; also, Arch. Dermat. u. Syph. 1906, 78, 335-368)

48. LANDSTEINER, K., AND REICH, M. Ueber die Verbindungen der Immunkörper. (Centr. Bakt., Orig., 1905, 39, 83-93)

49. LANDSTEINER, K., AND REICH, M. Ueber Unterschiede zwischen normalen und Immunagglutininen. (Wien. klin. Rundschau 1905, 19, 568-569)

50. LANDSTEINER, K., AND VON EISLER, M. Ueber Agglutinin- und Lysinwirkung. (Centr. Bakt. Orig. 1905, 39, 309-319)

VON EISLER, M. Ueber Antihämolysine. (Wien. klin. Woch. 1905, 18, 721-725)

51. LANDSTEINER, K. Darmverschluss durch eingedicktes Meconium; Pankreatitis. (Centr. allg. Path. 1905, 16, 903-907)

52. LANDSTEINER, K., AND REICH, M. Ueber Unterschiede zwischen normalen und durch Immunisierung entstandenen Stoffen des Blutserums. (Centr. Bakt. Orig. 1905, 39, 712-717)

53. FINGER, E. A report of work of Finger, E., and Landsteiner, K. (Fortsetzung der Untersuchungen über Syphilisimpfungen an Affen.) (K. Akad. der Naturwissenschaften, Wien. Math.-naturwissensch. Cl., Anzeiger, 1905, 42, 448-449)

1906

54. LANDSTEINER, K., AND UHLIRZ, R. Ueber die Adsorption von Eiweisskörpern. I. Mitteilung. (Centr. Bakt. Orig. 1906, 40, 265-270)

55. LANDSTEINER, K., AND STANKOVIĆ, R. Ueber die Adsorption von Eiweisskörpern und über Agglutininverbindungen. II. Mitteilung. (Centr. Bakt. Orig. 1906, 41, 108-117)

56. DONATH, J., AND LANDSTEINER, K. Ueber paroxysmale Hämoglobinurie. (Z. klin. Med. 1906, 58,173-189)

57. LANDSTEINER, K. Bemerkungen zu der vorläufigen Mitteilung über Hämolysinbildung von Bang und Forssman. (Centr. Bakt. Orig. 1906, 40, 723)

58. FINGER, E., AND LANDSTEINER, K. Untersuchungen über Syphilis an Affen. II. Mitteilung. (Sitzungsber. k. Akad. Wissensch., Math.-naturwissensch. Cl., Wien. Abt. 3, 1906, 115, 179-200; also, Arch. Dermat. u. Syph. 1906, 81, 147-166)

59. LANDSTEINER, K. Ueber Tumoren der Schweissdrüsen. (Beitr. path. Anat. 1906, 39, 316-332)

60. LANDSTEINER, K. Ueber Adsorptionsverbindungen. (Abstract.) (Centr. Bakt. Ref. 1906, 38, Beil., 25-26)

61. LANDSTEINER, K. Ueber den Immunisierungsprozess. (Abstract.) (Centr. Bakt. Ref. 1906, 38, Beil., 26-27)

62. LANDSTEINER, K., AND FINGER, E. Ueber Immunität bei Syphilis. (Abstract.) (Centr. Bakt. Ref. 1906, 38, Beil., 107-108)

63. LANDSTEINER, K. Beobachtungen über das Virus der Hühnerpest. (Centr. Bakt. Ref. 1906, 38, 540-542)

YEAR

64. LANDSTEINER, K., AND STANKOVIĆ, R.   Ueber die Bindung von Komplement durch suspendierte und kolloid gelöste Substanzen.   III. Mitteilung über die Adsorptionsverbindungen.   (Centr. Bakt. Orig. 1906, **42**, 353–356)

65. LANDSTEINER, K., AND BOTTERI, A.   Ueber Verbindungen von Tetanustoxin mit Lipoiden.   IV. Mitteilung über Adsorptionsverbindungen.   (Centr. Bakt. Orig. 1906, **42**, 562–566)

66. LANDSTEINER, K., AND MUCHA, V.   Zur Technik der Spirochaetenuntersuchung.   (Wien. klin. Woch. 1906, **19**, 1349–50)

67. LANDSTEINER, K.   Bemerkungen zu der Mitteilung von U. Friedemann und H. Friedenthal: Beziehungen der Kernstoffe zu den Immunkörpern.   (Centr. Physiol. 1906/07, **20**, 657–658)

1907

68. FINGER, E.   Untersuchungen über Immunität bei Syphilis [Finger and Landsteiner].   (Verhandl. deut. dermat. Gesell. 9th Kongress, 1906.   1907, 251–255)

69. LANDSTEINER, K.   Zu der Erwiderung von Friedemann und Friedenthal.   (Centr. Physiol. 1906/07, **20**, 806)

70. MUCHA, V., AND LANDSTEINER, K.   (Fortsetzung ihrer Untersuchungen über die Spirochaete pallida (mit Hilfe der Dunkelfeldbeleuchtung)).   (Wien. klin. Woch. 1907, **20**, 209)

71. LANDSTEINER, K., AND MUCHA, V.   Beobachtungen über Spirochaete pallida.   (Abstract.)   (Centr. Bakt. Ref. 1907, **39**, 540–541)

72. LANDSTEINER, K.   Ueber das Carcinom der Leber.   (Sitzungsber. k. Akad. Wissensch., Math.-naturwissensch. Cl., Wien. Abt. 3, 1907, **116**, 175–246)

73. DAUTWITZ, F., AND LANDSTEINER, K.   Ueber Beziehungen der Lipoide zur Serumhämolyse.   (Beitr. chem. Physiol. u. Path. 1907, **9**, 431–452)

74. LANDSTEINER, K.   Plattenepithelkarzinom und Sarkom der Gallenblase in einem Falle von Cholelithiasis.   (Z. klin. Med. 1907, **62**, 427–433)

MÜLLER, R.   (Komplementbindung durch die Zerebrospinalflüssigkeit von Kranken mit progressiver Paralyse).   (Wien. klin. Woch. 1907, **20**, 514–515)

75. LANDSTEINER, K., AND EHRLICH, H.   Ueber Lipoide bakterizide Zellstoffe.   (Wien. klin. Rundschau 1907, **21**, 526)

76. HEYROVSKY, H., AND LANDSTEINER, K.   Ueber Hämotoxine des Milzbrandbacillus und verwandter Bakterien.   (Centr. Bakt. Orig. 1907, **44**, 150–160)

77. LANDSTEINER, K., AND RAUBITSCHEK, H.   Beobachtungen über Hämolyse und Hämagglutination.   (Wien. klin. Rundschau 1907, **21**, 748)

78. DONATH, J., AND LANDSTEINER, K.   Weitere Beobachtungen über paroxysmale Hämoglobinurie.   (Centr. Bakt. Orig. 1907, **45**, 205–213)

79. LANDSTEINER, K., AND EHRLICH, H.   Ueber bakterizide Wirkungen von Lipoiden und ihre Beziehung zur Komplementwirkung.   (Centr. Bakt. Orig. 1907, **45**, 247–257)

80. LANDSTEINER, K., MÜLLER, R., AND PÖTZL, O.   Ueber Komplementbindungsreaktionen mit dem Serum von Dourinetieren.   (Wien. klin. Woch. 1907, **20**, 1421–1422)

YEAR

1907   81. LANDSTEINER, K., MÜLLER, R., AND PÖTZL, O.   Zur Frage der Komplementbindungsreaktion bei Syphilis. (Wien. klin. Woch. 1907, **20**, 1565–1567)

82. LANDSTEINER, K., AND RAUBITSCHEK, H.   Beobachtungen über Hämolyse und Hämagglutination. (Centr. Bakt. Orig. 1907, **45**, 660–667)

83. LANDSTEINER, K., AND REICH, M.   Ueber den Immunisierungsprozess.   (Z. Hyg. 1908, **58**, 213–232)

1908   84. LANDSTEINER, K., AND MÜLLER, R.   Bemerkungen zu der Mitteilung: "Ueber die Beeinflussung von Antistoffen durch alkoholische Organextrakte."   (Wien. klin. Woch. 1908, **21**, 230)

85. LANDSTEINER, K., AND PAULI, W.   Elektrische Wanderung der Immunstoffe.   (Wien. med. Woch. 1908, **58**, 1010–1011; also, Verhandl. XXV Kong. inn. Med. 1908, **25**, 571–574).   (Abstract in Wien. klin. Woch. 1908, **21**, 1201)

86. LANDSTEINER, K.   Immunität und Serodiagnostik bei menschlicher Syphilis.   (XIV Intern. Kong. für Hyg. u. Demographie, Berlin, 23–29 Sept. 1907, Section I)   (Centr. Bakt. Ref. 1908, **41**, 232)

87. LANDSTEINER, K.   Bemerkungen zu dem Artikel von Emmerich und Löw.   [Sind die bakteriziden Bestandteile der Pyozyanase Lipoide? -pp. 839–842]   (Wien. klin. Woch. 1908, **21**, 843)

88. LANDSTEINER, K.   Bemerkungen zur Kenntnis der übertragbaren tierischen Tumoren.   (Wien. klin. Woch. 1908, **21**, 1549–50)

89. LANDSTEINER, K., AND DONATH, J.   Zu dem Artikel von W. Czernecki, in der Wiener klin. Wochenschr. 1908, Nr. **42**, Hämoglobinurie und Hämolyse.   (Wien. klin. Woch. 1908, **21**, 1565–66)

90. LANDSTEINER, K.   Bemerkungen zu dem Aufsatze von Robertson über die Theorie der Adsorption.   (Z. Chem. u. Ind. d. Koll. 1908, **3**, 221–224)

91. LANDSTEINER, K.   Immunität und Serodiagnostik bei menschlicher Syphilis.   (Centr. Bakt. Ref. 1908, **41**, 785–795)

92. LANDSTEINER, K., AND RAUBITSCHEK, H.   (Demonstriert mikroskopische Präparate von einem menschlichen und zwei Affenrückenmarken (Poliomyelitis))   (Wien. klin. Woch. 1908, **21**, 1830).

93. LANDSTEINER, K., AND RAUBITSCHEK, H.   Ueber die Adsorption von Immunstoffen. V. Mitteilung. (Biochem. Z. 1908/09, **15**, 33–51)

94. LANDSTEINER, K.   Ueber das Streptokokkenlysin.   (Centr. Bakt. Ref. 1908/1909, **42**, 785)

1909   95. GRAF, R., AND LANDSTEINER, K.   Versuche über die Giftigkeit des Blutserums bei Eklampsie.   (Centr. Gynak. 1909, **33**, 142–152)

96. LANDSTEINER, K.   Zu der Mitteilung von C. Bezzola: "Ueber die bakteriolytischen Eigenschaften des Paratyphus-B-Immunserums."   (Centr. Bakt. Orig. 1909, **51**, 432)

97. LANDSTEINER, K., AND VON RAUCHENBICHLER, R.   Ueber das Verhalten des Staphylolysins beim Erwärmen.   (Z. Immunitätsforsch. Orig. 1908/09, **1**, 439–448)

MÜLLER, R.   (Report on experiments of Müller and Landsteiner on use of alcoholic extract of normal organs as antigen in Wassermann reaction.)   (Wien. med. Woch. 1909, **59**, 286)

98. LANDSTEINER, K., AND FÜRTH, J.   Ueber die Reaktivierung von hämolytischen Immunserum durch Lösungen von Hämotoxinen und durch Kaltblutersera.   (Wien. klin. Woch. 1909, **22**, 231–232)

YEAR

99. LANDSTEINER, K., AND FÜRTH, J. Nachträgliche Bemerkungen zu unserer Mitteilung: Ueber die Reaktivierung von hämolytischen Immunserum etc. (Wien. klin. Woch. 1909, 22, 606–607).

100. LANDSTEINER, K., AND POPPER, E. Uebertragung der Poliomyelitis acuta auf Affen. (Z. Immunitätsf. Orig. 1909, 2, 377–390)

101. LANDSTEINER, K. Bemerkungen zu der Mitteilung von P. Krause und E. Meinicke: Zur Aetiologie der akuten epidemischen Kinderlähmung. (Deut. med. Woch. 1909, 35, 1975–76)

102. LANDSTEINER, K. Die Theorien der Antikörperbildung. (Wien. klin. Woch. 1909, 22, 1623–1631; also, Ergebn. wiss. Med. 1909/10, 1, 185–207)

103. LANDSTEINER, K., AND LEVADITI, C. La transmission de la paralysie infantile aux singes. (Compt. rend. soc. biol. 1909, 67, 592–594)

104. LEVADITI, C., AND LANDSTEINER, K. La transmission de la paralysie infantile au chimpanzé. (Compt. rend. acad. sci. 1909, 149, 1014–1016)

105. LANDSTEINER, K., AND LEVADITI, C. La paralysie infantile expérimentale (Deuxième note). (Compt. rend. soc. biol. 1909, 67, 787–789)

1910

106. LANDSTEINER, K. Hämagglutination und Hämolyse. (Oppenheimer, C., ed., Handbuch der Biochemie, Jena, Fischer, 1910, 2, T.1, 395–541)

107. LANDSTEINER, K. Immunität gegen Körperzellen und Neubildungen (Cytotoxine). (Oppenheimer, C., ed., Handbuch der Biochemie, Jena, Fischer, 1910, 2, T.1, 542–551)

108. LEVADITI, C., AND LANDSTEINER, K. La paralysie infantile expérimentale. (Compt. rend. acad. sci. 1910, 150, 55–57)

109. LEVADITI, C., AND LANDSTEINER, K. Recherches sur la paralysie infantile expérimentale. (Compt. rend. acad. sci. 1910, 150, 131–132)

110. LANDSTEINER, K., AND PRÁŠEK, E. Uebertragung der Poliomyelitis acuta auf Affen. II. Mitteilung. (Z. Immunitätsf. Orig. 1909/10, 4, 584–589)

111. LEVADITI, C., AND LANDSTEINER, K. La poliomyélite expérimentale (Cinquième note). (Compt. rend. soc. biol. 1910, 68, 311–313)

112. LEVADITI, C., AND LANDSTEINER, K. Etude expérimentale de la poliomyélite aigue (Sixième note). (Compt. rend. soc. biol. 1910, 68, 417–418)

113. LANDSTEINER, K. Wirken Lipoide als Antigene? (Jahresber. Ergebn. Immunitätsf. 1910, 6, Abt. 1, 209–226)

114. LEVADITI, C., AND LANDSTEINER, K. Action exercée par le thymol, le permanganate de potasse et l'eau oxygénée sur le virus de la poliomyélite aigue. (Compt. rend. soc. biol. 1910, 68, 740–741)

115. HERZ, A., AND LANDSTEINER, K. Ueber das Verhalten pathologischer Sera zur Saponinhämolyse. (Med. Klin. 1910, 6, 1062–1063)

116. LANDSTEINER, K., AND LEVADITI, C. Etude expérimentale de la poliomyélite aigue (maladie de Heine-Medin). (Ann. Inst. Pasteur 1910, 24, 833–876)

117. LANDSTEINER, K., AND WELECKI, S. Ueber den Einfluss konzentrierter Lösungen von Salzen und Nicht-elektrolyten auf die Agglutination und Agglutininbildung. (Z. Immunitätsf. Orig. 1910, 8, 397–403)

YEAR

1911 118. LANDSTEINER, K. Technik der Untersuchungen über Poliomyelitis acuta (Heine-Medinsche Krankheit). (Kraus, R. and Levaditi, C., eds. Handbuch der Technik und Methodik der Immunitätsforschung, Jena, Fischer, 1911, Ergänzungsbd. 1, 458-464).

119. LANDSTEINER, K., LEVADITI, C., AND PRÁŠEK, E. Tentatives de transmission de la scarlatine au chimpanzé. (Compt. rend. soc. biol. 1911, 70, 641-643).

120. LANDSTEINER, K., LEVADITI, C., AND PRÁŠEK, E. Etude expérimentale du pemphigus infectieux aigu. (Compt. rend. soc. biol. 1911, 70, 643-645).

121. LANDSTEINER, K., LEVADITI, C., AND PRÁŠEK, E. Tentatives de transmission de la scarlatine au chimpanzé. (Compt. rend. acad. sci. 1911, 152, 1190-1192).

122. LANDSTEINER, K. Bemerkungen zu der Abhandlung von Traube: Die Resonanztheorie, eine physikalische Theorie der Immunitätserscheinungen. (Z. Immunitätsf. Orig. 1911, 9, 779-786).

123. LANDSTEINER, K., LEVADITI, C., AND PASTIA, C. Recherche du virus dans les organes d'un enfant atteint de poliomyélite aigue. (Compt. rend. acad. sci. 1911, 152, 1701-1702).

124. LANDSTEINER, K., LEVADITI, C., AND PRÁŠEK, E. Contribution a l'étiologie du Pemphigus infectieux aigu. (Compt. rend. soc. biol. 1911, 70, 1026-1028).

125. LANDSTEINER, K., AND PRÁŠEK, E. Ueber die Beziehung der Antikörper zu der präzipitablen Substanz des Serums. (Z. Immunitätsf. Orig. 1911, 10, 68-102).

126. LANDSTEINER, K., LEVADITI, C., AND PRÁŠEK, E. Essais de transmission de la scarlatine aux singes. (Ann. Inst. Pasteur 1911, 25, 754-775).

127. LANDSTEINER, K., LEVADITI, C., AND PASTIA, M. Etude expérimentale de la poliomyélite aigue (maladie de Heine-Medin). (Second mémoire). (Ann. Inst. Pasteur, 1911, 25, 805-829).

128. LANDSTEINER, K., AND DANULESCO. Présence du virus de la poliomyélite dans l'amygdale des singes paralysés et son élimination par le mucus nasal. (Compt. rend. soc. biol. 1911, 71, 558-560).

1912 129. LANDSTEINER, K. Experimentelle Syphilis. (Finger, E., and others, eds., Handbuch der Geschlechtskrankheiten, Wien. Hölder, 1912, 2, 873-895).

130. LANDSTEINER, K., LEVADITI, C., AND DANULESCO. Contribution à l'étude de la scarlatine expérimentale. (Compt. rend. soc. biol. 1912, 72, 358-360).

131. LANDSTEINER, K., AND PRÁŠEK, E. Ueber die bindenden und immunisierenden Substanzen der Blutkörperchen. (Z. Immunitätsforsch. Orig. 1912, 13, 403-420).

132. LANDSTEINER, K., AND ROCK, H. Untersuchungen über Komplementwirkung. Hämolyse durch Kieselsäure und Komplement. (Z. Immunitätsforsch. Orig. 1912, 14, 14-32).

133. LANDSTEINER, K., AND BERLINER, M. Ueber die Kultivierung des Virus der Hühnerpest. (Centr. Bakt. Orig. 1912, 67, 165-168).

DONATH, J. (Comments on cases of paroxysmal hemoglobinuria, and use of complement-fixation). (Mitt. Gesell. f. inn. Med. u. Kinderheilk. Wien, 1912, 11, 47; 1913, 12, 308).

1913 134. LANDSTEINER, K. Aetiologie der Poliomyelitis und die Möglichkeit ihrer Verhütung. (Trans. 15th Internat. Congr. Hyg. and Demography, Washington, D. C. Sept. 23-28, 1912. 1913, 1, 582).

YEAR

135. LANDSTEINER, K. Kolloide und Lipoide in der Immunitätslehre. (Kolle, W., u. von Wassermann, A. Handbuch der pathogenen Mikroorganismen, Jena, Fischer, 1913, ed. 2, 2, 1241–1300)

136. LANDSTEINER, K. Poliomyelitis acuta. (Kolle, W., u. von Wassermann, A. Handbuch der pathogenen Mikroorganismen, Jena, Fischer, 1913, ed. 2, 8, 427–462)

137. LANDSTEINER, K. Zur Frage der Spezifität der Immunreaktionen und ihrer kolloidchemischen Erklärbarkeit. (Biochem. Z. 1913, 50, 176–184)

138. LANDSTEINER, K., AND PRÁŠEK, E. Ueber die bindenden und immunisierenden Substanzen der roten Blutkörperchen. II. Mitteilung über Blutantigene. (Z. Immunitätsf. Orig. 1913, 17, 363–377)

139. LANDSTEINER, K. Zu der Mitteilung über die Bildung bakteriolytischer Immunkörper von Bail und Rotky. (Z. Immunitätsforsch. Orig. 1913, 18, 220–222)

GUSSENBAUER, R. Ueber eine zu Komplementbindung führende, durch Temperaturerniedrigung beförderte Reaktion. (Z. Immunitätsf. 1913, 18, 616–621)

140. DONATH, J., AND LANDSTEINER, K. Die Serumreaktion bei paroxysmaler Hämoglobinurie. Bemerkungen zu der Mitteilung von C. H. Browning und H. F. Watson im Journal of Pathology and Bacteriology, vol. 17, 1912, p. 117. (Z. Immunitätsf. Orig. 1913, 18, 701–704)

141. LANDSTEINER, K., AND PRÁŠEK, E. Ueber Säureflockung der Blutstromata. III. Mitteilung über Blutantigene. (Z. Immunitätsf. Orig. 1913/14, 20, 137–145)

PRÁŠEK, E. Ueber die Wärmeresistenz von normalen und Immunagglutininen. (Z. Immunitätsf. 1913, 20, 146–159)

142. LANDSTEINER, K., AND PRÁŠEK, E. Ueber die Aufhebung der Artspezifität von Serumeiweiss. IV. Mitteilung über Antigene. (Z. Immunitätsf. Orig. 1913/14, 20, 211–237)

143. LANDSTEINER, K. Ueber einige Eiweissderivate. (Biochem. Z. 1913/14, 58, 362–364)

1914 144. LANDSTEINER, K., SCHLAGENHAUFER, F., AND WAGNER VON JAUREGG, J. Experimentelle Untersuchungen über die Aetiologie des Kropfes. (Sitzungsber. k. Akad. Wissensch. Math.-naturwissensch. Cl. Wien. Abt. 3, 1914, 123, 35–76)

145. LANDSTEINER, K., AND JABLONS, B. Ueber die Bildung von Antikörpern gegen verändertes arteigenes Serumeiweiss. V. Mitteilung über Antigene. (Z. Immunitätsforsch. Orig. 1913/14, 20, 618–621)

146. LANDSTEINER, K., AND JABLONS, B. Ueber die Antigeneigenschaften von acetyliertem Eiweiss. VI. Mitteilung über die Antigene. (Z. Immunitätsforsch. Orig. 1914, 21, 193–201)

147. LANDSTEINER, K., AND PRÁŠEK, E. Notiz zu der Mitteilung über Immunisierungsversuche mit Lipoproteinen. (Biochem. Z. 1914, 61, 191–192)

148. HERZIG, J., AND LANDSTEINER, K. Ueber die Methylierung von Eiweisstoffen. (Biochem. Z. 1914, 61, 458–463)

149. LANDSTEINER, K. Ueber die Aetiologie der Pest. (Vorträge über Epidemiologie, abgehalten am 21–26 September 1914 in Wien). (Wien. med. Woch. 1914, 64, 2221–2224; Oesterr. Sanitätswesen. 1914, 26, Beil. zur Nr. 46, p. 60–63)

150. LANDSTEINER, K. Ueber Serumbehandlung. (Vorträge über Epidemiologie, abgehalten am 21–26 September 1914 in Wien) (Oesterr. Sanitätswesen. 1914, 26, Beil. zur Nr. 46, p. 21–24)

YEAR

1914

151. LANDSTEINER, K. Die Serotherapie der Seuchen. [Formal presentation of #150.] (Klin.-therap. Woch. 1914, 21, 1060-1071; abstracted in Wien. med. Woch. 1914, 64, 2157-2159)

152. LANDSTEINER, K. Die Cytotoxine. (completed for Kraus, R., and Levaditi, C., eds. Handb. d. Immunitätsf. und experimentellen Therapie, Bd. I, 1914 (Zweite Auflage des Handbuches der Technik und Methodik der Immunitätsf.), unpublished due to intervention of war)

153. LANDSTEINER, K. Pflanzliche Hämagglutinine. (completed for Kraus, R., and Levaditi, C., eds. Handb. d. Immunitätsf. u. exp. Therapie, Bd. IV, 1914 (Zweite Auflage des Handbuches der Technik und Methodik der Immunitätsf.), unpublished due to intervention of war)

154. HERZIG, J., AND LANDSTEINER, K. Ueber die Einwirkung von alkoholischen Säuren auf Eiweisstoffe. (Biochem. Z. 1914, 67, 334-337)

1915

155. LANDSTEINER, K., AND LAMPL, H. Untersuchung der Spezifizität von Serumreaktionen durch Einführung verschiedenartiger Gruppen in Eiweiss. (Centr. Physiol. 1915, 30, 329-330)

1916

156. LANDSTEINER, K. Serumdiagnostik der Syphilis. (Finger, E., and others, eds. Handbuch der Geschlechtskrankheiten, Wien. Hölder, 1916, 3, 2357-2405)

157. LANDSTEINER, K., AND PRÁŠEK, E. Ueber acetylierte Eiweisskörper. (Biochem. Z. 1916, 74, 388-393)

158. LANDSTEINER, K. (bacteriological investigation of polymyositis) (Wien. klin. Woch. 1916, 29, 930)

159. LANDSTEINER, K. Ueber knötchenförmige Infiltrate der Niere bei Scharlach. (Beitr. path. Anat. 1916, 62, 227-232)

160. HAUSMANN, W., AND LANDSTEINER, K. Ueber das Vorkommen hämorrhagischer Nephritis bei Infektion mit Paratyphusbazillen A und B. (Wien. med. Woch. 1916, 66, 1247)

1917

161. LANDSTEINER, K. Ueber die Aetiologie der Polymyositis. (Svenska Läkaresällakapets. 1917, 43, 759-774)

162. LANDSTEINER, K. Ueber die Antigeneigenschaften von methyliertem Eiweiss. VII. Mitteilung über Antigene. (Z. Immunitätsf. Orig. 1917, 26, 122-133)

163. LANDSTEINER, K., AND LAMPL, H. Ueber die Einwirkung von Formaldehyd auf Eiweissantigen. VIII. Mitteilung über Antigene. (Z. Immunitätsf. Orig. 1917, 26, 133-141)

164. LANDSTEINER, K., AND BARRON, C. Ueber die Einwirkung von Säure und Lauge auf Serumeiweissantigen (Restitution der Antigeneigenschaft). IX. Mitteilung über Antigene. (Z. Immunitätsf. Orig. 1917, 26, 142-153)

165. LAMPL, H., AND LANDSTEINER, K. Quantitative Untersuchungen über die Einwirkung von Komplement auf Präzipitate. (Z. Immunitätsf. Orig. 1917, 26, 193-198)

166. LANDSTEINER, K., AND LAMPL, H. Ueber Antigene mit verschiedenartigen Acylgruppen. X. Mitteilung über Antigene. (Z. Immunitätsf. Orig. 1917, 26, 258-276)

167. LANDSTEINER, K., AND LAMPL, H. Ueber die Antigeneigenschaften von Azoproteinen. XI. Mitteilung über Antigene. (Z. Immunitätsf. Orig. 1917, 26, 293-304)

YEAR

1918 168. HERZIG, J., AND LANDSTEINER, K.   Ueber die Methylierung der Eiweisstoffe.   (Sitzungsber. k. Akad. Wissensch., Math.-naturwissensch. Cl., Wien.Abt. 2B. 1918, 127, 71–86; also, Monatsh. Chem. 1918, 39, 269–284)

169. LANDSTEINER, K., AND LAMPL, H.   Ueber die Abhängigkeit der serologischen Spezifität von der chemischen Struktur. (Darstellung von Antigenen mit bekannter chemischer Konstitution der spezifischen Gruppen).   XII. Mitteilung über Antigene.   (Biochem. Z. 1918, 86, 343–394)

170. LANDSTEINER, K., AND HAUSMANN, W.   Einige Beobachtungen über das Fleckfiebervirus.   (Med. Klin. 1918, 14, 515–518)

LAMPL, H.   Ueber einen neuen Typus von Dysenteriebazillen (Bact. dysenteriae Schmitz).   (Wien. klin. Woch. 1918, 30, 835–837); (see also remarks by Landsteiner, K., in Wien. klin. Woch. 1915, 28, 485)

1919 171. LANDSTEINER, K.   Ueber die Bedeutung der Proteinkomponente bei den Präcipitinreaktionen der Azoproteine.   XIII. Mitteilung über Antigene.   (Biochem. Z. 1918/19, 93, 106–118)

1920 172. LANDSTEINER, K.   Spezifische Serumreaktionen mit einfach zusammengesetzten Substanzen von bekannter Konstitution (organischen Säuren).   XIV. Mitteilung über Antigene und serologische Spezifität.   (Biochem. Z. 1920, 104, 280–299)

173. HERZIG, J., AND LANDSTEINER, K.   Zur Einwirkung von Diazomethan auf Aminosäuren.   (Biochem. Z. 1920, 105, 111–114)

174. LANDSTEINER, K., AND EDELMANN, A.   Beitrag zur Kenntnis der anatomischen Befunde bei polyglandulärer Erkrankung (Insuffisance pluriglandulaire).   (Frankf. Z. Path. 1920, 24, 339–353)

1921 175. LANDSTEINER, K.   Over de serologische Specificiteit van het haemoglobine van verschillende diersoorten.   (K. Akad. Wetenschappen te Amsterdam. Verslag van de gewone vergaderingen der wis-en natuurkundige. Afdeeling. 1920/21, 29, 1029–1034)

176. LANDSTEINER, K.   Over heterogenetisch antigeen.   (K. Akad. Wetenschappen te Amsterdam. Verslag van de gewone vergaderingen der wis-en natuurkundige. Afdeeling. 1920/21, 29, 1118–1121)

177. LANDSTEINER, K.   Bemerkungen über Isoagglutination anlässlich einer Mitteilung von R. Zimmermann.   (Zentr. Gynäk. 1921, 45, 662–665, Nr. 19)

178. LANDSTEINER, K.   Nachtrag zu der Mitteilung in diesem Zentralblatt Nr. 19.   (Zentr. Gynäk. 1921, 45, 1153)

179. LANDSTEINER, K.   Ueber heterogenetisches Antigen und Hapten.   XV. Mitteilung über Antigene.   (Biochem. Z. 1921, 119, 294–306)

180. LANDSTEINER, K.   Over het samenstellen van heterogenetisch antigeen uit hapteen en proteïne.   (K. Akad. Wetenschappen te Amsterdam. Verslag van de gewone vergaderingen der wis-en natuurkundige. Afdeeling. 1921, 30, 329–330)

1922 181. LANDSTEINER, K.   Onderzoekingen over anaphylaxie door azoproteïnen.   (K. Akad. Wetenschappen te Amsterdam. Verslag van de gewone vergaderingen der wis-en natuurkundige. Afdeeling. 1922, 31, 54–55)

182. VAN BOUWDIJK BASTIAANSE, F. S., AND LANDSTEINER, K.   Een familiaire vorm van tubereuse sclerose.   (Nederl. Tijdschr. v. Geneesk. 1922, 66, pt. 2, 248–257)

YEAR

1922   183.   LANDSTEINER, K.   On the serological specificity of haemoglobin in different species of animals.   (Proc. K. Akad. Wetenschappen. Amsterdam. Section of sciences. 1922, 23, 1161–1165)

184.   LANDSTEINER, K.   On heterogenetic antigen.   (Proc. K. Akad. Wetenschappen. Amsterdam. Section of sciences. 1922, 23, 1166–1169)

185.   LANDSTEINER, K.   On the formation of heterogenetic antigen by combination of hapten and protein.   (Proc. K. Akad. Wetenschappen. Amsterdam. Section of sciences. 1922, 24, 237–238)

1923   186.   LANDSTEINER, K.   Experiments on anaphylaxis with azoproteins.   (Proc. K. Akad. Wetenschappen. Amsterdam. Section of sciences. 1923, 25, 34–35)

187.   LANDSTEINER, K., AND SIMMS, S.   Production of heterogenetic antibodies with mixtures of the binding part of the antigen and protein.   (J. Exp. Med. 1923, 38, 127–138)

188.   HEIDELBERGER, M., AND LANDSTEINER, K.   On the antigenic properties of hemoglobin.   (J. Exp. Med. 1923, 38, 561–571)

189.   LANDSTEINER, K., AND HEIDELBERGER, M.   Differentiation of oxyhemoglobins by means of mutual solubility tests.   (J. Gen. Physiol. 1923/24, 6, 131–135)

1924   190.   LANDSTEINER, K.   Darstellungsmethoden von Antigenen und Antikörpern für immunchemische Untersuchungen.   (Abderhalden, E., ed. Handbuch d. biol. Arbeitsmeth. Abt. 13, Teil 2, Heft 3, Lfg. 137, p. 547, 1924)

191.   LANDSTEINER, K., AND VAN DER SCHEER, J.   Serological examination of a species-hybrid.   (Proc. Soc. Exp. Biol. and Med. 1923/24, 21, 252)

192.   LANDSTEINER, K., AND WITT, D. H.   Observations on human isoagglutinins.   (Proc. Soc. Exp. Biol. and Med. 1923/24, 21, 389–391)

193.   LANDSTEINER, K.   Experiments on anaphylaxis to azoproteins.   (J. Exp. Med. 1924, 39, 631–637)

194.   LANDSTEINER, K., AND VAN DER SCHEER, J.   Serological examination of a species-hybrid.   I. On the inheritance of species-specific qualities.   (J. Immunol. 1924, 9, 213–219)

195.   LANDSTEINER, K., AND VAN DER SCHEER, J.   Serological examination of a species-hybrid.   II. Tests with normal agglutinins.   (J. Immunol. 1924, 9, 221–226)

196.   LANDSTEINER, K., AND VAN DER SCHEER, J.   On the specificity of agglutinins and precipitins.   (J. Exp. Med. 1924, 40, 91–107)

197.   LANDSTEINER, K., AND VAN DER SCHEER, J.   On the antigens of red blood corpuscles.   (Proc. Soc. Exp. Biol. and Med. 1924/25, 22, 98–99)

198.   LANDSTEINER, K., AND MILLER, C. P., JR.   On individual differences of the blood of chickens and ducks.   (Proc. Soc. Exp. Biol. and Med. 1924/25, 22, 100–102)

199.   LANDSTEINER, K., AND VAN DER SCHEER, J.   Flocculation reactions with hemolytic immune sera.   (Proc. Soc. Exp. Biol. and Med. 1924/25, 22, 170)

1925   200.   LANDSTEINER, K., VAN DER SCHEER, J., AND WITT, D. H.   Group specific flocculation reactions with alcoholic extracts of human blood.   (Proc. Soc. Exp. Biol. and Med. 1924/25, 22, 289–291)

201.   LANDSTEINER, K., AND VAN DER SCHEER, J.   On the antigens of red blood corpuscles.   The question of lipoid antigens.   (J. Exp. Med. 1925, 41, 427–437)

YEAR

202. LANDSTEINER, K., AND MILLER, C. P., JR. Serological observations on the relationship of the bloods of man and the anthropoid apes. (Science 1925, 61, 492-493)

203. MURPHY, J. B., AND LANDSTEINER, K. Experimental production and transmission of tar sarcomas in chickens. (J. Exp. Med. 1925, 41, 807-816)

204. LANDSTEINER, K., AND LEVENE, P. A. Observations on the specific part of the heterogenetic antigen. (J. Immunol. 1925, 10, 731-733)

VAN DER SCHEER, J. Flocculation reactions with immune sera produced by injections of organ emulsions. (J. Immunol. 1925, 10, 735-739)

205. LANDSTEINER, K., AND VAN DER SCHEER, J. On the antigens of red blood corpuscles. II. Flocculation reactions with alcoholic extracts of erythrocytes. (J. Exp. Med. 1925, 42, 123-142)

206. DONATH, J., AND LANDSTEINER, K. Ueber Kältehämoglobinurie. (Ergeb. Hyg. 1925, 7, 184-228)

207. LANDSTEINER, K., AND MILLER, C. P., JR. Serological studies on the blood of the primates. I. The differentiation of human and anthropoid bloods. (J. Exp. Med. 1925, 42, 841-852)

208. LANDSTEINER, K., AND MILLER, C. P., JR. Serological studies on the blood of the primates. II. The blood groups in anthropoid apes. (J. Exp. Med. 1925, 42, 853-862)

209. LANDSTEINER, K., AND MILLER, C. P., JR. Serological studies on the blood of the primates. III. Distribution of serological factors related to human isoagglutinogens in the blood of lower monkeys. (J. Exp. Med. 1925, 42, 863-872)

1926 210. LANDSTEINER, K., AND LEVENE, P. A. On the heterogenetic haptene. (Proc. Soc. Exp. Biol. and Med. 1925/26, 23, 343-344)

211. LANDSTEINER, K., AND WITT, D. H. Observations on the human blood groups. Irregular reactions. Isoagglutinins in sera of group IV. The factor A¹. (J. Immunol. 1926, 11, 221-247)

212. LANDSTEINER, K. Studies on anaphylaxis with the products of peptic digestion of proteins. (Proc. Soc. Exp. Biol. and Med. 1925/26, 23, 540-541)

213. LANDSTEINER, K., AND VAN DER SCHEER, J. Experiments with trypanosomes in relation to the Wassermann reaction. (Proc. Soc. Exp. Biol. and Med. 1925/26, 23, 641-644)

214. LANDSTEINER, K., AND LEVINE, P. On group specific substances in human spermatozoa. (J. Immunol. 1926, 12, 415-418)

215. LANDSTEINER, K., AND LEVINE, P. On the cold agglutinins in human serum. (J. Immunol. 1926, 12, 441-460)

216. LANDSTEINER, K., AND LEVINE, P. On a specific substance of the cholera vibrio. (Proc. Soc. Exp. Biol. and Med. 1926/27, 24, 248-249)

1927 217. LANDSTEINER, K. Ueber komplexe Antigene. (Klin. Woch. 1927, 6, 103-107)

218. LANDSTEINER, K., AND FURTH, J. Extraction of precipitable substances of bacilli with dilute alcohol. (Proc. Soc. Exp. Biol. and Med. 1926/27, 24, 379)

219. LANDSTEINER, K., AND VAN DER SCHEER, J. Experiments on the production of Wassermann reagins by means of trypanosomes. (J. Exp. Med. 1927, 45, 465-482)

YEAR

1927 220. LANDSTEINER, K., AND LEVINE, P. A new agglutinable factor differentiating individual human bloods. (Proc. Soc. Exp. Biol. and Med. 1926/27, 24, 600-602)
      FURTH, J. Further observations on the extraction of precipitable substances of bacilli. (Proc. Soc. Exp. Biol. and Med. 1927, 24, 602-603)
      221. LANDSTEINER, K., AND VAN DER SCHEER, J. Influence of organic salts on the action of tyrosinase. (Proc. Soc. Exp. Biol. and Med. 1926/27, 24, 692-693)
      222. LANDSTEINER, K., AND LEVENE, P. A. Further studies on the heterogenetic haptene. (Proc. Soc. Exp. Biol. and Med. 1926/27, 24, 693-695)
      223. FURTH, J., AND LANDSTEINER, K. Further observations on the precipitable substances of B. typhosus and B. paratyphosus B. (Proc. Soc. Exp. Biol. and Med. 1926/27, 24, 771-772)
      224. LANDSTEINER, K., AND VAN DER SCHEER, J. On the influence of acid groups on the serological specificity of azoproteins. (J. Exp. Med. 1927, 45, 1045-1056; also in Leopoldina, 1929, 4, 201-208)
      225. LANDSTEINER, K., AND LEVINE, P. Further observations on individual differences of human blood. (Proc. Soc. Exp. Biol. and Med. 1926/27, 24, 941-942)
      226. LANDSTEINER, K. Serologische Individualdifferenzen und die menschlichen Blutgruppen. (Wien. med. Woch. 1927, 77, 744-745)
      227. LEVENE, P. A., LANDSTEINER, K., AND VAN DER SCHEER, J. Immunization experiments with lecithin. (J. Exp. Med. 1927, 46, 197-204)
      228. LANDSTEINER, K., AND LEVENE, P. A. On the heterogenetic haptene. Fourth communication. (J. Immunol. 1927, 14, 81-83)
      229. LANDSTEINER, K., AND LEVINE, P. On a specific substance of the cholera vibrio. (J. Exp. Med. 1927, 46, 213-221)
      230. LEVENE, P. A., AND LANDSTEINER, K. On some new lipoids. (J. Biol. Chem. 1927, 75, 607-612)
      231. LANDSTEINER, K., AND VAN DER SCHEER, J. On the production of immune sers for tissues. (Proc. Soc. Exp. Biol. and Med. 1927/28, 25, 140-141)

1928 232. LANDSTEINER, K. The human blood groups. (Jordan, E. O., and Falk, I. S., Eds., The newer knowledge of bacteriology and immunology, Chicago, University of Chicago press, 1928, 892-908)
      233. FURTH, J., AND LANDSTEINER, K. On precipitable substances derived from Bacillus typhosus and Bacillus paratyphosus B. (J. Exp. Med. 1928, 47, 171-184)
      234. LANDSTEINER, K. Zur Frage der Gruppenbestimmung bei Transfusionen. (Klin. Woch. 1928, 7, 112-113)
      235. LANDSTEINER, K., AND FURTH, J. Precipitable substances prepared from bacilli of the Salmonella group. (Proc. Soc. Exp. Biol. and Med. 1927/28, 25, 565-567)
      236. LANDSTEINER, K., AND LEVINE, P. On individual differences in human blood. (J. Exp. Med. 1928, 47, 757-775)
      237. LANDSTEINER, K. Studies on anaphylaxis with the products of peptic digestion of proteins. II. (Proc. Soc. Exp. Biol. and Med. 1927/28, 25, 666-667)
      238. LANDSTEINER, K., LEVINE, P., AND JANE, M. L. On the development of isoagglutinins following transfusions. (Proc. Soc. Exp. Biol. and Med. 1928, 25, 672-674)

YEAR

239. LANDSTEINER, K.   Bemerkung zu dem Vortrage von Prof. K. Meixner.   Die Blutgruppen in der gerichtlichen Medizin.   (Wien. klin. Woch. 1928, 41, 638)

240. LANDSTEINER, K.   Sur les propriétés sérologiques du sang des anthropoïdes.   (Compt. Rend. Soc. Biol. 1928, 99, 658-660)

241. LANDSTEINER, K., AND VAN DER SCHEER, J.   Serological differentiation of steric isomers.   (J. Exp. Med. 1928, 48, 315-320)

242. LANDSTEINER, K., AND LEVINE, P.   On the inheritance of agglutinogens of human blood demonstrable by immune agglutinins.   (J. Exp. Med. 1928, 48, 731-749)

LEVINE, P.   Menschliche Blutgruppen und individuelle Blutdifferenzen.   (Erg. inn. Med. u. Kinderheilk. 1928, 34, 111-153)

LEVINE, P.   On pseudo-agglutination and cold agglutination.   (Ukrainisches Zentralblatt für Blutgruppenforsch. 1928, 2, 1-13)

243. LANDSTEINER, K.   On hemagglutination by tumor extracts.   (Proc. Soc. Exp. Biol. and Med. 1928/29, 26, 134-135)

244. LANDSTEINER, K.   Cell antigens and individual specificity.   [Presidential address delivered to the American Assoc. of Immunologists, Apr. 30, 1928, at Washington.]   (J. Immunol. 1928, 15, 589-600)

1929

245. LANDSTEINER, K.   Zur Frage der Untergruppen der Blutgruppe A und der Agglutinine in Gruppe AB.   (Deut. Z. ges. gerichtl. Med. 1929, 13, 1-4)

246. LANDSTEINER, K., AND LEVINE, P.   On the racial distribution of some agglutinable structures of human blood.   (J. Immunol. 1929, 16, 123-131)

247. LANDSTEINER, K.   Ueber einfache chemische Verbindungen enthaltende Antigene.   (Z. Immunitätsf. 1929, 62, 178-184)

248. FURTH, J., AND LANDSTEINER, K.   Studies on the precipitable substances of bacilli of the Salmonella group.   (J. Exp. Med. 1929, 49, 727-743)

249. LANDSTEINER, K.   Die Lipoide in der Immunitätslehre.   (Kolle, W., Kraus, R., and Uhlenhuth, P., Eds., Handbuch der pathogenen Mikroorganismen, Jena, Fischer, ed. 3, 1929, 1, 1069-1108)

250. LANDSTEINER, K., AND LEVINE, P.   On isoagglutinin reactions of human blood other than those defining the blood groups.   (J. Immunol. 1929, 17, 1-28)

251. LANDSTEINER, K., AND VAN DER SCHEER, J.   Serological differentiation of steric isomers (antigens containing tartaric acids).   Second paper.   (J. Exp. Med. 1929, 50, 407-417)

252. LANDSTEINER, K.   Poliomyelitis acuta.   (Kolle, W., Kraus, R., and Uhlenhuth, P., Eds., Handbuch der pathogenen Mikroorganismen, Jena, Fischer, ed. 3, 1929, 8, 777-820)

253. LEVINE, P., AND LANDSTEINER, K.   On immune isoagglutinins in rabbits.   (J. Immunol. 1929, 17, 559-565)

254. LANDSTEINER, K., AND LEVINE, P.   On the racial distribution of certain agglutinable structures in human blood.   (Profilakticheskaia Meditsina, Supplément, Rec. dédié Prof. s. I. Zlatogoroff en l'honneur du XXX-ème ann. de son travail, sci., med. et péd. 1897-1927, Kharkov, 1929, 63) (also in Russian)

YEAR

1930

VAN DER SCHEER, J.   On the action of serum on the fibrins of various species.   (J. Immunol. 1930, 18, 17-22)

WORMALL, A.   The immunological specificity of chemically altered proteins.   Halogenated and nitrated proteins.   (J. Exp. Med., 1930, 51, 295-317)

255. LANDSTEINER, K., AND LEVINE, P.   On the inheritance and racial distribution of agglutinable properties of human blood.   (J. Immunol. 1930, 18, 87-94)

256. LANDSTEINER, K., LEVINE, P., AND VAN DER SCHEER, J.   Anaphylactic reactions produced by azodyes in animals sensitized with azoproteins.   (Proc. Soc. Exp. Biol. and Med. 1929/30, 27, 811-812)

257. LANDSTEINER, K., AND VAN DER SCHEER, J.   Antigens containing peptides of known structure and antigenic properties of azoalbumoses.   (Proc. Soc. Exp. Biol. and Med. 1929/30, 27, 812-813)

258. LANDSTEINER, K.   Further observations on hemagglutination by tumor extracts.   (Proc. Soc. Exp. Biol. and Med. 1929/30, 27, 813-814)

259. LANDSTEINER, K.   Ueber einige neuere Ergebnisse der Serologie.   (Naturwissenschaften, 1930, 18, 653-659)

260. LANDSTEINER, K., AND LEVINE, P.   Experiments on anaphylaxis to azoproteins.   Third paper.   (J. Exp. Med. 1930, 52, 347-359)

261. NIGG, C., AND LANDSTEINER, K.   Growth of rickettsia of typhus fever (Mexican type) in the presence of living tissue.   (Proc. Soc. Exp. Biol. and Med. 1930/31, 28, 3-5)

262. LANDSTEINER, K., AND LEVINE, P.   Note on individual differences in human blood.   (Proc. Soc. Exp. Biol. and Med. 1930/31, 28, 309-310)

1931   263   LANDSTEINER, K.   Die menschlichen Blutgruppen mit Rücksicht auf die Transfusionstherapie.   (Wolff-Eisner, A., Ed., Handbuch der experimentellen Therapie, Serum- und Chemotherapie, München, Lehmann, 1931, ed. 2, Ergänzungsband, p. 32-42)

264. LANDSTEINER, K., AND LEVINE, P.   The differentiation of a type of human blood by means of normal animal serum.   (J. Immunol. 1931, 20, 179-185)

265. LANDSTEINER, K.   Individual differences in human blood.   (Translation of Nobel lecture read in German at Stockholm, December 11, 1930; see 266).   (Science, 1931, 73, 403-409)

VAN DER SCHEER, J.   Zur Frage der quantitativen Bedingungen bei der Lipoidantikörperbildung durch Kombinationsimmunisierung.   (Z. Immunitätsforsch. 1931, 71, 190-192)

266. LANDSTEINER, K.   Ueber individuelle Unterschiede des menschlichen Blutes.   Nobel-Vortrag, gehalten in Stockholm am 11. Dezember 1930.   Stockholm, Norstedt, 1931, pp. 1-13.

267. LANDSTEINER, K.   Serological tests with the blood of Cavia porcellus and Cavia rufescens.   (Proc. Soc. Exp. Biol. and Med. 1930/31, 28, 981-982)

268. LANDSTEINER, K., AND VAN DER SCHEER, J.   Observations on serological reactions with albumose preparations.   (Proc. Soc. Exp. Biol. and Med. 1930/31, 28, 983-984)

269. LANDSTEINER, K.   Die Blutgruppen und ihre praktische Anwendung, besonders für die Bluttransfusion.   [Auszug aus einem Vortrag, gehalten anlässlich der Verleihung des Nobelpreises im Dezember 1930].   (Forsch. u. Fortschr. 1931, 7, 311-312; also Fortschr. Med. 1932, 50, 1061-1062)

270. LANDSTEINER, K., AND VAN DER SCHEER, J.   On the specificity of serological reactions with simple chemical compounds (inhibition reactions).   (J. Exp. Med. 1931, 54, 295-305)

YEAR

271. LANDSTEINER, K., AND VAN DER SCHEER, J. Beobachtungen über Präcipitinreaktionen mit Azoalbumosen. [Enlargement of Nos. 257 and 268]. (Z. Hyg. Infektionskrankh, 1931/32, **113**, 1–8)

272. LEVINE, P., AND LANDSTEINER, K. On immune isoagglutinins in rabbits. Second paper. (J. Immunol. 1931, **21**, 513–515)

1932

273. LANDSTEINER, K. Die Blutgruppen und ihre praktische Anwendung besonders für Bluttransfusion. (Congrès international de microbiologie, 1st, Paris, 1930. 1932, **2**, 149–163)

274. LANDSTEINER, K., AND VAN DER SCHEER, J. Ueber die Spezifität serologischer Reaktionen mit einfachen chemischen Verbindungen (Hemmungs-Reaktionen). [Translation of No. 270]. (Acta Soc. Med. Fenn. Duodecim. (Ser. A, art. 3). 1932, **15**, 1–13)

275. LANDSTEINER, K., AND JACOBS, J. L. Experiments on immunization with haptens. (Proc. Soc. Exp. Biol. and Med. 1931/32, **29**, 570–571)

276. LANDSTEINER, K., AND LEVINE, P. On the Forssman antigens in B. paratyphosus B and B. dysenteriae Shiga. (J. Immunol. 1932, **22**, 75–82)

277. LANDSTEINER, K., AND VAN DER SCHEER, J. Precipitin reactions of immune sera with simple chemical substances. (Proc. Soc. Exp. Biol. and Med. 1931/32, **29**, 747–748)

278. NIGG, C., AND LANDSTEINER, K. Studies on the cultivation of the typhus fever rickettsia in the presence of live tissue. (J. Exp. Med. 1932, **55**, 563–576)

279. LANDSTEINER, K., AND VAN DER SCHEER, J. On the serological specificity of peptides. (J. Exp. Med. 1932, **55**, 781–796)

LEVINE, P. The application of blood groups in forensic medicine. (Am. J. Police Sci. 1932, **3**, 157–168)

280. LANDSTEINER, K., AND LEVINE, P. Immunization of chimpanzees with human blood. (J. Immunol. 1932, **22**, 397–400)

281. LANDSTEINER, K., AND VAN DER SCHEER, J. Note on the serological differentiation of steric isomers. (Proc. Soc. Exp. Biol. and Med. 1931/32, **29**, 1261–1263)

282. NIGG, C., AND LANDSTEINER, K. Note on the cultivation of the typhus fever Rickettsia. (Proc. Soc. Exp. Biol. and Med. 1931/32, **29**, 1291)

283. LANDSTEINER, K., AND VAN DER SCHEER, J. Serological reactions with simple chemical compounds (precipitin reactions). (J. Exp. Med. 1932, **56**, 399–409)

284. LANDSTEINER, K. Note on the group specific substance of horse saliva. (Science 1932, **76**, 351–352)

285. LANDSTEINER, K., AND LEVINE, P. On individual differences in chicken blood. (Proc. Soc. Exp. Biol. and Med. 1932/33, **30**, 209–212)

1933

286. LANDSTEINER, K., AND VAN DER SCHEER, J. Anaphylactic shock by azodyes. (J. Exp. Med. 1933, **57**, 633–636)

287. LANDSTEINER, K., AND JACOBS, J. L. II. Experiments on the immunization with haptens. (Proc. Soc. Exp. Biol. and Med. 1932/33, **30**, 1055–1057)

YEAR

1933 288. LANDSTEINER, K., AND CHASE, M. W. Observations on serological reactions with albumose preparations. II. (Proc. Soc. Exp. Biol. and Med. 1932/33, **30**, 1413–1415)
FINKELSTEIN, M. H. On the specificity of brominated and iodinated proteins. (J. Immunol. 1933, **25**, 179–182)

289. LANDSTEINER, K. Die Spezifizität der serologischen Reaktionen, Berlin, Springer, 1933

1934 290. LANDSTEINER, K. Immunchemische Spezifizität. (Atti del III convegno "Fondazione Alessandro Volta", Roma, 25 Settembre–1 Ottobre, 1933. 1934, 61–79)

291. LANDSTEINER, K., STRUTTON, W. R., AND CHASE, M. W. On agglutination reactions observed with some human bloods, chiefly among negroes. (J. Immunol. 1934, **27**, 469–472)
VAN DER SCHEER, J. The preparation of γ-(p-aminophenyl)-butyric and ε-(p-aminophenyl)-caproic acids. (J. Am. Chem. Soc. 1934, **56**, 744–745)
JACOBS, J. On the use of adsorbents in immunizations with haptens. (J. Exp. Med. 1934, **59**, 479–490)

292. LANDSTEINER, K., AND JACOBS, J. L. Experiments on sensitization of guinea pigs with simple chemical compounds. (Proc. Soc. Exp. Biol. and Med. 1933/34, **31**, 790–791)

293. LANDSTEINER, K., AND JACOBS, J. L. Supplementary note on skin sensitization with simple chemical compounds. (Proc. Soc. Exp. Biol. and Med. 1933/34, **31**, 1079–1080)

294. LANDSTEINER, K., AND VAN DER SCHEER, J. Serological studies on azoproteins. Antigens containing azo-components with aliphatic side chains. (J. Exp. Med. 1934, **59**, 751–768)

295. LANDSTEINER, K., AND VAN DER SCHEER, J. On the serological specificity of peptides. II. (J. Exp. Med. 1934, **59**, 769–780)

296. LANDSTEINER, K. Forensic application of serologic individuality tests. (J. Am. Med. Assn. 1934, **103**, 1041–1044)

1935 297. LANDSTEINER, K., AND VAN DER SCHEER, J. Ueber die serologische Spezifizität von Peptiden. (Festschrift, Heinrich Zangger. Zürich, Rascher, 1935, Teil 2, 600. This is a translation of No. 295)
NIGG, C. On the preservation of typhus fever Rickettsiae in cultures. (J. Exp. Med. 1935, **61**, 17–26)
HOLZER, F. J. Ueber die serologische Differenzierung zweier Meerschweinchenarten. (Z. Immunitätsforsch. 1935, **84**, 170–176)

298. LANDSTEINER, K., AND CHASE, M. W. Decomposition of the group A substance in horse saliva by a myxobacterium. (Proc. Soc. Exp. Biol. and Med. 1934/35, **33**, 713–714)

299. LANDSTEINER, K., AND CHASE, M. W. Additional note on decomposition of the group A substances. (Proc. Soc. Exp. Biol. and Med. 1934/35, **32**, 1208)

300. LANDSTEINER, K., AND JACOBS, J. L. Studies on the sensitization of animals with simple chemical compounds. (J. Exp. Med. 1935, **61**, 643–656)
NIGG, C. On the presence of typhus virus in wild rats in New York City. (J. Infect. Dis., 1935, **57**, 252–254)

301. VAN DER SCHEER, J., AND LANDSTEINER, K. Serological tests with amino acids. (J. Immunol. 1935, **29**, 371–376)

1936 302. LANDSTEINER, K. On the group specific A substance in horse saliva, II. (J. Exp. Med. 1936, **63**, 185–190)

YEAR

303. LANDSTEINER, K., AND VAN DER SCHEER, J. On cross reactions of immune sera to azoproteins. (J. Exp. Med. 1936, 63, 325-339)

NIGG, C. Studies on culture strains of European and murine typhus. (J. Exp. Med., 1936, 63, 341-351)

304. LANDSTEINER, K., AND JACOBS, J. L. Sensitization of guinea pigs with methyl heptine carbonate. (J. Am. Med. Assn. 1936, 106, 1112)

304a LANDSTEINER, K. Foreword. (In Dujarric de la Rivière, R. and Kossovitch, N. Les groupes sanguins. Paris, Baillière et fils, 1936, p. 1)

HOLZER, F. J. Zum Nachweis der Bluteigenschaften M und N. (Deutsch. Z. ges. gerichtl. Med. 1936, 26, 515-518)

305. LANDSTEINER, K., AND CHASE, M. W. On group specific A substances. III. The substance in commercial pepsin. (J. Exp. Med. 1936, 63, 813-817)

306. LANDSTEINER, K., AND JACOBS, J. L. Studies on the sensitization of animals with simple chemical compounds. II. (J. Exp. Med. 1936, 64, 625-639)

307. LANDSTEINER, K., AND JACOBS, J. L. Studies on the sensitization of animals with simple chemical compounds. III. Anaphylaxis induced by arsphenamine. (J. Exp. Med. 1936, 64, 717-721)

308. LANDSTEINER, K. Serological and allergic reactions with simple chemical compounds. (New England J. Med. 1936, 215, 1199-1204)

309. LANDSTEINER, K. The specificity of serological reactions, Springfield, Ill., Thomas, 1936.

1937

JACOBS, J. Serological reactions of azoproteins derived from aromatic hydrocarbons and diaryl compounds. (J. Gen. Physiol. 1937, 20, 353-361)

PARKER, RAYMOND C. Studies on the production of antibodies in vitro. (Science 1937, 85, 292-294)

310. LANDSTEINER, K., AND CHASE, M. W. Immunochemistry. [Survey of the literature during 1935 and 1936.] (Ann. Rev. Biochem. 1937, 6, 621-643)

311. LANDSTEINER, K., AND WIENER, A. S. On the presence of M agglutinogens in the blood of monkeys. (J. Immunol. 1937, 33, 19-25)

312. LANDSTEINER, K. Comment on article by Straus, H. W., and Cocs, A. F.: Studies in experimental hypersensitiveness in the rhesus monkey. III. On the manner of development of the hypersensitiveness in contact dermatitis. (J. Immunol. 1937, 33, 224-225)

313. LANDSTEINER, K., AND CHASE, M. W. Studies on the sensitization of animals with simple chemical compounds. IV. Anaphylaxis induced by picryl chloride and 2:4 dinitrochlorobenzene. (J. Exp. Med. 1937, 66, 337-351)

NIGG, C. Serological behavior of heated protein mixtures. (J. Immunol. 1937, 33, 229-234)

314. LANDSTEINER, K., AND PIRIE, N. W. Serological specificity in pyridine derivatives. (J. Immunol. 1937, 33, 265-270)

1938

315. LANDSTEINER, K., AND VAN DER SCHEER, J. Anaphylactic shock by azodyes. II. (J. Exp. Med. 1938, 67, 79-87)

YEAR

1938 316. LANDSTEINER, K., AND VAN DER SCHEER, J. On cross reactions of immune sera to azoproteins. II. Antigens with azocomponents containing two determinant groups. (J. Exp. Med. 1938, 67, 709–723)

317. HARTE, ROBERT A. Serological tests with pyrazolone compounds. (J. Immunol., 1938, 34, 433–439)

LANDSTEINER, K., LONGSWORTH, L. G., AND VAN DER SCHEER, J. Electrophoresis experiments with egg albumins and hemoglobins. (Science 1938, 88, 83–85)

318. LANDSTEINER, K., AND DISOMMA, A. A. Studies on the sensitization of animals with simple chemical compounds. V. Sensitization to diazomethane and mustard oil. (J. Exp. Med. 1938, 68, 505–512)

1939 319. LANDSTEINER, K., ROSTENBERG, A., JR., AND SULZBERGER, M. B. Individual differences in susceptibility to eczematous sensitization with simple chemical substances. (J. Invest. Dermat. 1939, 2, 25–29)

CHASE, M. W. A microorganism decomposing group-specific A substances. (J. Bact. 1938, 36, 383–390)

320. CHASE, M. W., AND LANDSTEINER, K. Immunochemistry. [Survey of the literature during 1937 and 1938]. (Ann. Rev. Biochem. 1939, 8, 579–610)

321. LANDSTEINER, K., AND VAN DER SCHEER, J. On the serological specificity of peptides. III. (J. Exp. Med. 1939, 69, 705–719)

322. LANDSTEINER, K., AND CHASE, M. W. Studies on the sensitization of animals with simple chemical compounds. VI. Experiments on the sensitization of guinea pigs to poison ivy. (J. Exp. Med. 1939, 69, 767–784)

323. ROTHEN, A., AND LANDSTEINER, K. Adsorption of antibodies by egg albumin films. (Science 1939, 90, 65–66)

1940 324. LANDSTEINER, K., AND CHASE, M. W. Breeding experiments in reference to drug allergy in animals. (Proc. 3rd. Internat. Cong. Microbiol., New York, Sept. 2–9, 1939. 1940, 772–773)

325. LANDSTEINER, K., AND WIENER, A. S. An agglutinable factor in human blood recognized by immune sera for Rhesus blood. (Proc. Soc. Exp. Biol. and Med. 1940, 43, 223–224)

326. LANDSTEINER, K., AND PARKER, R. C. Serological tests for homologous serum proteins in tissue cultures maintained on a foreign medium. (J. Exp. Med. 1940, 71, 231–236)

327. LANDSTEINER, K., AND CHASE, M. W. Studies on the sensitization of animals with simple chemical compounds. VII. Skin sensitization by intraperitoneal injections. (J. Exp. Med. 1940, 71, 237–245)

328. LANDSTEINER, K., AND VAN DER SCHEER, J. On cross reactions of egg albumin sera. (J. Exp. Med. 1940, 71, 445–454)

329. LANDSTEINER, K., AND HARTE, R. A. On group specific A substances. IV. The substance from hog stomach. (J. Exp. Med. 1940, 71, 551–562)

330. LANDSTEINER, K., AND CHASE, M. W. Skin sensitization to a simple compound by injections of conjugates. (Proc. Soc. Exp. Biol. and Med. 1940, 44, 559)

331. LANDSTEINER, K., AND DISOMMA, A. A. Studies on the sensitization of animals with simple chemical compounds. VIII. Sensitization to picric acid; subsidiary agents and mode of sensitization. (J. Exp. Med. 1940, 72, 361–366)

1941 332. LANDSTEINER, K., AND CHASE, M. W. Quinine hypersensitivity in guinea pigs. (Proc. Soc. Exp. Biol. and Med. 1941, 46, 223)

YEAR

333. LANDSTEINER, K., AND CHASE, M. W.   Studies on the sensitization of animals with simple chemical compounds. IX. Skin sensitization induced by injection of conjugates.   (J. Exp. Med. 1941, 73, 431–438.)

CHASE, M. W.   Inheritance in guinea pigs of the susceptibility to skin sensitization with simple chemical compounds.   (J. Exp. Med. 1941, 73, 711–726)

334. LANDSTEINER, K., AND HARTE, R. A.   Group-specific substances in human saliva.   (J. Biol. Chem. 1941, 140, 673–674)

335. LANDSTEINER, K., AND WIENER, A. S.   Studies on an agglutinogen (Rh) in human blood reacting with anti-rhesus sera and with human isoantibodies.   (J. Exp. Med. 1941, 74, 309–320)

336. LANDSTEINER, K.   Versuche über Hautallergie gegen einfache chemische Substanzen.   (Schweiz. med. Woch. 1941, 71, 1359–1360)

1942

337. LANDSTEINER, K.   Foreword.   (In Schiff, F. and Boyd, W. C.   Blood grouping technic, New York, Interscience, 1942, p. iii)

338. LANDSTEINER, K.   Serological reactivity of hydrolytic products from silk.   (J. Exp. Med. 1942, 75, 269–276)

339. LANDSTEINER, K., AND CHASE, M. W.   Experiments on transfer of cutaneous sensitivity to simple compounds.   (Proc. Soc. Exp. Biol. and Med. 1942, 49, 688–690)

340. LANDSTEINER, K., WIENER, A. S., AND MATSON, G. A.   Distribution of the Rh factor in American Indians.   (J. Exp. Med. 1942, 76, 73–78)

341. ROTHEN, A., AND LANDSTEINER, K.   Serological reactions of protein films and denatured proteins.   (J. Exp. Med. 1942, 76, 437–450)

342. LANDSTEINER, K., AND WIENER, A. S.   Tests for the Rh factor with guinea pig immune sera.   (Proc. Soc. Exp. Biol. and Med. 1942, 51, 313)

CHASE, M. W.   Production of local skin reactivity by passive transfer of anti-protein sera.   (Proc. Soc. Exp. Biol. and Med. 1943, 52, 238–240)

1943

343. WIENER, A. S., AND LANDSTEINER, K.   Heredity of variants of the Rh type.   (Proc. Soc. Exp. Biol. and Med. 1943, 53, 167–170)

344. CHASE, M. W., AND LANDSTEINER, K.   Studies on the sensitization of animals with simple chemical compounds. X. Antibodies producing early skin reactions (to be published)

345. CHASE, M. W., AND LANDSTEINER, K.   Studies on the sensitization of animals with simple chemical compounds. XI. Transfer of skin reactivity of the delayed type (in preparation)

346. LANDSTEINER, K.   The Specificity of Serological Reactions. A Survey of Immunochemistry.   [2nd ed. of No. 309] (in press)

INDEX

(Italic figures indicate the principal reference)

Drosophila, 81
Dyes, specifically precipitable, *see* Azo-dyes
Dysenteriae Shiga, *see* Shigella dysenteriae

Echinoderms, 14, 82
Edestin, 27
Egg proteins, 25, 28, 29, 61. *See also* Ovalbumin
Ehrlich, hypothesis of, 148
Electrophoresis, 21, 24, 53, 133
Endotoxins, 7, *222*
Enzymes, 23, 29, *38*, 109, 142, 148, 183, 230, 266; decomposing bacterial polysaccharides, 234, 235; purification of antitoxins by, 143.
Equivalence zone, 241, 242, 243, 251, 259, 260
Ergosterol, *see* Sterols
Erythrocytes, as adsorbents, 144. *See also* Blood cells
Esterification, effect on reactivity, 51, 59, 168, 178, 192, 263
Ethylbenzene, 264
E. typhosa: complex antigens from, 224; O-specific polysaccharides of, 215
Euglobulin, 24, 25, 104, 136
Euxanthic acid, 175
Exotoxins, *see* Toxins

Factors (receptors), 93, 113, 114, 148, 266 ff.
Fats, 111; of tubercle bacilli, 225
Fatty acids, 192, 225; halogenated, 172, 173
Ferritin, 26
Fertilizin, 82
Fibrinogen, 21, 24
Films, 43, 57, 149, 260, 261
Flagellar antigens, antibodies, *91* ff., 141, 143
Flocculation, of emulsions, 81, 99, 102, 130; of toxin, *see* Ramon's flocculation test; velocity of, 248–250. *See also* Antigen-antibody reactions

Flower pigments, 95
Fluorescent conjugates, 142
Forensic tests, 66, 87, 88
Formaldehyde, as allergen, 202; effect on antibodies, 142, 263; effect on antigens, toxins, viruses, 46, 47. *See also* Toxoid
Forssman antigens (haptens, antisera), 76, 79, 81, *95*, 98, 99, 103, 105, 115, 223, 229, 231, 271; in bacteria, 97, 100, 109, 131, 215, 225; interpretation of, 99, 115
Friedländer bacillus, *see* Bact. pneumoniae Friedländer
Fumaric acid, 174, 191, 192
Fungi, 35

Galactose, 174, 175
Galacturonic acid, 175
Gelatin, 59, 62; azo-, 62, 158; derivatives, 158
Gentiobiose, 174, 176
Gentiobiuronic acid, 176
Gliadin, 27, 55
Globin, 20, 24, 25
Globoglycoid, 24
Globulins, 24, 25, 80, *133* ff., 141, 146, 149, 150
Glucosamine, *see* Hexosamine
Glucose, 174, 175
Glucosides, 175, 185, 186, 193, 267
Glucosido-carbobenzyloxytyrosyl protein, 174
Glucosido conjugate, 162
Glucuronic acid, 175, 176, 272
Glutaric acid, 177, 179, 190, 191
Glutathione, 178
Glycogen, 110
Glycoproteins, 26, 27
Gonococcus, antigens from, 217
Griffith's phenomenon, 235
Group specific substances, 87, 104, 116, *231* ff.
Gums 174, 227, 250, 271, 272; antigenicity of, 110; polysaccharides of, 218, 220

Hapten, 7, *76*, 81, 106, 110, 156, 161,

A CATALOG OF SELECTED
DOVER BOOKS
IN SCIENCE AND MATHEMATICS

DOVER BOOKS
IN SCIENCE AND MATHEMATICS

QUALITATIVE THEORY OF DIFFERENTIAL EQUATIONS, V.V. Nemytskii and V.V. Stepanov. Classic graduate-level text by two prominent Soviet mathematicians covers classical differential equations as well as topological dynamics and erqodic theory. Bibliographies. 523pp. 5⅜ × 8½.   65954-2 Pa. $10.95

MATRICES AND LINEAR ALGEBRA, Hans Schneider and George Phillip Barker. Basic textbook covers theory of matrices and its applications to systems of linear equations and related topics such as determinants, eigenvalues and differential equations. Numerous exercises. 432pp. 5⅜ × 8½.   66014-1 Pa. $8.95

QUANTUM THEORY, David Bohm. This advanced undergraduate-level text presents the quantum theory in terms of qualitative and imaginative concepts, followed by specific applications worked out in mathematical detail. Preface. Index. 655pp. 5⅜ × 8½.   65969-0 Pa. $10.95

ATOMIC PHYSICS (8th edition), Max Born. Nobel laureate's lucid treatment of kinetic theory of gases, elementary particles, nuclear atom, wave-corpuscles, atomic structure and spectral lines, much more. Over 40 appendices, bibliography. 495pp. 5⅜ × 8½.   65984-4 Pa. $11.95

ELECTRONIC STRUCTURE AND THE PROPERTIES OF SOLIDS: The Physics of the Chemical Bond, Walter A. Harrison. Innovative text offers basic understanding of the electronic structure of covalent and ionic solids, simple metals, transition metals and their compounds. Problems. 1980 edition. 582pp. 6⅛ × 9¼.   66021-4 Pa. $14.95

BOUNDARY VALUE PROBLEMS OF HEAT CONDUCTION, M. Necati Özisik. Systematic, comprehensive treatment of modern mathematical methods of solving problems in heat conduction and diffusion. Numerous examples and problems. Selected references. Appendices. 505pp. 5⅜ × 8½.   65990-9 Pa. $11.95

A SHORT HISTORY OF CHEMISTRY (3rd edition), J.R. Partington. Classic exposition explores origins of chemistry, alchemy, early medical chemistry, nature of atmosphere, theory of valency, laws and structure of atomic theory, much more. 428pp. 5⅜ × 8½. (Available in U.S. only)   65977-1 Pa. $10.95

A HISTORY OF ASTRONOMY, A. Pannekoek. Well-balanced, carefully reasoned study covers such topics as Ptolemaic theory, work of Copernicus, Kepler, Newton, Eddington's work on stars, much more. Illustrated. References. 521pp. 5⅜ × 8½.   65994-1 Pa. $11.95

PRINCIPLES OF METEOROLOGICAL ANALYSIS, Walter J. Saucier. Highly respected, abundantly illustrated classic reviews atmospheric variables, hydrostatics, static stability, various analyses (scalar, cross-section, isobaric, isentropic, more). For intermediate meteorology students. 454pp. 6⅛ × 9¼. 65979-8 Pa. $12.95

RELATIVITY, THERMODYNAMICS AND COSMOLOGY, Richard C. Tolman. Landmark study extends thermodynamics to special, general relativity; also applications of relativistic mechanics, thermodynamics to cosmological models. 501pp. 5⅜ × 8½. 65383-8 Pa. $11.95

APPLIED ANALYSIS, Cornelius Lanczos. Classic work on analysis and design of finite processes for approximating solution of analytical problems. Algebraic equations, matrices, harmonic analysis, quadrature methods, much more. 559pp. 5⅜ × 8½. 65656-X Pa. $11.95

SPECIAL RELATIVITY FOR PHYSICISTS, G. Stephenson and C.W. Kilmister. Concise elegant account for nonspecialists. Lorentz transformation, optical and dynamical applications, more. Bibliography. 108pp. 5⅜ × 8½. 65519-9 Pa. $3.95

INTRODUCTION TO ANALYSIS, Maxwell Rosenlicht. Unusually clear, accessible coverage of set theory, real number system, metric spaces, continuous functions, Riemann integration, multiple integrals, more. Wide range of problems. Undergraduate level. Bibliography. 254pp. 5⅜ × 8½. 65038-3 Pa. $7.00

INTRODUCTION TO QUANTUM MECHANICS With Applications to Chemistry, Linus Pauling & E. Bright Wilson, Jr. Classic undergraduate text by Nobel Prize winner applies quantum mechanics to chemical and physical problems. Numerous tables and figures enhance the text. Chapter bibliographies. Appendices. Index. 468pp. 5⅜ × 8½. 64871-0 Pa. $9.95

ASYMPTOTIC EXPANSIONS OF INTEGRALS, Norman Bleistein & Richard A. Handelsman. Best introduction to important field with applications in a variety of scientific disciplines. New preface. Problems. Diagrams. Tables. Bibliography. Index. 448pp. 5⅜ × 8½. 65082-0 Pa. $10.95

MATHEMATICS APPLIED TO CONTINUUM MECHANICS, Lee A. Segel. Analyzes models of fluid flow and solid deformation. For upper-level math, science and engineering students. 608pp. 5⅜ × 8½. 65369-2 Pa. $12.95

ELEMENTS OF REAL ANALYSIS, David A. Sprecher. Classic text covers fundamental concepts, real number system, point sets, functions of a real variable, Fourier series, much more. Over 500 exercises. 352pp. 5⅜ × 8½. 65385-4 Pa. $8.95

PHYSICAL PRINCIPLES OF THE QUANTUM THEORY, Werner Heisenberg. Nobel Laureate discusses quantum theory, uncertainty, wave mechanics, work of Dirac, Schroedinger, Compton, Wilson, Einstein, etc. 184pp. 5⅜ × 8½. 60113-7 Pa. $4.95

INTRODUCTORY REAL ANALYSIS, A.N. Kolmogorov, S.V. Fomin. Translated by Richard A. Silverman. Self-contained, evenly paced introduction to real and functional analysis. Some 350 problems. 403pp. 5⅜ × 8½. 61226-0 Pa. $7.95

PROBLEMS AND SOLUTIONS IN QUANTUM CHEMISTRY AND PHYSICS, Charles S. Johnson, Jr. and Lee G. Pedersen. Unusually varied problems, detailed solutions in coverage of quantum mechanics, wave mechanics, angular momentum, molecular spectroscopy, scattering theory, more. 280 problems plus 139 supplementary exercises. 430pp. 6½ × 9¼. 65236-X Pa. $10.95

ASYMPTOTIC METHODS IN ANALYSIS, N.G. de Bruijn. An inexpensive, comprehensive guide to asymptotic methods—the pioneering work that teaches by explaining worked examples in detail. Index. 224pp. 5⅜ × 8½. 64221-6 Pa. $5.95

OPTICAL RESONANCE AND TWO-LEVEL ATOMS, L. Allen and J.H. Eberly. Clear, comprehensive introduction to basic principles behind all quantum optical resonance phenomena. 53 illustrations. Preface. Index. 256pp. 5⅜ × 8½. 65533-4 Pa. $6.95

COMPLEX VARIABLES, Francis J. Flanigan. Unusual approach, delaying complex algebra till harmonic functions have been analyzed from real variable viewpoint. Includes problems with answers. 364pp. 5⅜ × 8½. 61388-7 Pa. $7.95

ATOMIC SPECTRA AND ATOMIC STRUCTURE, Gerhard Herzberg. One of best introductions; especially for specialist in other fields. Treatment is physical rather than mathematical. 80 illustrations. 257pp. 5⅜ × 8½. 60115-3 Pa. $4.95

APPLIED COMPLEX VARIABLES, John W. Dettman. Step-by-step coverage of fundamentals of analytic function theory—plus lucid exposition of 5 important applications: Potential Theory; Ordinary Differential Equations; Fourier Transforms; Laplace Transforms; Asymptotic Expansions. 66 figures. Exercises at chapter ends. 512pp. 5⅜ × 8½. 64670-X Pa. $10.95

ULTRASONIC ABSORPTION: An Introduction to the Theory of Sound Absorption and Dispersion in Gases, Liquids and Solids, A.B. Bhatia. Standard reference in the field provides a clear, systematically organized introductory review of fundamental concepts for advanced graduate students, research workers. Numerous diagrams. Bibliography. 440pp. 5⅜ × 8½. 64917-2 Pa. $8.95

UNBOUNDED LINEAR OPERATORS: Theory and Applications, Seymour Goldberg. Classic presents systematic treatment of the theory of unbounded linear operators in normed linear spaces with applications to differential equations. Bibliography. 199pp. 5⅜ × 8½. 64830-3 Pa. $7.00

LIGHT SCATTERING BY SMALL PARTICLES, H.C. van de Hulst. Comprehensive treatment including full range of useful approximation methods for researchers in chemistry, meteorology and astronomy. 44 illustrations. 470pp. 5⅜ × 8½. 64228-3 Pa. $9.95

CONFORMAL MAPPING ON RIEMANN SURFACES, Harvey Cohn. Lucid, insightful book presents ideal coverage of subject. 334 exercises make book perfect for self-study. 55 figures. 352pp. 5⅜ × 8¼. 64025-6 Pa. $8.95

OPTICKS, Sir Isaac Newton. Newton's own experiments with spectroscopy, colors, lenses, reflection, refraction, etc., in language the layman can follow. Foreword by Albert Einstein. 532pp. 5⅜ × 8½. 60205-2 Pa. $8.95

GENERALIZED INTEGRAL TRANSFORMATIONS, A.H. Zemanian. Graduate-level study of recent generalizations of the Laplace, Mellin, Hankel, K. Weierstrass, convolution and other simple transformations. Bibliography. 320pp. 5⅜ × 8½. 65375-7 Pa. $7.95

THE ELECTROMAGNETIC FIELD, Albert Shadowitz. Comprehensive undergraduate text covers basics of electric and magnetic fields, builds up to electromagnetic theory. Also related topics, including relativity. Over 900 problems. 768pp. 5⅜ × 8¼. 65660-8 Pa. $15.95

FOURIER SERIES, Georgi P. Tolstov. Translated by Richard A. Silverman. A valuable addition to the literature on the subject, moving clearly from subject to subject and theorem to theorem. 107 problems, answers. 336pp. 5⅜ × 8½. 63317-9 Pa. $7.95

THEORY OF ELECTROMAGNETIC WAVE PROPAGATION, Charles Herach Papas. Graduate-level study discusses the Maxwell field equations, radiation from wire antennas, the Doppler effect and more. xiii + 244pp. 5⅜ × 8½. 65678-0 Pa. $6.95

DISTRIBUTION THEORY AND TRANSFORM ANALYSIS: An Introduction to Generalized Functions, with Applications, A.H. Zemanian. Provides basics of distribution theory, describes generalized Fourier and Laplace transformations. Numerous problems. 384pp. 5⅜ × 8½. 65479-6 Pa. $8.95

THE PHYSICS OF WAVES, William C. Elmore and Mark A. Heald. Unique overview of classical wave theory. Acoustics, optics, electromagnetic radiation, more. Ideal as classroom text or for self-study. Problems. 477pp. 5⅜ × 8½. 64926-1 Pa. $10.95

CALCULUS OF VARIATIONS WITH APPLICATIONS, George M. Ewing. Applications-oriented introduction to variational theory develops insight and promotes understanding of specialized books, research papers. Suitable for advanced undergraduate/graduate students as primary, supplementary text. 352pp. 5⅜ × 8½. 64856-7 Pa. $8.50

A TREATISE ON ELECTRICITY AND MAGNETISM, James Clerk Maxwell. Important foundation work of modern physics. Brings to final form Maxwell's theory of electromagnetism and rigorously derives his general equations of field theory. 1,084pp. 5⅜ × 8½. 60636-8, 60637-6 Pa., Two-vol. set $19.00

AN INTRODUCTION TO THE CALCULUS OF VARIATIONS, Charles Fox. Graduate-level text covers variations of an integral, isoperimetrical problems, least action, special relativity, approximations, more. References. 279pp. 5⅜ × 8½. 65499-0 Pa. $6.95

HYDRODYNAMIC AND HYDROMAGNETIC STABILITY, S. Chandrasekhar. Lucid examination of the Rayleigh-Benard problem; clear coverage of the theory of instabilities causing convection. 704pp. 5⅜ × 8¼. 64071-X Pa. $12.95

CALCULUS OF VARIATIONS, Robert Weinstock. Basic introduction covering isoperimetric problems, theory of elasticity, quantum mechanics, electrostatics, etc. Exercises throughout. 326pp. 5⅜ × 8½. 63069-2 Pa. $7.95

DYNAMICS OF FLUIDS IN POROUS MEDIA, Jacob Bear. For advanced students of ground water hydrology, soil mechanics and physics, drainage and irrigation engineering and more. 335 illustrations. Exercises, with answers. 784pp. 6⅛ × 9¼. 65675-6 Pa. $19.95

NUMERICAL METHODS FOR SCIENTISTS AND ENGINEERS, Richard Hamming. Classic text stresses frequency approach in coverage of algorithms, polynomial approximation, Fourier approximation, exponential approximation, other topics. Revised and enlarged 2nd edition. 721pp. 5⅜ × 8½.
65241-6 Pa. $14.95

THEORETICAL SOLID STATE PHYSICS, Vol. I: Perfect Lattices in Equilibrium; Vol. II: Non-Equilibrium and Disorder, William Jones and Norman H. March. Monumental reference work covers fundamental theory of equilibrium properties of perfect crystalline solids, non-equilibrium properties, defects and disordered systems. Appendices. Problems. Preface. Diagrams. Index. Bibliography. Total of 1,301pp. 5⅜ × 8½. Two volumes. Vol. I 65015-4 Pa. $12.95
Vol. II 65016-2 Pa. $12.95

OPTIMIZATION THEORY WITH APPLICATIONS, Donald A. Pierre. Broad-spectrum approach to important topic. Classical theory of minima and maxima, calculus of variations, simplex technique and linear programming, more. Many problems, examples. 640pp. 5⅜ × 8½. 65205-X Pa. $12.95

THE MODERN THEORY OF SOLIDS, Frederick Seitz. First inexpensive edition of classic work on theory of ionic crystals, free-electron theory of metals and semiconductors, molecular binding, much more. 736pp. 5⅜ × 8½.
65482-6 Pa. $14.95

ESSAYS ON THE THEORY OF NUMBERS, Richard Dedekind. Two classic essays by great German mathematician: on the theory of irrational numbers; and on transfinite numbers and properties of natural numbers. 115pp. 5⅜ × 8½.
21010-3 Pa. $4.95

THE FUNCTIONS OF MATHEMATICAL PHYSICS, Harry Hochstadt. Comprehensive treatment of orthogonal polynomials, hypergeometric functions, Hill's equation, much more. Bibliography. Index. 322pp. 5⅜ × 8½. 65214-9 Pa. $8.95

NUMBER THEORY AND ITS HISTORY, Oystein Ore. Unusually clear, accessible introduction covers counting, properties of numbers, prime numbers, much more. Bibliography. 380pp. 5⅜ × 8½. 65620-9 Pa. $8.95

THE VARIATIONAL PRINCIPLES OF MECHANICS, Cornelius Lanczos. Graduate level coverage of calculus of variations, equations of motion, relativistic mechanics, more. First inexpensive paperbound edition of classic treatise. Index. Bibliography. 418pp. 5⅜ × 8½. 65067-7 Pa. $10.95

MATHEMATICAL TABLES AND FORMULAS, Robert D. Carmichael and Edwin R. Smith. Logarithms, sines, tangents, trig functions, powers, roots, reciprocals, exponential and hyperbolic functions, formulas and theorems. 269pp. 5⅜ × 8½. 60111-0 Pa. $5.95

THEORETICAL PHYSICS, Georg Joos, with Ira M. Freeman. Classic overview covers essential math, mechanics, electromagnetic theory, thermodynamics, quantum mechanics, nuclear physics, other topics. First paperback edition. xxiii + 885pp. 5⅜ × 8½. 65227-0 Pa. $17.95

HANDBOOK OF MATHEMATICAL FUNCTIONS WITH FORMULAS, GRAPHS, AND MATHEMATICAL TABLES, edited by Milton Abramowitz and Irene A. Stegun. Vast compendium: 29 sets of tables, some to as high as 20 places. 1,046pp. 8 × 10½. 61272-4 Pa. $21.95

MATHEMATICAL METHODS IN PHYSICS AND ENGINEERING, John W. Dettman. Algebraically based approach to vectors, mapping, diffraction, other topics in applied math. Also generalized functions, analytic function theory, more. Exercises. 448pp. 5⅜ × 8¼. 65649-7 Pa. $8.95

A SURVEY OF NUMERICAL MATHEMATICS, David M. Young and Robert Todd Gregory. Broad self-contained coverage of computer-oriented numerical algorithms for solving various types of mathematical problems in linear algebra, ordinary and partial, differential equations, much more. Exercises. Total of 1,248pp. 5⅜ × 8½. Two volumes. Vol. I 65691-8 Pa. $13.95
Vol. II 65692-6 Pa. $13.95

TENSOR ANALYSIS FOR PHYSICISTS, J.A. Schouten. Concise exposition of the mathematical basis of tensor analysis, integrated with well-chosen physical examples of the theory. Exercises. Index. Bibliography. 289pp. 5⅜ × 8½. 65582-2 Pa. $7.95

INTRODUCTION TO NUMERICAL ANALYSIS (2nd Edition), F.B. Hildebrand. Classic, fundamental treatment covers computation, approximation, interpolation, numerical differentiation and integration, other topics. 150 new problems. 669pp. 5⅜ × 8½. 65363-3 Pa. $13.95

INVESTIGATIONS ON THE THEORY OF THE BROWNIAN MOVEMENT, Albert Einstein. Five papers (1905–8) investigating dynamics of Brownian motion and evolving elementary theory. Notes by R. Fürth. 122pp. 5⅜ × 8½. 60304-0 Pa. $3.95

NUMERICAL METHODS FOR SCIENTISTS AND ENGINEERS, Richard Hamming. Classic text stresses frequency approach in coverage of algorithms, polynomial approximation, Fourier approximation, exponential approximation, other topics. Revised and enlarged 2nd edition. 721pp. 5⅜ × 8½. 65241-6 Pa. $14.95

AN INTRODUCTION TO STATISTICAL THERMODYNAMICS, Terrell L. Hill. Excellent basic text offers wide-ranging coverage of quantum statistical mechanics, systems of interacting molecules, quantum statistics, more. 523pp. 5⅜ × 8½. 65242-4 Pa. $10.95

ELEMENTARY DIFFERENTIAL EQUATIONS, William Ted Martin and Eric Reissner. Exceptionally, clear comprehensive introduction at undergraduate level. Nature and origin of differential equations, differential equations of first, second and higher orders. Picard's Theorem, much more. Problems with solutions. 331pp. 5⅜ × 8½. 65024-3 Pa. $8.95

STATISTICAL PHYSICS, Gregory H. Wannier. Classic text combines thermodynamics, statistical mechanics and kinetic theory in one unified presentation of thermal physics. Problems with solutions. Bibliography. 532pp. 5⅜ × 8½. 65401-X Pa. $10.95

ORDINARY DIFFERENTIAL EQUATIONS, Morris Tenenbaum and Harry Pollard. Exhaustive survey of ordinary differential equations for undergraduates in mathematics, engineering, science. Thorough analysis of theorems. Diagrams. Bibliography. Index. 818pp. 5⅜ × 8½. 64940-7 Pa. $15.95

STATISTICAL MECHANICS: Principles and Applications, Terrell L. Hill. Standard text covers fundamentals of statistical mechanics, applications to fluctuation theory, imperfect gases, distribution functions, more. 448pp. 5⅜ × 8½. 65390-0 Pa. $9.95

ORDINARY DIFFERENTIAL EQUATIONS AND STABILITY THEORY: An Introduction, David A. Sánchez. Brief, modern treatment. Linear equation, stability theory for autonomous and nonautonomous systems, etc. 164pp. 5⅜ × 8¼. 63828-6 Pa. $4.95

THIRTY YEARS THAT SHOOK PHYSICS: The Story of Quantum Theory, George Gamow. Lucid, accessible introduction to influential theory of energy and matter. Careful explanations of Dirac's anti-particles, Bohr's model of the atom, much more. 12 plates. Numerous drawings. 240pp. 5⅜ × 8½. 24895-X Pa. $5.95

ORDINARY DIFFERENTIAL EQUATIONS, I.G. Petrovski. Covers basic concepts, some differential equations and such aspects of the general theory as Euler lines, Arzel's theorem, Peano's existence theorem, Osgood's uniqueness theorem, more. 45 figures. Problems. Bibliography. Index. xi + 232pp. 5⅜ × 8½. 64683-1 Pa. $6.00

GREAT EXPERIMENTS IN PHYSICS: Firsthand Accounts from Galileo to Einstein, edited by Morris H. Shamos. 25 crucial discoveries: Newton's laws of motion, Chadwick's study of the neutron, Hertz on electromagnetic waves, more. Original accounts clearly annotated. 370pp. 5⅜ × 8½. 25346-5 Pa. $8.95

INTRODUCTION TO PARTIAL DIFFERENTIAL EQUATIONS WITH APPLICATIONS, E.C. Zachmanoglou and Dale W. Thoe. Essentials of partial differential equations applied to common problems in engineering and the physical sciences. Problems and answers. 416pp. 5⅜ × 8½. 65251-3 Pa. $9.95

BURNHAM'S CELESTIAL HANDBOOK, Robert Burnham, Jr. Thorough guide to the stars beyond our solar system. Exhaustive treatment. Alphabetical by constellation: Andromeda to Cetus in Vol. 1; Chamaeleon to Orion in Vol. 2; and Pavo to Vulpecula in Vol. 3. Hundreds of illustrations. Index in Vol. 3. 2,000pp. 6⅛ × 9¼. 23567-X, 23568-8, 23673-0 Pa., Three-vol. set $38.85

ASYMPTOTIC EXPANSIONS FOR ORDINARY DIFFERENTIAL EQUATIONS, Wolfgang Wasow. Outstanding text covers asymptotic power series, Jordan's canonical form, turning point problems, singular perturbations, much more. Problems. 384pp. 5⅜ × 8½. 65456-7 Pa. $8.95

AMATEUR ASTRONOMER'S HANDBOOK, J.B. Sidgwick. Timeless, comprehensive coverage of telescopes, mirrors, lenses, mountings, telescope drives, micrometers, spectroscopes, more. 189 illustrations. 576pp. 5⅜ × 8¼. 24034-7 Pa. $8.95

SPECIAL FUNCTIONS, N.N. Lebedev. Translated by Richard Silverman. Famous Russian work treating more important special functions, with applications to specific problems of physics and engineering. 38 figures. 308pp. 5⅜ × 8½.
60624-4 Pa. $6.95

OBSERVATIONAL ASTRONOMY FOR AMATEURS, J.B. Sidgwick. Mine of useful data for observation of sun, moon, planets, asteroids, aurorae, meteors, comets, variables, binaries, etc. 39 illustrations 384pp. 5⅜ × 8¼. (Available in U.S. only)
24033-9 Pa. $5.95

INTEGRAL EQUATIONS, F.G. Tricomi. Authoritative, well-written treatment of extremely useful mathematical tool with wide applications. Volterra Equations, Fredholm Equations, much more. Advanced undergraduate to graduate level. Exercises. Bibliography. 238pp. 5⅜ × 8½.
64828-1 Pa. $6.95

CELESTIAL OBJECTS FOR COMMON TELESCOPES, T.W. Webb. Inestimable aid for locating and identifying nearly 4,000 celestial objects. 77 illustrations. 645pp. 5⅜ × 8½.
20917-2, 20918-0 Pa., Two-vol. set $12.00

MODERN NONLINEAR EQUATIONS, Thomas L. Saaty. Emphasizes practical solution of problems; covers seven types of equations. ". . . a welcome contribution to the existing literature. . . ."—*Math Reviews.* 490pp. 5⅜ × 8½. 64232-1 Pa. $9.95

FUNDAMENTALS OF ASTRODYNAMICS, Roger Bate et al. Modern approach developed by U.S. Air Force Academy. Designed as a first course. Problems, exercises. Numerous illustrations. 455pp. 5⅜ × 8½.
60061-0 Pa. $8.95

INTRODUCTION TO LINEAR ALGEBRA AND DIFFERENTIAL EQUATIONS, John W. Dettman. Excellent text covers complex numbers, determinants, orthonormal bases, Laplace transforms, much more. Exercises with solutions. Undergraduate level. 416pp. 5⅜ × 8½.
65191-6 Pa. $8.95

INCOMPRESSIBLE AERODYNAMICS, edited by Bryan Thwaites. Covers theoretical and experimental treatment of the uniform flow of air and viscous fluids past two-dimensional aerofoils and three-dimensional wings; many other topics. 654pp. 5⅜ × 8½.
65465-6 Pa. $14.95

INTRODUCTION TO DIFFERENCE EQUATIONS, Samuel Goldberg. Exceptionally clear exposition of important discipline with applications to sociology, psychology, economics. Many illustrative examples; over 250 problems. 260pp. 5⅜ × 8½.
65084-7 Pa. $6.95

LAMINAR BOUNDARY LAYERS, edited by L. Rosenhead. Engineering classic covers steady boundary layers in two- and three-dimensional flow, unsteady boundary layers, stability, observational techniques, much more. 708pp. 5⅜ × 8½.
65646-2 Pa. $15.95

LECTURES ON CLASSICAL DIFFERENTIAL GEOMETRY, Second Edition, Dirk J. Struik. Excellent brief introduction covers curves, theory of surfaces, fundamental equations, geometry on a surface, conformal mapping, other topics. Problems. 240pp. 5⅜ × 8½.
65609-8 Pa. $6.95

ROTARY-WING AERODYNAMICS, W.Z. Stepniewski. Clear, concise text covers aerodynamic phenomena of the rotor and offers guidelines for helicopter performance evaluation. Originally prepared for NASA. 537 figures. 640pp. 6⅛ × 9¼.
64647-5 Pa. $14.95

DIFFERENTIAL GEOMETRY, Heinrich W. Guggenheimer. Local differential geometry as an application of advanced calculus and linear algebra. Curvature, transformation groups, surfaces, more. Exercises. 62 figures. 378pp. 5⅜ × 8½.
63433-7 Pa. $7.95

INTRODUCTION TO SPACE DYNAMICS, William Tyrrell Thomson. Comprehensive, classic introduction to space-flight engineering for advanced undergraduate and graduate students. Includes vector algebra, kinematics, transformation of coordinates. Bibliography. Index. 352pp. 5⅜ × 8½. 65113-4 Pa. $8.00

A SURVEY OF MINIMAL SURFACES, Robert Osserman. Up-to-date, in-depth discussion of the field for advanced students. Corrected and enlarged edition covers new developments. Includes numerous problems. 192pp. 5⅜ × 8½.
64998-9 Pa. $8.00

ANALYTICAL MECHANICS OF GEARS, Earle Buckingham. Indispensable reference for modern gear manufacture covers conjugate gear-tooth action, gear-tooth profiles of various gears, many other topics. 263 figures. 102 tables. 546pp. 5⅜ × 8½. 65712-4 Pa. $11.95

SET THEORY AND LOGIC, Robert R. Stoll. Lucid introduction to unified theory of mathematical concepts. Set theory and logic seen as tools for conceptual understanding of real number system. 496pp. 5⅜ × 8¼. 63829-4 Pa. $8.95

A HISTORY OF MECHANICS, René Dugas. Monumental study of mechanical principles from antiquity to quantum mechanics. Contributions of ancient Greeks, Galileo, Leonardo, Kepler, Lagrange, many others. 671pp. 5⅜ × 8½.
65632-2 Pa. $14.95

FAMOUS PROBLEMS OF GEOMETRY AND HOW TO SOLVE THEM, Benjamin Bold. Squaring the circle, trisecting the angle, duplicating the cube: learn their history, why they are impossible to solve, then solve them yourself. 128pp. 5⅜ × 8½. 24297-8 Pa. $3.95

MECHANICAL VIBRATIONS, J.P. Den Hartog. Classic textbook offers lucid explanations and illustrative models, applying theories of vibrations to a variety of practical industrial engineering problems. Numerous figures. 233 problems, solutions. Appendix. Index. Preface. 436pp. 5⅜ × 8½. 64785-4 Pa. $8.95

CURVATURE AND HOMOLOGY, Samuel I. Goldberg. Thorough treatment of specialized branch of differential geometry. Covers Riemannian manifolds, topology of differentiable manifolds, compact Lie groups, other topics. Exercises. 315pp. 5⅜ × 8½. 64314-X Pa. $6.95

HISTORY OF STRENGTH OF MATERIALS, Stephen P. Timoshenko. Excellent historical survey of the strength of materials with many references to the theories of elasticity and structure. 245 figures. 452pp. 5⅜ × 8½. 61187-6 Pa. $9.95

GEOMETRY OF COMPLEX NUMBERS, Hans Schwerdtfeger. Illuminating, widely praised book on analytic geometry of circles, the Moebius transformation, and two-dimensional non-Euclidean geometries. 200pp. 5⅜ × 8¼.
63830-8 Pa. $6.95

MECHANICS, J.P. Den Hartog. A classic introductory text or refresher. Hundreds of applications and design problems illuminate fundamentals of trusses, loaded beams and cables, etc. 334 answered problems. 462pp. 5⅜ × 8½. 60754-2 Pa. $8.95

TOPOLOGY, John G. Hocking and Gail S. Young. Superb one-year course in classical topology. Topological spaces and functions, point-set topology, much more. Examples and problems. Bibliography. Index. 384pp. 5⅜ × 8¼.
65676-4 Pa. $7.95

STRENGTH OF MATERIALS, J.P. Den Hartog. Full, clear treatment of basic material (tension, torsion, bending, etc.) plus advanced material on engineering methods, applications. 350 answered problems. 323pp. 5⅜ × 8½. 60755-0 Pa. $7.50

ELEMENTARY CONCEPTS OF TOPOLOGY, Paul Alexandroff. Elegant, intuitive approach to topology from set-theoretic topology to Betti groups; how concepts of topology are useful in math and physics. 25 figures. 57pp. 5⅜ × 8½.
60747-X Pa. $2.95

ADVANCED STRENGTH OF MATERIALS, J.P. Den Hartog. Superbly written advanced text covers torsion, rotating disks, membrane stresses in shells, much more. Many problems and answers. 388pp. 5⅜ × 8½. 65407-9 Pa. $8.95

COMPUTABILITY AND UNSOLVABILITY, Martin Davis. Classic graduate-level introduction to theory of computability, usually referred to as theory of recurrent functions. New preface and appendix. 288pp. 5⅜ × 8½. 61471-9 Pa. $6.95

GENERAL CHEMISTRY, Linus Pauling. Revised 3rd edition of classic first-year text by Nobel laureate. Atomic and molecular structure, quantum mechanics, statistical mechanics, thermodynamics correlated with descriptive chemistry. Problems. 992pp. 5⅜ × 8½. 65622-5 Pa. $18.95

AN INTRODUCTION TO MATRICES, SETS AND GROUPS FOR SCIENCE STUDENTS, G. Stephenson. Concise, readable text introduces sets, groups, and most importantly, matrices to undergraduate students of physics, chemistry, and engineering. Problems. 164pp. 5⅜ × 8½. 65077-4 Pa. $5.95

THE HISTORICAL BACKGROUND OF CHEMISTRY, Henry M. Leicester. Evolution of ideas, not individual biography. Concentrates on formulation of a coherent set of chemical laws. 260pp. 5⅜ × 8½. 61053-5 Pa. $6.00

THE PHILOSOPHY OF MATHEMATICS: An Introductory Essay, Stephan Körner. Surveys the views of Plato, Aristotle, Leibniz & Kant concerning propositions and theories of applied and pure mathematics. Introduction. Two appendices. Index. 198pp. 5⅜ × 8½. 25048-2 Pa. $5.95

THE DEVELOPMENT OF MODERN CHEMISTRY, Aaron J. Ihde. Authoritative history of chemistry from ancient Greek theory to 20th-century innovation. Covers major chemists and their discoveries. 209 illustrations. 14 tables. Bibliographies. Indices. Appendices. 851pp. 5⅜ × 8½. 64235-6 Pa. $15.95

THE FOUR-COLOR PROBLEM: Assaults and Conquest, Thomas L. Saaty and Paul G. Kainen. Engrossing, comprehensive account of the century-old combinatorial topological problem, its history and solution. Bibliographies. Index. 110 figures. 228pp. 5⅜ × 8½.   65092-8 Pa. $6.00

CATALYSIS IN CHEMISTRY AND ENZYMOLOGY, William P. Jencks. Exceptionally clear coverage of mechanisms for catalysis, forces in aqueous solution, carbonyl- and acyl-group reactions, practical kinetics, more. 864pp. 5⅜ × 8½.   65460-5 Pa. $18.95

PROBABILITY: An Introduction, Samuel Goldberg. Excellent basic text covers set theory, probability theory for finite sample spaces, binomial theorem, much more. 360 problems. Bibliographies. 322pp. 5⅜ × 8½.   65252-1 Pa. $7.95

LIGHTNING, Martin A. Uman. Revised, updated edition of classic work on the physics of lightning. Phenomena, terminology, measurement, photography, spectroscopy, thunder, more. Reviews recent research. Bibliography. Indices. 320pp. 5⅜ × 8¼.   64575-4 Pa. $7.95

PROBABILITY THEORY: A Concise Course, Y.A. Rozanov. Highly readable, self-contained introduction covers combination of events, dependent events, Bernoulli trials, etc. Translation by Richard Silverman. 148pp. 5⅜ × 8¼.
63544-9 Pa. $4.50

THE CEASELESS WIND: An Introduction to the Theory of Atmospheric Motion, John A. Dutton. Acclaimed text integrates disciplines of mathematics and physics for full understanding of dynamics of atmospheric motion. Over 400 problems. Index. 97 illustrations. 640pp. 6 × 9.   65096-0 Pa. $16.95

STATISTICS MANUAL, Edwin L. Crow, et al. Comprehensive, practical collection of classical and modern methods prepared by U.S. Naval Ordnance Test Station. Stress on use. Basics of statistics assumed. 288pp. 5⅜ × 8½.
60599-X Pa. $6.00

WIND WAVES: Their Generation and Propagation on the Ocean Surface, Blair Kinsman. Classic of oceanography offers detailed discussion of stochastic processes and power spectral analysis that revolutionized ocean wave theory. Rigorous, lucid. 676pp. 5⅜ × 8½.   64652-1 Pa. $14.95

STATISTICAL METHOD FROM THE VIEWPOINT OF QUALITY CONTROL, Walter A. Shewhart. Important text explains regulation of variables, uses of statistical control to achieve quality control in industry, agriculture, other areas. 192pp. 5⅜ × 8½.   65232-7 Pa. $6.00

THE INTERPRETATION OF GEOLOGICAL PHASE DIAGRAMS, Ernest G. Ehlers. Clear, concise text emphasizes diagrams of systems under fluid or containing pressure; also coverage of complex binary systems, hydrothermal melting, more. 288pp. 6½ × 9¼.   65389-7 Pa. $8.95

STATISTICAL ADJUSTMENT OF DATA, W. Edwards Deming. Introduction to basic concepts of statistics, curve fitting, least squares solution, conditions without parameter, conditions containing parameters. 26 exercises worked out. 271pp. 5⅜ × 8½.   64685-8 Pa. $7.95

DE RE METALLICA, Georgius Agricola. The famous Hoover translation of greatest treatise on technological chemistry, engineering, geology, mining of early modern times (1556). All 289 original woodcuts. 638pp. 6¾ × 11.
60006-8 Clothbd. $15.95

SOME THEORY OF SAMPLING, William Edwards Deming. Analysis of the problems, theory and design of sampling techniques for social scientists, industrial managers and others who find statistics increasingly important in their work. 61 tables. 90 figures. xvii + 602pp. 5⅜ × 8½. 64684-X Pa. $14.95

THE VARIOUS AND INGENIOUS MACHINES OF AGOSTINO RAMELLI: A Classic Sixteenth-Century Illustrated Treatise on Technology, Agostino Ramelli. One of the most widely known and copied works on machinery in the 16th century. 194 detailed plates of water pumps, grain mills, cranes, more. 608pp. 9 × 12.
25497-6 Clothbd. $34.95

LINEAR PROGRAMMING AND ECONOMIC ANALYSIS, Robert Dorfman, Paul A. Samuelson and Robert M. Solow. First comprehensive treatment of linear programming in standard economic analysis. Game theory, modern welfare economics, Leontief input-output, more. 525pp. 5⅜ × 8½. 65491-5 Pa. $12.95

ELEMENTARY DECISION THEORY, Herman Chernoff and Lincoln E. Moses. Clear introduction to statistics and statistical theory covers data processing, probability and random variables, testing hypotheses, much more. Exercises. 364pp. 5⅜ × 8½. 65218-1 Pa. $8.95

THE COMPLEAT STRATEGYST: Being a Primer on the Theory of Games of Strategy, J.D. Williams. Highly entertaining classic describes, with many illustrated examples, how to select best strategies in conflict situations. Prefaces. Appendices. 268pp. 5⅜ × 8½. 25101-2 Pa. $5.95

MATHEMATICAL METHODS OF OPERATIONS RESEARCH, Thomas L. Saaty. Classic graduate-level text covers historical background, classical methods of forming models, optimization, game theory, probability, queueing theory, much more. Exercises. Bibliography. 448pp. 5⅜ × 8¼. 65703-5 Pa. $12.95

CONSTRUCTIONS AND COMBINATORIAL PROBLEMS IN DESIGN OF EXPERIMENTS, Damaraju Raghavarao. In-depth reference work examines orthogonal Latin squares, incomplete block designs, tactical configuration, partial geometry, much more. Abundant explanations, examples. 416pp. 5⅜ × 8¼.
65685-3 Pa. $10.95

THE ABSOLUTE DIFFERENTIAL CALCULUS (CALCULUS OF TENSORS), Tullio Levi-Civita. Great 20th-century mathematician's classic work on material necessary for mathematical grasp of theory of relativity. 452pp. 5⅜ × 8½.
63401-9 Pa. $9.95

VECTOR AND TENSOR ANALYSIS WITH APPLICATIONS, A.I. Borisenko and I.E. Tarapov. Concise introduction. Worked-out problems, solutions, exercises. 257pp. 5⅜ × 8¼. 63833-2 Pa. $6.95

TENSOR CALCULUS, J.L. Synge and A. Schild. Widely used introductory text covers spaces and tensors, basic operations in Riemannian space, non-Riemannian spaces, etc. 324pp. 5⅜ × 8¼. 63612-7 Pa. $7.00

A CONCISE HISTORY OF MATHEMATICS, Dirk J. Struik. The best brief history of mathematics. Stresses origins and covers every major figure from ancient Near East to 19th century. 41 illustrations. 195pp. 5⅜ × 8½. 60255-9 Pa. $7.95

A SHORT ACCOUNT OF THE HISTORY OF MATHEMATICS, W.W. Rouse Ball. One of clearest, most authoritative surveys from the Egyptians and Phoenicians through 19th-century figures such as Grassman, Galois, Riemann. Fourth edition. 522pp. 5⅜ × 8½. 20630-0 Pa. $9.95

HISTORY OF MATHEMATICS, David E. Smith. Non-technical survey from ancient Greece and Orient to late 19th century; evolution of arithmetic, geometry, trigonometry, calculating devices, algebra, the calculus. 362 illustrations. 1,355pp. 5⅜ × 8½. 20429-4, 20430-8 Pa., Two-vol. set $21.90

THE GEOMETRY OF RENÉ DESCARTES, René Descartes. The great work founded analytical geometry. Original French text, Descartes' own diagrams, together with definitive Smith-Latham translation. 244pp. 5⅜ × 8½.
60068-8 Pa. $6.00

THE ORIGINS OF THE INFINITESIMAL CALCULUS, Margaret E. Baron. Only fully detailed and documented account of crucial discipline: origins; development by Galileo, Kepler, Cavalieri; contributions of Newton, Leibniz, more. 304pp. 5⅜ × 8½. (Available in U.S. and Canada only) 65371-4 Pa. $7.95

THE HISTORY OF THE CALCULUS AND ITS CONCEPTUAL DEVELOPMENT, Carl B. Boyer. Origins in antiquity, medieval contributions, work of Newton, Leibniz, rigorous formulation. Treatment is verbal. 346pp. 5⅜ × 8½.
60509-4 Pa. $6.95

THE THIRTEEN BOOKS OF EUCLID'S ELEMENTS, translated with introduction and commentary by Sir Thomas L. Heath. Definitive edition. Textual and linguistic notes, mathematical analysis. 2500 years of critical commentary. Not abridged. 1,414pp. 5⅜ × 8½. 60088-2, 60089-0, 60090-4 Pa., Three-vol. set $26.85

A HISTORY OF VECTOR ANALYSIS: The Evolution of the Idea of a Vectorial System, Michael J. Crowe. The first large-scale study of the history of vector analysis, now the standard on the subject. Unabridged republication of the edition published by University of Notre Dame Press, 1967, with second preface by Michael C. Crowe. Index. 278pp. 5⅜ × 8½. 64955-5 Pa. $7.00

THE HISTORICAL ROOTS OF ELEMENTARY MATHEMATICS, Lucas N.H. Bunt, Phillip S. Jones, and Jack D. Bedient. Fundamental underpinnings of modern arithmetic, algebra, geometry and number systems derived from ancient civilizations. 320pp. 5⅜ × 8½. 25563-8 Pa. $7.95

CALCULUS REFRESHER FOR TECHNICAL PEOPLE, A. Albert Klaf. Covers important aspects of integral and differential calculus via 756 questions. 566 problems, most answered. 431pp. 5⅜ × 8½. 20370-0 Pa. $7.95

CHALLENGING MATHEMATICAL PROBLEMS WITH ELEMENTARY SOLUTIONS, A.M. Yaglom and I.M. Yaglom. Over 170 challenging problems on probability theory, combinatorial analysis, points and lines, topology, convex polygons, many other topics. Solutions. Total of 445pp. 5⅜ × 8½. Two-vol. set.
Vol. I 65536-9 Pa. $5.95
Vol. II 65537-7 Pa. $5.95

FIFTY CHALLENGING PROBLEMS IN PROBABILITY WITH SOLUTIONS, Frederick Mosteller. Remarkable puzzlers, graded in difficulty, illustrate elementary and advanced aspects of probability. Detailed solutions. 88pp. 5⅜ × 8½.
65355-2 Pa. $3.95

EXPERIMENTS IN TOPOLOGY, Stephen Barr. Classic, lively explanation of one of the byways of mathematics. Klein bottles, Moebius strips, projective planes, map coloring, problem of the Koenigsberg bridges, much more, described with clarity and wit. 43 figures. 210pp. 5⅜ × 8½.
25933-1 Pa. $4.95

RELATIVITY IN ILLUSTRATIONS, Jacob T. Schwartz. Clear non-technical treatment makes relativity more accessible than ever before. Over 60 drawings illustrate concepts more clearly than text alone. Only high school geometry needed. Bibliography. 128pp. 6⅛ × 9¼.
25965-X Pa. $5.95

AN INTRODUCTION TO ORDINARY DIFFERENTIAL EQUATIONS, Earl A. Coddington. A thorough and systematic first course in elementary differential equations for undergraduates in mathematics and science, with many exercises and problems (with answers). Index. 304pp. 5⅜ × 8¼.
65942-9 Pa. $7.95

FOURIER SERIES AND ORTHOGONAL FUNCTIONS, Harry F. Davis. An incisive text combining theory and practical example to introduce Fourier series, orthogonal functions and applications of the Fourier method to boundary-value problems. 570 exercises. Answers and notes. 416pp. 5⅜ × 8½.
65973-9 Pa. $8.95

THE THOERY OF BRANCHING PROCESSES, Theodore E. Harris. First systematic, comprehensive treatment of branching (i.e. multiplicative) processes and their applications. Galton-Watson model, Markov branching processes, electron-photon cascade, many other topics. Rigorous proofs. Bibliography. 240pp. 5⅜ × 8½.
65952-6 Pa. $6.95

AN INTRODUCTION TO ALGEBRAIC STRUCTURES, Joseph Landin. Superb self-contained text covers "abstract algebra": sets and numbers, theory of groups, theory of rings, much more. Numerous well-chosen examples, exercises. 247pp. 5⅜ × 8½.
65940-2 Pa. $6.95

GAMES AND DECISIONS: Introduction and Critical Survey, R. Duncan Luce and Howard Raiffa. Superb non-technical introduction to game theory, primarily applied to social sciences. Utility theory, zero-sum games, n-person games, decision-making, much more. Bibliography. 509pp. 5⅜ × 8½.   65943-7 Pa. $10.95

Prices subject to change without notice.
Available at your book dealer or write for free Mathematics and Science Catalog to Dept. GI, Dover Publications, Inc., 31 East 2nd St., Mineola, N.Y. 11501. Dover publishes more than 175 books each year on science, elementary and advanced mathematics, biology, music, art, literary history, social sciences and other areas.